# Adhering to Green and Open Cooperation to Advance Modernization

China Council for International Cooperation on Environment and Development (CCICED) Secretariat

# Adhering to Green and Open Cooperation to Advance Modernization

## CCICED Annual Policy Report 2024

Springer

China Council for International
Cooperation on Environment and
Development (CCICED) Secretariat
Beijing, China



ISBN 978-981-95-5546-8 ISBN 978-981-95-5547-5 (eBook)
https://doi.org/10.1007/978-981-95-5547-5

Jointly published with China Environment Publishing Group Co., Ltd.

The print edition is not for sale in China (Mainland). Customers from China (Mainland) please order the print book from: China Environment Publishing Group Co., Ltd.



This Springer imprint is published by the registered company Springer Nature Singapore Pte Ltd.
The registered company address is: 152 Beach Road, #21-01/04 Gateway East, Singapore 189721, Singapore

# Preface

At present, the global economy is facing a general downward trend, and the world is in a period of turbulence and change. A century of change is accelerating as the geopolitical landscape undergoes profound adjustments, technological change reshapes the industrial ecosystem, and climate change intersects with biodiversity loss. 2024 is the year when the construction of a beautiful China will be advanced in all aspects, and it is also a key year for the realization of the goals and tasks of the 14th Five-Year Plan. At the same time, the global environmental and climate governance process is also accelerating, and the realization of the United Nations Sustainable Development Goals by 2030 is still a long way off, so the international community urgently needs to build consensus, deepen cooperation, and jointly respond to the systemic challenges of mankind's sustainable development with innovative thinking.

As a high-level international advisory body and two-way exchange platform dedicated to promoting sustainable development, the China Council for International Cooperation on Environment and Development (CCICED) continues to aim to develop into an innovative platform for international cooperation on the environment and development that embraces global inclusiveness, open cooperation, and mutual benefits and development. CCICED has carried out the plan as approved by the executive members under the annual theme of "Adhering to Green and Open Cooperation to Advance Modernization." It has also provided advice and recommendations on the environment and development that embraces open cooperation, mutual benefits and development. The aim of this period is to provide suggestions for promoting China's environment and development undertakings and contribute wisdom and strength to building a community with a shared future for mankind and a clean and beautiful world.

The CCICED Secretariat publishes CCICED Annual Policy Report every year, in order to better reach a consensus on green development and promote green transition and transformation. The Annual Policy Report is a flagship publication that presents research outputs, papers, and reports by Chinese and international CCICED teams, as well as outlining policy recommendations that the CCICED has proposed to the Chinese government. It shares the observations and thinking of the Chinese and international members, special advisors, and experts on hot issues in the environment and development, for the reference of decision-makers at all levels, scholars, and the public.

The 2024 CCICED Annual Policy Report incorporates the policy recommendation proposed to the Chinese government "Further Deepen Comprehensive Reforms, Seize New Opportunities Brought by the Technological Revolution and Industrial Transformation, and Accelerate the Comprehensive Green Transition of the Economy and Society"; the 2024 issues paper which is written by CCICED's chief advisors; research reports from nine Special Policy Studies, including "China's Pathways for Achieving Carbon Neutrality and Global Climate Governance," "Sustainable Oceans Management under the Vision of Carbon Neutrality," "Biodiversity Conservation and Implementation of the Kunming-Montreal Global Biodiversity Framework (GBF)," "Collaborative Mechanism for Carbon Reduction, Pollution Reduction, Green Expansion and Growth," "Green Development and Climate Adaptation for Urban and Rural Areas," "Digitalization and Green Technologies for Sustainable Development," "Innovative Mechanisms of Sustainable Investment in Environment and Climate," "Sustainable Trade and Supply Chains," and "Green Opening-up and South-South Cooperation"; five Scoping Study reports on "Technology Innovation for Green and Low-Carbon Transition," "Beyond GDP," "Carbon Pricing," "Circular Economy," "Low-Carbon Transition of Traditional Energy Regions of China"; three Discussion Papers on "CCS-based Carbon Dioxide Removal (CDR)," "Green Consumption," "Nature Markets"; Progress on Environment and Development Policies in China and Impact of CCICED's Policy Recommendations (2023–2024) and Report on Gender Mainstreaming in SPS Research for the Period 2023–2024.

Beijing, China

China Council for International Cooperation on Environment and Development (CCICED) Secretariat

# Contents

# Chapter 1
# China's Pathways for Achieving Carbon Neutrality and Global Climate Governance

## 1 Foreword

As China is preparing for the next round of Nationally Determined Contributions (NDCs), integrating the 15th Five-Year Plan with NDC renewal is essential. A robust 2035 target will clarify the transition pathway and accelerate action in the coming years, laying a solid foundation for the transition to carbon neutrality.

## 2 Analysis of Global Climate Governance Situation

### *2.1 COP28 Has Resulted in a Strengthened, Balanced and Historic Package of Climate Actions, and Future Climate Governance Needs to Emphasize a Balance of Ambition and Pragmatic Solutions*

At COP28, parties agreed for the first time on a roadmap for a "transition away from fossil fuels in energy systems, in a just, orderly and equitable manner". In the *UAE Consensus* agreed at the end of the conference, the UAE, Azerbaijan and Brazil Presidencies will take forward the 1.5 °C Mission Roadmap and encourage countries to submit economy-wide NDCs in 2025. However, more and more countries are realizing the need for pragmatic climate action, rather than simply setting ambitious targets.

---

The content of this chapter is extracted from the main findings of the 2023–2024 CCICED Special Policy Study (SPS) on "China's Pathways for Achieving Carbon Neutrality and Global Climate Governance".



*Adhering to Green and Open Cooperation to Advance Modernization*,
https://doi.org/10.1007/978-981-95-5547-5_1

## 2.2 Complexity and Uncertainty in Climate Governance Increase Due to Intersection with Trade, Green Supply Chains, and Other Issues

The European Union's Carbon Border Adjustment Mechanism (CBAM) will enter a full implementation phase from 2026. In the medium to long term, it will exert continued pressure on the trade systems of major manufacturing countries like China. The varied economic structures, energy endowments, and policies across different countries complicate the establishment and development of uniform carbon markets, making it challenging to accurately reflect advanced emission reduction efforts through CBAM. This could result in even the most advanced industries facing additional carbon taxes, which contradicts the Just Transition principle. Furthermore, developing countries, which often rely on high-carbon product exports, may be disproportionately burdened by unilateral carbon tariff policies.

## 2.3 Decoupling and Breaking of Green and Low-Carbon Industrial Technologies Is Intensifying, and the Pace of the Global Low-Carbon Transition Is Under Attack

International cooperation in the low-carbon field has significantly reduced climate governance costs. Achieving global climate goals requires the rapid and extensive deployment of renewable energy installations, but reliance on domestic production alone will drive up costs and delay this process. According to the scenario studied, if countries gradually shift to domestic production of photovoltaic modules between 2020 and 2030, the production cost may increase by 20% to 30% by 2030, directly slowing down deployment [1].

## 2.4 Global Climate Leadership Landscape Is Changing Due to Key National and Regional Elections

Major political elections will create an uncertain outlook, becoming a crucial variable in global climate governance. The outcome of the U.S. presidential election will be particularly decisive for climate policy. If the Republican candidate is elected, the United States is expected to reduce its focus on climate issues and increase its support for fossil fuels. Internationally, the U.S. would reduce participation in climate cooperation, potentially withdrawing from the *Paris Agreement* again. In EU parliamentary elections, the Conservatives remained the biggest winners. Climate policy will most likely remain stable and continuous for the next five years. At the same time, the deeply polarized national interests makes consensus harder to achieve. Various groups and alliances are exploring alternative cooperation mechanisms outside the formal treaty framework to address their diverse interests.

### *2.5 Climate Risk Has Become One of the Most Important Risks Affecting Human Societies, and the Shift from a "High-Carbon System" to a "Low-Carbon/ Zero-Carbon System" Is Irreversible. The Fight against Climate Change Represents a Vital Avenue for International Cooperation*

The political will of major countries and groups in climate governance and international cooperation remains significant. Europe and the United States continue to seek collaboration with the Global South countries, including emerging markets such as China.

Investment in global climate governance and clean energy transition has shown strong growth momentum. In 2023, global investment in the low-carbon energy transition increased by 17% to a new record of $1.77 trillion [2]. Of these, electrified transportation, the largest area of energy transition spending, grew by 36% to $634 billion; investment in the renewable energy sector grew by 8% to $623 billion; and grid investment, the third largest area of energy transition investment, reached $310 billion. Emerging sectors such as hydrogen and Carbon Capture and Storage (CCS) technologies achieved investment growth of 200% and 100% respectively, while energy storage investment grew by 76% (Fig. 1).

## 3 Progress in China's Comprehensive Green and Low-Carbon Transition and Challenges Ahead

### *3.1 High-Quality Economic and Social Development and "Dual Carbon" Efforts Are Synergistically Promoted, and Green and Low-Carbon Industries Have Become a New Driving Force for Economic Growth*

China remains a key engine of global growth. In 2023, China's gross domestic product (GDP) exceeded 126 trillion yuan, reflecting a 5.2% increase from 2022 in constant prices. It ranks among the world's major economies (Fig. 2) and is expected to contribute over 30% to global economic growth.

China's industrial structure is continuously being optimized towards greening and decarbonization. From 2014 to 2023, the share of China's tertiary industry value added in GDP has increased annually. In 2023, the value added by China's

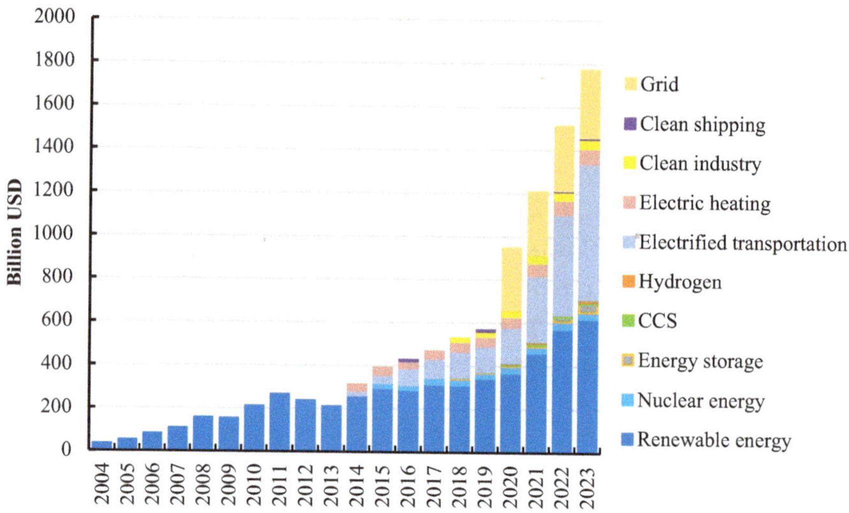


**Fig. 1** Global energy transition investment profile by sector, 2004–2023. *Source* Bloomberg NEF. (Starting years vary by industry, but all industries have statistics starting in 2020, with nuclear data starting in 2015 and grid data starting in 2021)

tertiary industry amounted to 68.8 trillion yuan, a year-on-year growth rate of 5.8%, accounting for 54.6% of the country's total GDP—an increase of 6.3% points compared with 2014 (Fig. 3).

Strategic emerging industries in China are experiencing robust growth, becoming a new force driving industrial transformation and long-term economic stability. Between 2014 and 2023, the share of strategic emerging industries value added in China's GDP increased from 7.6% to over 13% [3].

## 3.2 *The Dual Carbon "1 + N" Policy Framework Has Been Completed and Implemented, and a New Mechanism for the Comprehensive Transition, from Controlling both the Amount and Intensity of Energy Consumption to Controlling both the Amount and Intensity of Carbon Emissions, Is Established*

During the 14th Five-Year Plan period, China proposes to further refine the dual control system for energy consumption and steadily shift to a dual control system for carbon emissions (carbon amount and carbon intensity control).

**Table 1** Policy lineage of dual control of carbon emissions

| Policy documents | Issuance date | Policy content |
|---|---|---|
| *Working guidance for carbon dioxide peaking and carbon neutrality in full and faithful implementation of the new development philosophy* | October 2021 | Strictly control energy consumption and $CO_2$ emission intensity, appropriately control total energy consumption, and establish a system to control the amount of $CO_2$ emissions |
| *An action plan for peaking carbon emissions before 2030* | October 2021 | Implementing a system that based on carbon intensity control, supplemented by carbon amount control and gradually establishing a systematic and comprehensive evaluation and assessment system for carbon peaking and carbon neutrality |
| The central economic work conference | December 2021 | Creating conditions for the early realization of shift from dual control for energy consumption to dual control for carbon emissions |
| The report to 20th National Congress of the Communist Party of China | October 2022 | We will exercise better control over the amount and intensity of energy consumption, particularly of fossil fuels, and transition gradually toward controlling both the amount and intensity of carbon emissions |
| *Guidelines on promoting the gradual transition from duel control for energy consumption to duel control for carbon emissions* | July 2023 | Promoting the shift from dual control for energy consumption to dual control for carbon emissions in a planned and step-by-step manner |
| *Resolution of the central committee of the communist party of China on further deepening reform comprehensively to advance Chinese modernization* | July 2024 | New mechanisms will be put in place to facilitate the transition from controlling the total amount and intensity of energy consumption to controlling the total amount and intensity of carbon emissions |
| *Work plan to accelerate the establishment of a dual control system for carbon emissions* | August 2024 | By 2025, the statistical accounting system for carbon emissions will be further improved. During the 15th Five-Year Plan period, a dual control system for carbon emissions will be implemented with intensity control as the main focus and total amount control as a supplement. A comprehensive evaluation and assessment system for carbon neutrality will be set up for carbon peaks. After the carbon peaks, a dual control system for carbon emissions will be implemented with amount control as the main focus and intensity control as a supplement, and a system for evaluating and assessing the carbon neutrality targets will be set up. Establish a carbon neutrality target evaluation and assessment system |

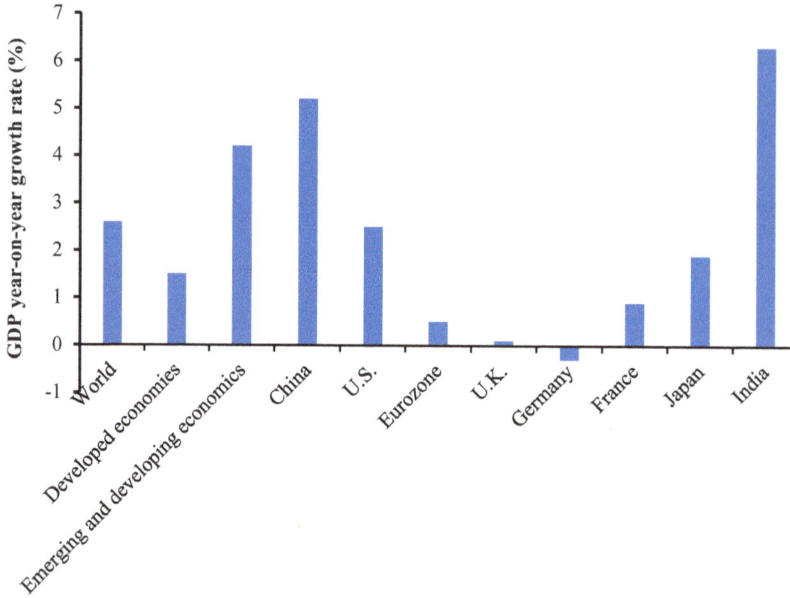


**Fig. 2** Comparison of economic growth rates among major global economies in 2023. *Sources* Global and Indian data from the World Bank report released in January, other data from the National Bureau of Statistics of China, U.S. Department of Commerce, UK Office for National Statistics, Japan's Cabinet Office, Germany's Federal Statistical Office, France's National Institute of Statistics and Economic Studies, and Eurostat

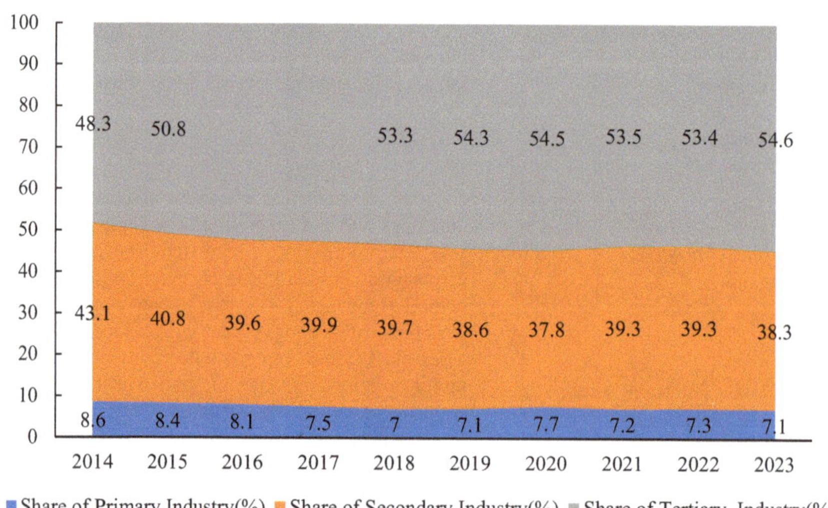


**Fig. 3** Changes in the value added by China's three major industries as a proportion of GDP, 2014–2023. *Source* National Bureau of Statistics of China

## 3.3 *Progress in the Energy Transition Has Been Remarkable, and the Development of a New Power System Is Advancing Steadily*

The proportion of clean energy consumption has increased significantly, accelerating the green and low-carbon transition. According to the Statistical Communique of the People's Republic of China on the 2023 National Economic and Social Development published by the National Bureau of Statistics, China's total energy consumption in 2023 was 5.72 billion tons of standard coal, reflecting a 5.7% increase from the previous year. Among the components of this consumption, coal consumption increased by 5.6%, crude oil consumption by 9.1%, natural gas consumption by 7.2%, and electric power consumption by 6.7%. Over the decade from 2014 to 2023, China's coal consumption share declined from 65.8 to 55.5%, while the share of non-fossil energy consumption rose from 11.3 to 17.7% (Fig. 4).

The development of renewable energy has attracted worldwide attention and is increasingly becoming the mainstay of meeting new electricity demand. By the end of 2023, China's total installed capacity for renewable energy generation is 1.45 billion kilowatts, accounting for more than 50% of the total installed power capacity, historically exceeding thermal power installed capacity. In the first three years of the 14th Five-Year Plan period, China's new grid-connected installed capacity for photovoltaic and wind power grew at an average annual rate of 93.9% and 26.3%, respectively (Fig. 5).

China's power system is evolving into a new type of power system characterized by being clean and low-carbon, safe and controllable, flexible and efficient, open and interactive, and intelligent and convenient. During the first three years of the

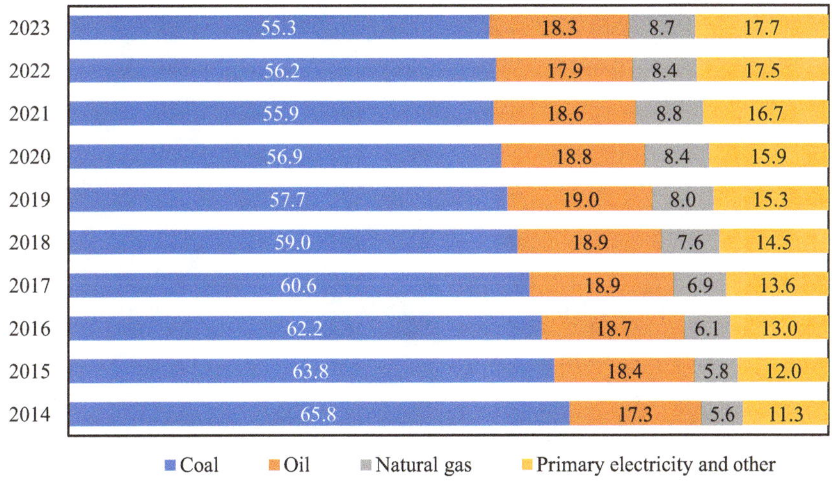


**Fig. 4** Structure of energy consumption in China, 2014–2023 (%). *Source* National Bureau of Statistics of China

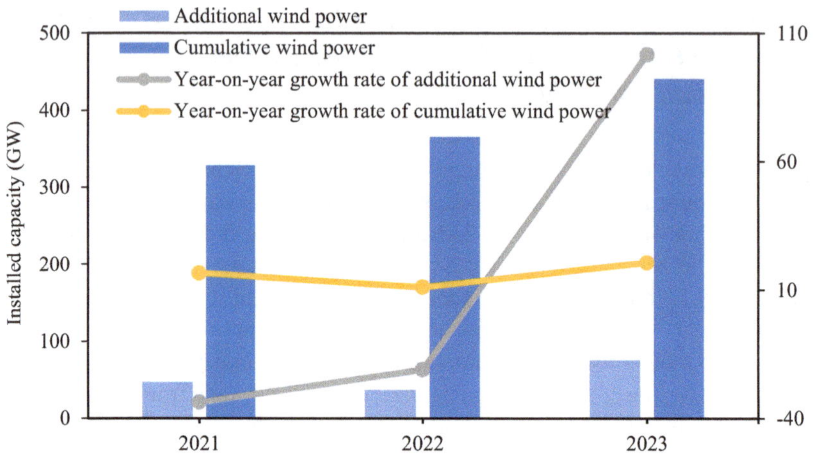


**Fig. 5** New and total installed capacity of grid-connected PV and wind power in China and its Growth Rate, 2021–2023. *Source* National Bureau of Statistics of China

14th Five-Year Plan period, new installed energy storage capacity grew rapidly. In 2023, with a 22.6 million kilowatt increase, representing a more than 260% rise from the end of 2022. (Fig. 6). Second, power grid enterprises play a fundamental and pioneering role in accelerating the development of new power systems, serving major national strategic needs.

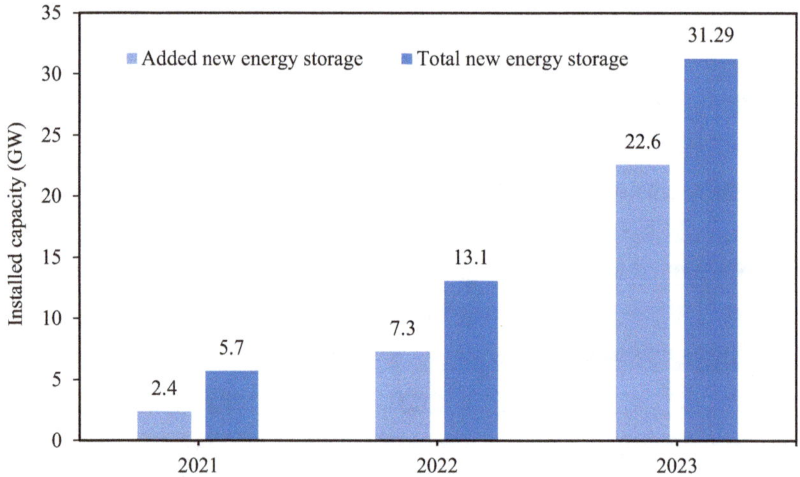


**Fig. 6** New and total installed new energy storage capacity and growth rate in China, 2021–2023. *Source* National Bureau of Statistics of China

## 3.4 *Energy-Saving and Carbon-Reducing Upgrades and Transformation of Industries, Products and Equipment in Key Areas Are Accelerating*

The effectiveness of energy conservation and consumption reduction in the industrial sector is gradually emerging. The green transition in the construction sector has been effective. By the end of 2023, a total of approximately 11.85 billion square meters of green building area had been constructed in urban areas across China, with more than 27,000 projects receiving Green Building Labels. In 2023 alone, approximately 2.07 billion square meters of new green building area was constructed, accounting for 94% of the newly built urban area [4]. The construction of a green and low-carbon transportation system is advancing rapidly. From 2018 to 2023, the proportion of freight transported by rail and waterway increased from 25.2 to 26.3%, while the share transported by road decreased from 74.8 to 73.7%. The upgrade of energy efficiency in energy-using products and equipment is steadily advancing. Data shows that China has over 5 billion units (sets) of major energy-using products and equipment in operation, accounting for approximately 80% of the country's total energy consumption [5].

## 3.5 *Carbon Market Are Improving and Will Play a Greater Role in the Dual Carbon Systems in the Future*

As of December 31, 2023, the cumulative trading volume of carbon emission allowances in the national carbon market reached 442 million tons, with a cumulative transaction value of 24.919 billion yuan.

## 3.6 *A Top-Level Design Document for Methane Emissions Control Issued, Setting Quantitative Targets for Methane Recovery and Utilization in Energy, Agriculture and Waste Sectors*

China emphasizes methane emission reduction under the dual carbon goals. In November 2023, eleven departments, including the Ministry of Ecology and Environment, jointly released the *Action Plan for Methane Emission Control*, which sets quantitative targets for methane recovery and utilization in the energy, agriculture, and waste treatment sectors, summarized in Table 2.

**Table 2** Some quantified targets in the action plan for Methane emission control

| Industry | Target |
|---|---|
| Energy | • By 2025, the annual utilization rate of coal mine gas will reach 6 billion cubic meters<br>• By 2030, the collection rate of oilfield associated gas will reach the international advanced level |
| Agriculture | • By 2025, the comprehensive utilization rate of livestock and poultry excrement will reach 80% or more, and by 2030, it will reach 85% or more |
| Waste and sewage treatment | • By 2025, the resource utilization rate of urban life waste in China will reach around 60%<br>• By 2025, the rate of safe disposal of urban sludge will reach 90% or more |

## *3.7 Strengthening Synergies with Biodiversity Conservation Efforts and Steady Increasing Ecosystem Carbon Sinks*

In April 2023, China completed the comprehensive delineation of the national ecological protection red line, encompassing the vast majority of important natural ecosystems. In the same month, the *Implementation Plan for Consolidation and Enhancement of Ecosystem Carbon Sequestration* was released, outlining the main objectives and key tasks for strengthening ecosystem carbon sinks by 2025 and 2030.Moreover, China has been a strong advocate for supporting developing countries in protecting and restoring biodiversity. At COP15, China committed to take the lead in contributing 1.5 billion yuan to establish the Kunming Biodiversity Fund and formally signed the cooperation agreement for the Kunming Biodiversity Fund in May 2024.

## *3.8 China's Green and Low-Carbon Transition Process Still Faces Multiple Challenges and Needs to Unlock Markets, Technologies and Policies Potential*

**The targets for energy consumption and carbon emissions in the 14th Five-Year Plan have not been accomplished as effectively as those in the 13th Five-Year Plan**. In the coming period, China's total energy consumption will continue to grow, posing challenges in meeting the binding targets of the 14th Five-Year Plan to reduce national energy consumption and carbon emissions intensity.

**Energy and electricity consumption remain in a phase of rigid growth**. In 2023, China's per capita primary energy consumption was 4.09 tons of standard coal, about one-third of Canada's, two-fifths of the U.S.'s, and 87.4% of German's per capita primary energy consumption, indicating potential for further energy consumption growth [6]. In 2023, China's energy and electricity consumption elasticity coefficients exceeded 1 (Fig. 7).

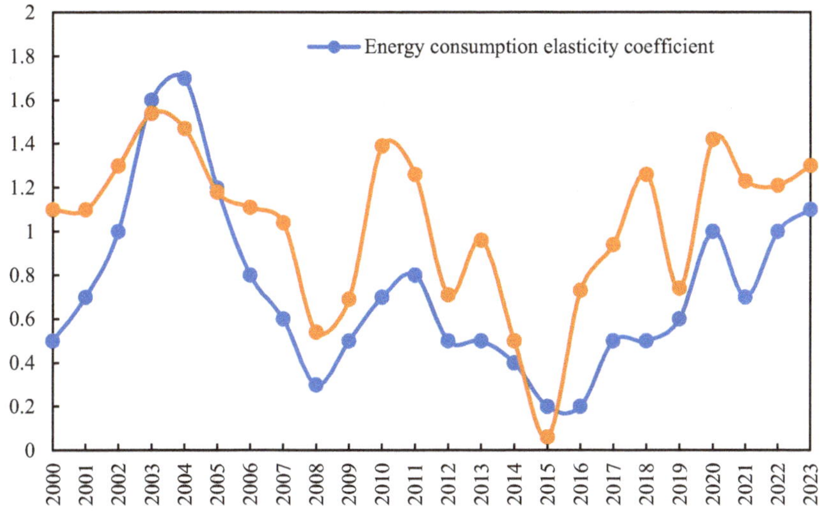


**Fig. 7** Trends in China's energy consumption and electricity consumption elasticity coefficients, 2000–2023. *Source* National Bureau of Statistics of China

**Renewable energy consumption will become the focal point of the low-carbon transition in the coming period.** China's the growth rate of renewable energy generation is much slower than the growth rate of installed capacity. Between 2015 and 2023, China's installed renewable energy capacity as a share of total installed capacity rose from about 30 to 50%, an increase of approximately 20% points over the decade. Meanwhile, the share of renewable energy generation in total electricity consumption increased from about 20% to about 30% (Fig. 8).

**The capacity for basic research and development needs to be strengthened, and there is an inadequate stock of key low-carbon, zero-carbon, and carbon-negative technologies.** Institutional mechanisms have hindered the transformation of scientific and technological achievements, and innovation factors have yet to be efficiently configured.

**The institutional system to support carbon peaking and post-peaking emissions needs further improvement.** First, a basic legal system to support carbon peaking and carbon neutrality is lacking. Second, carbon emissions trading has not yet been effectively linked to the overall goal of controlling total carbon emissions. Third, as the new energy industry enters a new stage of development, the financial and tax policy system needs to be optimized and improved.

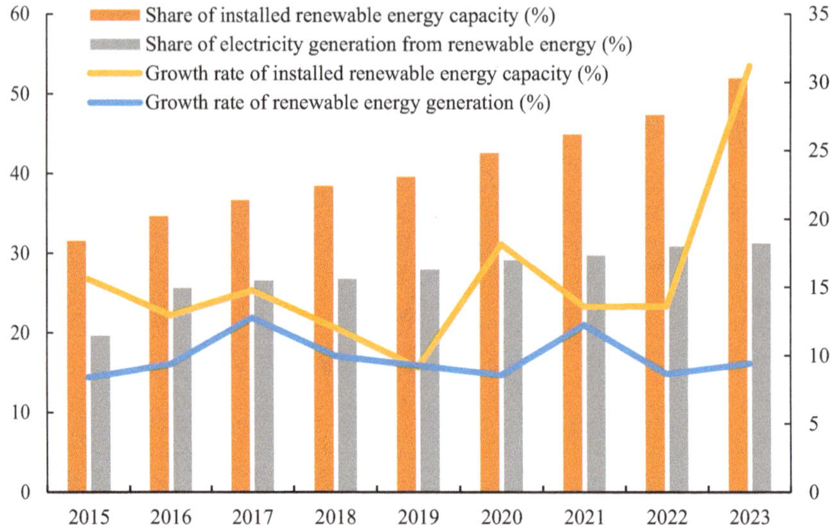


**Fig. 8** Renewable energy installed capacity and electricity generation in China, 2015–2023. *Source* National Energy Administration of China, IRENA

# 4 China's Contribution and Obstacles to Promoting a Global Green and Low-Carbon Transition

## *4.1 China's Clean Energy Products Contribute Significantly to Lowering Global Cost of Clean Energy Use, Accelerating Low-Carbon Technologies Deployment and Improving Energy Accessibility*

The positive externalities of China's green technology and products are evident, with continuous breakthroughs in clean energy technology innovation providing a critical boost for the global green and low-carbon transition. Over the past decade, the global installed capacities of photovoltaic, onshore wind, and offshore wind have grown, while the weighted average levelized cost of electricity (LCOE) has been decreasing (Fig. 9).

China's progress in international cooperation on renewable energy is closely linked to the high-quality advancement of the Belt and Road Initiative. In September 2021, China's President Xi Jinping stated at the General Debate of the 76th Session of the UN General Assembly that China will vigorously support the green and low-carbon energy development of developing countries and will no longer build new overseas coal-fired power projects. In terms of investment scale, major Chinese companies invested a total of $31.1 billion in overseas renewable energy projects between 2015 and 2023 [7].

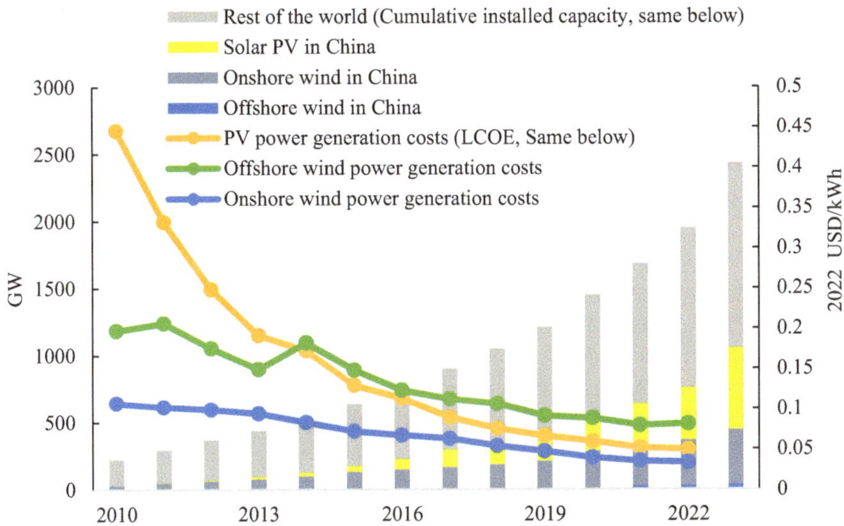


**Fig. 9** Trends in global installed capacity and power generation costs of solar PV and wind power, 2010–2023. *Source* International Renewable Energy Agency

## 4.2 China Actively Engages in the International Sustainable Trading System and both Participates in and Leads the Development of a Global Green Value Chain for Soft Commodities

China is actively involved in accelerating the green transformation of the soft commodities supply chain. At the strategic level, in October 2021, China and the European Union held the second China-EU High-Level Environment and Climate Dialogue, issuing a joint communiqué that highlighted the consensus on strengthening cooperation in areas such as forest resource protection, sustainable supply chains, and combating illegal logging and related trade. China also signed the *Glasgow Leaders' Declaration on Forests and Land Use* at COP26, committing to halting and reversing global forest loss and land degradation by 2030. Furthermore, China is also actively promoting green supply chain cooperation under the Belt and Road Initiative.

## *4.3 Challenges in International Cooperation on the Green Low-Carbon Transition*

### 4.3.1 Unilateral Trade Measures, Tariffs and Non-tariff Barriers Increase Uncertainty for the Global Green Low-Carbon Transition

Since the development of CBAM in the EU, trade and climate issues have become increasingly intertwined. On May 14, 2024, the White House announced that the Biden administration would continue to retain tariffs imposed by the previous administration and introduce additional tariffs on other Chinese goods. Notably, import tariffs on Chinese electric cars are set to surge from 25% to 100%. In addition, on December 18, 2023, the UK government officially announced that it would implement its own CBAM from 2027 onwards. Countries like Japan and Canada are also conducting CBAM-related research and discussions and are likely to introduce similar policies in the coming years. Non-tariff barriers are also on the rise. For lithium-ion batteries and photovoltaic products, the *Regulation Concerning Batteries and Waste Batteries* of EU also came into force on August 17, 2023.

### 4.3.2 Developing Countries Face Tight Fiscal Constraints and High Transition Costs in Meeting Climate and Sustainable Development Goals

Developing countries have difficulties in securing sufficient external financing to stimulate investment and growth, address climate-related risks, and accelerate the achievement of sustainable development goals. International investment in renewable energy has nearly tripled since the adoption of the *Paris Agreement* in 2015, but most of this investment is concentrated in developed countries. Furthermore, higher country risk ratings in developing countries make them less attractive to international private investment.

### 4.3.3 Challenges in Equitably Distributing Benefits from Global Green Value Chains

The greening of global value chains promises broader environmental and social benefits, but challenges remain in ensuring an equitable distribution of economic gains among and within countries. Developed countries, due to their early layout of green industries, occupy a relatively high position in the value chain with their technological advantages and capital strength, such as setting technical standards and producing core equipment. While developing and least developed countries are mostly limited to the low end of the value chain, such as the provision of raw materials and primary processing, making it difficult for them to capture corresponding high-value-added gains.

# 5 Pathways to Gender Mainstreaming in Climate Change Analysis

## *5.1 Importance of Gender Issues in Addressing Climate Change and Global Climate Governance*

Gender equality is a fundamental human right and an important prerequisite for realizing the full potential of humankind, advancing sustainable development and ultimately achieving peaceful societies. Unfortunately, women and their unique perspectives are frequently marginalized in public and decision-making forums. Therefore, it is essential to fully consider women's participation to ensure their perspectives are included.

## *5.2 Gender Issues in Global Climate Governance Continue to be Highlighted, and Multilateral Cooperation Needs to be Further Promoted to Ensure Women's Equal Rights and Interests in Climate Issues*

Promoting gender equality within multilateral frameworks for climate governance is critical in several ways: (1) international conferences such as the United Nations Climate Change Conference, should ensure proportional representation of women and enable them to increase their voice and influence in the decision-making process. (2) financing support and capacity-building provided by multilateral international organizations, can promote the integration of gender equality with climate change and energy transition issues in developing countries and poor regions. (3) it is possible to provide funding to support women's participation in climate action and access to relevant technical training through the establishment of a gender fund in the climate sector, which is dedicated to supporting gender equality and climate change response. Alternatively, existing climate funds could increase their allocations to gender equality projects. (4) Multilateral platforms can promote transnational and cross-regional technical cooperation to help developing countries enhance women's capacities and skills in the field of new energy and green technology. International training programs could also be established to upgrade women's technological and managerial capacities in climate change-related fields.

### *5.3 Promoting a Just Domestic Transition and Enhancing Green Jobs for Women in the Transition*

Currently, women are overrepresented in industries that are more susceptible to the impacts of the green transition. Therefore, when the coal industry is facing a transition, female employees are more vulnerable to the risks associated with the transition [8]. Specific recommendations include: First, providing retraining and skills upgrading opportunities to help women workers adapt to new green technologies and job requirements. Second, strengthening social security to provide better protection for female workers affected by the transition, to ensure their stability during the transition period. Third, creating appropriate policy mechanisms to encourage women's participation in emerging green sectors.

### *5.4 Promoting Gender Equality in the Green and Just Transition of Developing Countries, and Utilizing China's Leading Role in Global Climate Governance*

In recent years, developing countries have begun to recognize the importance of gender equality in energy and climate change efforts. Key practices and progress include: First, developing countries have successively formulated and introduced policies and legal frameworks related to gender equality. Second, providing education and training resources related to green industries for women in developing countries and impoverished regions; offering financial, technical, and market access support for women entrepreneurs.

Cross-border investment and assistance through bilateral countries must consider the comprehensive economic, social and environmental impact on recipient countries, particularly the vulnerability of local women. Specific recommendations include: First, developing standards and guidelines for overseas investment and assistance that incorporate gender equality considerations. Second, considering the integration of gender issues into the information disclosure processes of overseas investment projects through a self-regulatory convention on overseas investment. Third, providing women with training and technical support which is tailored to industrial and cultural contexts, and ensuring a significant proportion of female participation. Fourth, promoting knowledge sharing and gender mainstreaming through international cooperation, encouraging more stakeholders to address gender equality issues.

# 6 Policy Recommendations

## 6.1 *Taking the NDC Update as an Opportunity to Exercise Climate Leadership and Combine Ambitious Goals with Concrete Actions*

The rise of increasingly affordable and reliable non-fossil energy sources can meet new electricity demand and gradually replace existing fossil energy stocks. A strong, rapid push to generate power using non-fossil energy will be a decisive factor in shaping China's overall $CO_2$ emission trajectory. Specific recommendations include: **First**, proposing an absolute reduction target for $CO_2$ emissions by 2035. **Second**, updating NDC to reflect the reduction efforts across all greenhouse gases. **Third**, proposing a target of installed capacity of non-water renewable energy by 2035, contributing to achieving the goal of "tripling global renewable energy capacity by 2030". **Fourth**, proposing a target for the installed capacity of energy storage or new energy storage by 2035, ensuring the security, stability, and sustainability of the energy supply. **Fifth**, proposing infrastructure construction targets such as charging, replacing and hydrogenation by 2035 to provide guarantee for new energy vehicles to become the mainstream of new sales vehicles. **Sixth**, balancing ambitious goals and practical actions.

## 6.2 *Promoting the Consumption of Wind/PV Power Based on Local Conditions, and Supporting more Low-Carbon Technology Innovation and Industrialization through a Combination of Measures such as Green Industry Classification, Market Mechanisms, and Subsidies*

**First**, enhancing the research and development, investment, and innovations in business models and institutional mechanisms to address demand-side challenges. **Second**, implementing a classification of green industries. **Third**, deploying a new set of industrial policies for emerging low-carbon technologies. **Fourth**, placing emphasis on aligning new industrial policies with existing ones, converting revenue from carbon emissions trading and energy consumption as well as carbon emission control policies into competitive subsidies for emerging green and low-carbon industries. **Fifth**, linking assessments of carbon emission indicators at the provincial and municipal levels to transfer payments.

### 6.3 *Facilitating the Implementation of the "Dual Control" Policy for Carbon Emissions and Constantly Improving the "Dual Carbon" Policy System*

**First**, designing a system improvement roadmap within the total carbon emissions framework, integrating it with the NDC update. **Second**, improving the system of laws and regulations related to dual carbon efforts. **Third**, further advancing carbon market reforms, including promoting the integration of total carbon emissions with the carbon market. **Fourth**, the research of and response to the Carbon Border Adjustment Mechanism (CBAM) should be coordinated with the development of China's carbon market. This involves assessing CBAM's impact on key industries, aligning domestic carbon market rules with fair and reasonable international standard, and exploring mechanisms for cross-border linkage with the global carbon market.

### 6.4 *Promoting the Localization and Internationalization of the Industry Chain and Supply Chain for the "New Three" Products, and Actively Supporting and Promoting the Liberalization and Facilitation of Global Trade and Investment*

**First**, establishing an inter-departmental working group to study and advance the coordination between climate and trade policies. **Second**, balancing the domestic and international markets and effectively managing the domestic and international circulations. **Third**, internationalizing the emerging low-carbon industry chain and supply chain, actively participating in the deployment of global industry chain, and ensuring the safety and sustainability of the green and low-carbon supply chain globally. **Fourth**, facilitating free trade in low-carbon products and technologies. Existing multilateral mechanisms should be utilized to establish exchange platforms for investment and trade in low-carbon industries, fostering dialogue and seeking mutually acceptable solutions.

### 6.5 *Actively Participating in and Contributing to the Reform of the International Financial Architecture, and Incorporating Climate into Export Credit Policies and Overseas Investment and Financing Decisions*

**First**, climate finance negotiations should adhere to the UNFCCC and its *Paris Agreement* frameworks, with developed countries bearing primary responsibility for financing, to advance productive outcomes in New Collective Quantified Goal on

Climate Finance (NCQG) negotiations. **Second**, focusing on actively participating in discussions and practices related to the reform of the international financial architecture. **Third**, collaborating with the international community to explore governance models that incorporate multiple funding sources such as aid, philanthropy, and commercialization; enhancing fund utilization capabilities and developing feasible and sustainable business models for various energy transition scenarios and climate solutions. **Fourth**, integrating climate into export credit policies and decisions on overseas investment and financing and strengthening climate information disclosure requirements for overseas projects. **Fifth**, enhancing enterprise capabilities to assess climate risks, conducting climate stress tests on the balance sheets of overseas projects, and integrating climate change risks into credit ratings and insurance designs. **Sixth**, financial support and transparency for South-South cooperation on climate change should be reinforced.

## References


1. Helveston, J. P., He, G., & Davidson, M. R. (2022). Quantifying the cost savings of global solar photovoltaic supply chains. *Nature, 612*(7938), 83–87.
2. Bloomberg N. E. F. Global Clean Energy Investment Jumps 17%, Hits $1.8 Trillion in 2023[EB/OL]. (2024-01-30) [2024-08-15]. https://about.bnef.com/blog/global-clean-energy-investment-jumps-17-hits-1-8-trillion-in-2023-according-to-bloombergnef-report/.
3. National Bureau of Statistics. Statistical Communique of the People's Republic of China on the 2023 National Economic and Social Development. (2024-02-29) [2024-08-15]. https://www.gov.cn/lianbo/bumen/202402/content_6934935.htm.
4. China Construction News. Adhere to the people-centered philosophy of development and promote the high-quality development of green building. (2024-05-22) [2024-08-15]. https://www.mohurd.gov.cn/xinwen/jsyw/202405/20240522_778125.html.
5. National Development and Reform Commission. Responsible comrades of National Development and Reform Commission on the "Advanced Level, Energy-Saving Level and Access Level of Energy Efficiency of Key Energy-Using Products and Equipment (2024 Version)" to answer the reporter's question. (2024-02-09) [2024-08-15]. https://www.gov.cn/zhengce/202402/content_6951571.htm.
6. Our World in Data. Energy use per person. [2024-08-15]. https://ourworldindata.org/grapher/per-capita-energy-use?tab=chart.
7. China|BloombergNEF|Bloomberg Finance LP. [2024-08-16]. https://about.bnef.com/china/.
8. Institute of Energy of Peking University. Toward a Future of Just Transition: The Impact of China's Green Transition on Employment. (2023-06) [2024-08-16]. https://www.undp.org/sites/g/files/zskgke326/files/2023–08/paiban9gao0807_cn_zouxianggongzhengzhuanxingdeweilaizhongguolusezhuanxingduijiuyedeyingxiang.pdf.

# Chapter 2
# Sustainable Oceans Management Under the Vision of Carbon Neutrality

## 1 Framing the Issues

### *1.1 Global and Chinese Context*

The ocean is constantly being explored for new uses of its space and resources, leading to a steady increase in the economic value it provides. Existing and potential new economic activities related to oceans, seas, and coasts—the so-called ocean economy, or blue economy—thus cover a wide range of interlinked established and emerging sectors. The value of the global ocean economy is an estimated US$2.5trn annually, equivalent to the size of the world's seventh-largest economy and China's gross ocean product has been estimated to RMB 3.41 trillion in 2021, accounting for 8% of China's overall Gross Domestic Product (GDP). The long-term potential for innovation, employment creation and economic growth offered by the ocean economies in the future is impressive. It must be pointed out that the ocean economy is highly dependent on the marine spatial environment and resource endowment. Essentially, its economic value is determined by the value of the products and services provided by the marine space and its ecosystem.

### *1.2 Aiming for a Sustainable Blue Economy*

Sustainable development is important to maintain long-term economic development and social well-being globally. **To achieve a more ambitious transformation, Sustainable Blue Economy (SBE) should be included in the top-level policy**

---

The content of this chapter is extracted from the main findings of the 2023–2024 CCICED Special Policy Study (SPS) on "Sustainable Oceans Management under the Vision of Carbon Neutrality".



*Adhering to Green and Open Cooperation to Advance Modernization*,
https://doi.org/10.1007/978-981-95-5547-5_2

**framing with clear definitions and principles, as well as incorporated into the next 5-year plan to drive the necessary changes.**

The World Bank's definition of the blue economy is the "sustainable use of ocean resources for economic growth, improved livelihoods and jobs, and ocean ecosystem health." With the concept of SBE being widely disseminated globally, there is a growing international consensus on the development of an SBE. As blue economy being gradually incorporated into regional development strategy frameworks, some organizations have already proposed guidelines for SBE development, including guiding principles, focus and priority areas, initiatives, and recommendations.

### *1.3 Equity*

Gender equality is recognized as essential for the effective protection of oceans, the sustainable management of ocean and marine resources, and the accomplishment of the UN Sustainable Development Goals (SDGs).

Nevertheless, there remains a stark gap between men and women's participation in the blue economy, both in terms of numbers and the types of work. In line with this, special attention was given to the issue of gender equality for ocean sustainability at the 2024 UN Ocean Decade Conference, which put forward recommendations and call for actions to reduce the gender gap, provide educational opportunities for young female scientists, and acknowledge the contributions of women in marine conservation efforts.[1]

## 2 Ocean Economy and Blue Finance

### *2.1 Introduction*

The ocean is a treasure trove of resources and strategic space that supports future development, and the blue economy is an important engine for economic growth in coastal areas. After years of efforts, China's marine economy has achieved significant results in transitioning towards a quality- and efficiency-oriented approach: the construction of marine ecological civilization has been accelerated, and breakthrough results in marine scientific and technological innovation are achieved. Meanwhile, the development of blue economy also faces many challenges: decline in coastal resources and environmental carrying capacity, poor nearshore environmental quality, degradation of marine and so on.

[1] https://www.unesco.org/en/articles/gender-equality-ocean-sustainability

## 2.2 Challenges and Opportunities

### 2.2.1 Identifying Gaps in Chinese Frameworks and Policy Directions

In recent years, the concept of the "blue economy" has been widely discussed both at home and abroad and has gained increasing consensus. At the global level, the value and strategic significance of the blue economy have gradually attracted great attention from various countries and have been put on the international policy agenda. China attaches great importance to the development of the blue economy, it puts forward the "21st Century Maritime Silk Road" initiative and strengthens the "Blue Partnership". However, the development of the blue economy in China is still immature.

**There are various versions of the term of "sustainable blue economy"**. Official documents or reports use "ocean economy", while the blue economy of SBE mostly appears in provincial events or occasions held by institutes attached to ministries, usually referring to sustainable, modern and/or high-tech ocean economy. However, there is no consensus on the definition of the term "blue economy" in China. At the same time, there is still a lack of sufficient discussion on the boundary of the blue economy, which directly leads to the fact that the terms "blue economy", "green economy" and "ocean economy" are often confused. The first is to equate the "blue economy" with the "ocean economy", ignoring the environmental sustainability, equity and inclusiveness, and climate resilience that the blue economy emphasizes. Secondly, most people think that the "blue economy" is the "green economy", failing to recognize that the blue economy can achieve sustainable development through economic growth while promoting low-carbon, environmental protection and sustainable use of resources.

**China has not yet formulated a national or provincial-wide systematic strategic plan for the development of the blue economy**. At the international level, an SBE aims to promote low-carbon, environmentally friendly, and sustainable use of resources while also striving to protect and enhance the livelihoods of residents. However, China's major policies related to the blue economy emphasize more on the synergy between marine ecological protection and marine economic development, overlooking the protection and improvement of the population's livelihoods, including that of women and other marginalized groups, and lacking alignment with the global conversation on the blue economy.

**There are yet not enough policy instruments or pragmatic methodologies for relevant entities to practice SBE**. There should be a relatively clear definition of SBE—usually several principles, clear instruments, or screening tools (e.g. taxonomy) on what industries or practices are being considered as SBE should be provided. These tools can guide development departments to prepare policies to promote, central banks to design incentives, financial supervisory administrations to regulate the financial institutes, and statistical administrations to calculate the volume of SBE. Investors interested in SBE should have appropriate methodologies to identify them. These can be sustainable projects ready to be funded, screening tools on specific sub-industries in detail, and/or training tools that can build their capacity. Therefore, it is crucial to develop toolkits to assess the economic activities of the ocean. Conservation organizations and research institutes, both international and domestic, will be the key users.

**Another gap is a shared goal for the SBE**. China is renowned globally for its carbon peak and carbon neutrality goal for its climate agenda. This common goal propels all related administrations to prepare policies accordingly and enables enterprises to "do the right things". The indicator—carbon—is easy to break down and thus acts as a currency in the climate mitigation realm. Obviously, there is a gap between the ocean economy today and a common goal. Fortunately, several attempts at ocean accounting can somewhat serve as baselines. In the future, relatively simple and clear indicators and their measurements will be required.

**The incentive level for local governments or entities to develop an SBE is still low**. Currently, several provinces and cities are building their ocean pilot zones, each taking a specific sector to incubate their industrial expertise. In January 2024, Weihai City formulated and issued China's first catalogue of sustainable investment and financing support for blue industries. This initiative is an innovative sample of Weihai's development of the blue economy. However, more resources should be mobilized to match these policies.

**Promoting a sustainable economy related to the ocean is an important contribution to combatting climate change**. Thanks to the combination of policy and incentive to reduce carbon emissions, China has smashed the price of renewable energy and enabled small and medium-sized enterprises (SMEs) and even individuals to benefit from this movement. Similarly, policies to allow these groups to practice an SBE, subsidies for favourable activities to encourage innovation, organization of industrial clusters like industry parks, provision of appropriate capacity building will all help develop the SBE. Compared to climate, the SBE is equally, if not more, related to conservation in the field. Therefore, innovative solutions for mobilizing resources should also be further sought. These incentives should be built on the progress of the points mentioned above.

### 2.2.2 Governance System on Ocean Economic Sectors and Conservation Matters

Currently, at the national level, China has yet to distinguish between sustainable and unsustainable industries or products in its marine economic statistics. This status quo may not be conducive to integrating the allocation of resources. The lack of a leading governmental department may undermine the progress of SBE development. Existing marine legislation remains fragmented, showing a tendency towards fragmentation and sectionalization which poses an obstacle to the development of an SBE.

### 2.2.3 Inclusiveness and Livelihood of Communities During the Sustainable Blue Economy Transition Process

***Inclusiveness from industry perspective***

Small-scale (in some cases artisanal) fisheries are still the ocean's largest employer. Hence any policy or collective action needs to ensure that their rights and traditional uses are protected as part of an equitable blue economy.

Throughout the entire fishing industry supply chain, women's contributions are significant. Despite their significant role in ocean economy sectors, women often hold low-ranking, underpaid, or informal positions, instead of managerial roles and struggle to access resources. Thus, sustainable blue economy related policies should be inclusive to value women's contributions and recognize their distinct roles to assure their equitable livelihood opportunities.

***Engagement for broader public participation***

The successful implementation and sustainability of an SBE greatly depend on public participation and acceptance. The Department of Marine Ecology and Environment under the Ministry of Ecology and Environment (MEE) launched an online platform and an app to encourage public involvement in marine environmental protection and to gather coastal residents' views on coastal development and conservation. These platforms can serve as educational tools, as well as leveraging the collective community wisdom to inform policy making for developing sustainable blue economy at both national and local level.

### 2.2.4 Measurement and Accounting of Ocean's Value

Currently, the valuation of the ocean economy worldwide focuses mainly on ocean industries, whereas the value assessment of marine ecosystem services remains at a relatively early stage and is still defining specific approaches and parameters. In addition, ocean-specific data is often aggregated with terrestrial data (for example combining marine and freshwater fisheries), making it more difficult to build understanding of ocean health and economic performance. While China has established a relatively sophisticated statistical investigation and accounting system for its ocean industries, gaps remain in marine ecosystem accounting and more can be done to improve it. It is important to note that especially for deep ocean ecosystems, the current knowledge gaps remain too significant and thus impede the potential measuring or accounting of the values associated with these ecosystems. This further highlights the importance of the continuous deep-sea mapping and research efforts.

### 2.2.5 Marine Spatial Planning and Ecosystem-Based Management

Two key guidelines govern China's marine spaces usage: the national plan for functional zoning of major marine spaces and the regulation on coastal protection, utilization, and management. These guidelines establish a policy framework for allocating uses in different marine spaces based on three functions: industrial use and urbanization, agriculture and aquaculture, and ecosystem services provision. Coastal and marine spaces are then classified into four types, each with different natural capital values: optimized development zone, key development zone, restricted development zone, and prohibited development zone.

Recognizing the true value of natural capital and ecosystem services across all coastal and marine spaces is essential for reflecting their worth and maximizing their role in all types of development, including urban and industrial areas. This can also promote sustainable development by facilitating the integration of multiple sectors in shared spaces. Furthermore, the realization of an SBE necessitates the integration of marine and coastal management through ecosystem-based approaches.

### 2.2.6 Guidance to Developing a Blue Finance Mechanism Driving Towards Ocean Sustainability

China's current industry policies lack a five-year plan and explicit favourable policy signals for the advancement of the blue economy.

At the financing policy level, first, there are no technical standards or taxonomy at the national level to define and identify blue industries and activities. This not only makes it difficult to comprehensively cover all areas of sustainable marine development but also brings a higher risk of assessment bias to project investment and financing activities, which restricts the large-scale development of blue finance in China. Second, the lack of blue information disclosure makes the capital market lack of basis for assessing and ranking the environmental benefits or green performance of enterprises, which in turn restricts the enthusiasm of financial institutions to invest in blue enterprises and blue projects. Third, the de-risking mechanisms for blue financing and investment needs to be designed to attract the private sector, as part of a wider supportive enabling environment. Fourth, China currently faces a lack of incentive policies and mechanisms for financial institutions to support SBE activities. Specifically, financial institutions have yet to integrate blue economy activities into their marine protected area (MPA) assessments. Moreover, local governments lack incentive mechanisms for enterprises and financial institutions that endorse SBE projects.

### 2.2.7 International Collaboration of Sustainable Blue Economy Development

The global ocean covers over two thirds of the Earth's surface and by definition will require international collaboration to manage the effects of growing use and reduce conflict. China should champion global collaboration in managing ocean resources and encourage the worldwide adoption of the SBE. One aspect that has received attention is the role of international collaboration in helping to fund the transition to more sustainable ocean use, and particular public goods for this transition (e.g. development and implementation of conservation rules, cleanups, etc.). It is also important to join the agendas of SBE, climate actions and biodiversity conservation in partnership with the global community to meet the targets outlined in the Paris Agreement, the Convention on Biological Diversity (CBD), and Agenda 2030.

To further global ocean governance and the development of an SBE, China should maintain its proactive participation in international negotiations, promote greater international cooperation and collaboration in blue finance, advance international cooperation on maritime governance, advocate for the concept of ecological civilization, and support sustainable marine resource management and global climate change mitigation efforts.

## *2.3 Recommendations*

- and adopt SBE definitions and principles in China in the 15th Five Year Plan and develop policy drivers around it, including integrated governance across all levels and SBE transition planning. The agenda should be complementary to protecting, restoring and sustainably managing marine and coastal ecosystems as the bedrock of a sustainable blue economy, and therefore benefiting the long-term social, economic and environmental resilience.
- guidance must be given as to defining the blue finance taxonomy of available instruments and investment opportunities, and creating the enabling policy environment with incentives and disincentives to attract new and sustainable forms of finance for sustainable blue economy initiatives. This should be considered both for national purposes and for international cooperation programmes.
- of the expanding industries in the blue economy, such as coastal infrastructure, energy, shipping, blue foods, need to tie environmental sustainability considerations into their finance approval and monitoring strategies, establishing specific indicators to have a baseline, and tracking mechanism.

## 3 Industry Design and Transition

### *3.1 Introduction*

A key direction for China's future prosperity and the ongoing value of the ocean is the transition to sustainability of sectors of the blue economy, as well as the acceleration of low-carbon development of ocean industries. Many of these solutions are already well aligned with China's five-year plan for the ocean economy. It is important to understand the environmental impacts of different sectors across the SBE. It is essential in this context to promote an understanding of ecosystem services, to make industrial and commercial circles aware of the environmental impacts of their operations, and to build science-based targets for impact reduction and sustainable development across different sectors.

### *3.2 Ocean Renewable Energy*

#### 3.2.1 Introduction

Each Ocean Renewable Energy (ORE) technologies (wind, wave, current, tidal range, ocean thermal) are under consideration in different stages of development and will present its own unique challenges. China, is stepping up its push into offshore renewable energy, proposing higher green power consumption targets. ORE is a fast-growing ocean economy which is advancing the goals of low-carbon and circular economy. Enhancing knowledge of the ORE technologies potential impacts is crucial to inform future growth plans and inform effectively licensing for ORE activities. ORE technologies offer opportunities for China to develop a new industry, create jobs and take advantage of opportunities within its competency to global markets.

#### 3.2.2 Challenges and Opportunities

Challenges appearing as the ocean energy industry are completely high-tech and high-risk industry sectors, which need to overcome not only its natural risks and difficulties caused by the strong and tough marine circumstances, but also the new challenges which are brought by the industry transition.

Fortunately, the major challenges have already been recognized and new research works are just ongoing conducted by the industrial community to try to track and find the best solutions. The challenges which the international industrial community has identified and listed as the following:

- green transition of traditional offshore oil & gas industry: enhancing the electrification level for offshore oil & gas platform and large-scale connection with offshore green power
- cost increase due to the green power alternation
- complete change of technical infrastructure in offshore oil & gas sector
- lack of engineering facilities in this new direction
- offshore wind industry stepped into the deep and remote frontier: quick development of offshore floating wind turbine technologies and fast growth of floating wind farms
- increase of the water depth causes the rapid growth of installation costs for offshore wind turbines
- long-distance transportation makes it very difficult during the O&M process in deep and remote areas
- ability and reliability for offshore wind turbine in deep and remote areas
- cultivation and acceleration to the promising ocean energy industry: apparently improving the TRL (Technical Readiness Level) of offshore solar, wave/tidal current energy and natural gas hydrates
- limited demonstration scale of wave/tidal current energy
- cost and relative low reliability
- obstacles in technology exploration for new industry due to the lack of confidence and commercial perspective
- the industrial level production of offshore green hydrogen in generation scale and cost: significantly reducing the LCOE (Levelized Cost of Energy) for offshore renewable energy and creating the competitiveness of green hydrogen from offshore renewable energy
- understanding for the usage of green hydrogen and green fuels
- safety and stability control in hydrogen generation
- storage and transportation technologies

### 3.2.3 Recommendations

The ocean energy industry in China has just entered a very quick growth stage but is still under the old version of the industry design. The industry transition and new design seems very necessary to improve the current industrial level and enhance the economic output and contribution to blue economy during the following five or ten years, so the policy will be very sensitive and significant to ensure the industry stays in the right pathway to the industrial transition and shaping the new industry.

Based on the latest research and conclusive summary, we propose the following policy recommendations, which can act as research finding and guidance support to promote the SBE towards carbon neutrality:

- strengthen the top-level design of ocean energy industry, including a national strategic plan, comprehensive/integrating resources mapping, industry promotion action plan and financial support policy, to construct the infrastructure of ocean energy industry design/transition.

- enable the multiple-scenario utilization or demonstration of ocean energy, especially focus on the new promising industries like offshore photovoltaic, wave/tidal energy and offshore green hydrogen, to apparently make the contribution to SBE easily calculated from the ocean energy industry.
- enhance the leading effects of research & innovation & integrated ocean management in the process of making ocean energy the future major contributor to the Chinese/global carbon neutrality.
- enlarge the collaboration scope with the international academic or industrial partners to integrate the global knowledge and research outputs.

## *3.3 Deep-Sea Mining*

### 3.3.1 Introduction

Deep-sea mining is a complex process, it also has a wide range of environmental and social impacts. Environmentally, deep-sea mining can cause irreversible damage to seabed ecosystems, generate sediment plumes and chemical contamination with metals, and affect marine biodiversity and ecosystem function. Socially, deep-sea mining activities can raise numerous social and ethical issues, such as impacts on indigenous communities, inequitable access to benefits for women and other marginalised groups, and environmental justice. This further highlights the importance of the continuous deep-sea mapping and research efforts.

### 3.3.2 Challenges and Opportunities

***Calculation of carbon emissions from deep-sea mining***

Carbon accounting for deep-sea mining faces a variety of challenges, including data acquisition and accuracy issues, complexity of accounting methods and technical and policy barriers. In terms of data acquisition, the difficulty, cost and lack of data precision directly affect the accuracy of the accounting results. The accounting boundaries are not unified, and there is a lack of unified standards and norms, making it difficult to compare and synthesize the results of different studies.

Currently, there is very limited discussion of the potential carbon footprint and emissions from deep-sea mining processes, focusing on sediment plumes. According to existing research, it can be found that the potential carbon emissions from the mining process might have a small impact range and are in the controllable range.

However, these predictions are premised on very small-scale component tests of in situ prototype mining equipment carried out over a very short duration at a small location. Moreover, most studies have focused on only one resource type. Thus, in addition to the obvious carbon burial utility arising from fuel-powered processes, the potential carbon emissions from deep-sea mining need further careful assessment and study.

***Environmental impacts of deep-sea mining***

Although deep-sea mining is receiving renewed interest, it faces scientific, technological, regulatory and geopolitical challenges. Indeed, new research published in July 2024 suggests that some deep-sea resources (i.e. polymetallic nodules) may be capable of producing oxygen, which could prove to be a ground-breaking finding if verified by further research and capable of causing a paradigm shift in how we understand and appreciate the ecosystem functioning of the deep-sea [1]. Some groups have even argued that deep-sea mining activities if conducted in the near future and without sufficient safeguards would "represent the beginning of a large-scale uncontrolled experiment", while the legitimacy of deep-sea mining has been seriously called into question [2].

***Deep-sea green mining technology***

The development and application of responsible deep-sea mining technology faces multiple challenges, including

(1) resource potential and demand growth. The global demand for metal resources continues to grow. At the same time, it is important to acknowledge that large scale technological change in many mineral rich fields. As advances are made in finding substitutes or alternatives, and accompanied by growing investments to build a more circular economy, the profitability of deep-sea mining, which is already doubtful under current circumstances, will face further uncertainty.
(2) Deep-sea mineral resources could have potential in terms of reserves and quality However, it is important to note that processing capacities and development are still under development and questions remain with respect to the efficiency with respect to the further processing and refinement of these mineral resources. Finally, the growing number of financial bodies, global industries and leading businesses that have committed to not support or source for the metals they require from deep-sea mining activities could also affect the demand and profitability of such activities.
(3) Environmental challenges and risks. The potential impact of deep-sea mining activities on the marine environment is a difficult issue for current technology and scientific assessment. Deep-sea mining technology needs to minimize these environmental risks through technological innovation and scientific assessment to achieve a balance between ecological protection and resource development.
(4) Technical bottlenecks and R&D needs. The deep-sea environment is extremely special, characterized by high pressure, low temperature, darkness, etc., which puts forward extremely high requirements for mining equipment and technology. At the same time, due to the environmental specificity and complexity

of deep-sea mining, it is often necessary to combine a variety of advanced technologies, such as robotics, automation technology, remote control technology and intelligent systems, to cope with a variety of complex situations in the mining process.

(5) Jurisprudence and international norms. Deep-sea areas are usually defined as international public areas or "commons", and resource development needs to follow the provisions of the international law of the sea and relevant international treaties. The International Seabed Authority (ISA), an organization established under the United Nations Convention on the Law of the Sea, is responsible for managing and supervising mineral resource development activities in the international seabed area and to ensure the effective protection of the marine environment. Although there are some preliminary regulations and environmental protection guidelines, there are still loopholes and imperfections in the process of concrete implementation and enforcement.

***Deep-sea mining and "peak carbon, neutral" strategies***

While deep-sea mining could provide key metals, such as cobalt and nickel, contribute to the energy transition, it still faces multiple challenges. Deep-sea mining areas are remote from land, the ore bodies are found thousands of metres under the sea surface, the host environments are complex, and the mining technologies are immature and potentially expensive, with no examples of commercialized mining to date. The deep-sea hosts unique and fragile ecosystems, whereby deep-sea mining is likely to cause significant and irreversible harm. Moreover, deep-sea mining also could potentially cause significant human rights violations, it also has the potential to reduce land-based mining impacts. In addition, deep-sea mineral resources are often polymetallic ores, with long processing and metallurgical processes that consume large amounts of energy, resulting in high processing and metallurgical costs as well as large footprint for deep-sea minerals. According to the technical and economic evaluation of polymetallic nodule mining, the cost of beneficiation and metallurgy accounts for more than half of the total cost [3]. Mining activities may also lead to imbalances in the ocean carbon cycle, affecting carbon sequestration and storage and thus indirectly increasing carbon emissions [4].

Consequently, the United Nations Environmental Program Finance Initiative concluded that deep-sea mining as currently understood should not form part of the sustainable blue economy.

### 3.3.3 Recommendations

***Development of a comprehensive carbon emissions assessment framework for deep-sea mining***

(1) Establishment of a comprehensive methodology for assessing.
(2) Establishment of a comparative inventory of potential carbon emissions from marine protected areas and deep-sea mining areas.

***Promoting responsible technologies for deep-sea mining***

(1) Developing responsible mining technologies and methods.
(2) Innovating low-carbon deep-sea processing and metallurgical technologies and equipment.

***Enhanced monitoring and management systems***

(1) Establishment of a system for monitoring and continuous observation of environmental impacts, including carbon emissions from deep-sea mining.
(2) Strengthening scientific research and environmental management systems.

***Strengthening international cooperation and regulatory systems***

(1) Strengthening global international cooperation.
(2) Constructing a comprehensive system of international laws and regulations
(3) Ensuring coherence between future deep-sea mining activities with the recently adopted BBNJ agreement and the Kunming-Montreal Global Biodiversity Framework, among other internationally binding instruments that apply to the marine environment, and assessing how any potential deep-sea mining activities could affect the obligations, ambitions, goals and targets set by these other international agreements and processes.

***Promoting clean energy applications***

Developing non-fossil renewable and clean energy: Reducing the intensity of carbon emissions from deep-sea mining, actively adopting renewable and clean energy and other renewable energy sources, constructing a platform for the comprehensive supply of energy from the sea, realizing the large-scale storage and transportation of clean energy supplies and their efficient utilization, while ensuring the effective protection of the marine environment.

***Increased public awareness and industry transparency***

Raise public and industry awareness.
Implement gender mainstreaming policies.
Develop inclusive strategies and mechanisms.
Promote education and capacity building initiatives.

## 3.4 *Offshore Aquaculture*

### 3.4.1 Introduction

With the continuous rapid growth of the world's population and the improvement of people's living standards, people will put forward higher demands for high-quality animal protein in the foreseeable future. In the aquatic food production sector, capture fisheries are in the stage of reducing production capacity and optimizing production structure due to long-term overfishing, thus aquaculture is the main driving force of growth.

Excessively intensive nearshore aquaculture not only endangers the yield and safety of aquatic food through water pollution, disease transmission, etc., but also has profound negative impacts on coastal ecosystems and socioeconomic systems [5–7].

Promoting the expansion of mariculture from nearshore to offshore has become an urgent need at this stage. However, it should be noted that the offshore aquaculture industry is still in its early stages, and the global academic and industrial communities have not yet reached a sufficiently solid consensus on the definition, expansion potential, technical trends, and main risks of offshore aquaculture. Therefore, it is also necessary to collect more comprehensive scientific knowledge and organize more stakeholder-based decision-making processes to outline a clearer strategic path for the future development of offshore aquaculture.

### 3.4.2 Challenges and Opportunities

***Ecological risks***

The impact of aquaculture on natural ecological environment has been a widely studied topic in recent decades. Compared to nearshore mariculture, offshore aquaculture areas are less disturbed by human activities, have better water exchange conditions, and stronger pollution assimilation capacity. However, offshore aquaculture also poses considerable ecological risks, some of which are due to the inherent characteristics of aquaculture activities and others due to the special characteristics of the offshore environment.

(1) One of the most typical ecological risks of offshore aquaculture is the use of wild forage fish resources.
(2) In addition to the ecological risks caused by using external biological resources, the ecological risks posed by aquaculture infrastructures themselves to the surrounding environment are also noteworthy. At the same time, the risk of escape of farmed organisms due to potential causes also needs to be guarded against. In addition, it is necessary to strengthen research on the interaction between offshore aquaculture and threatened wildlife.

***Economic risks***

Overall, offshore aquaculture is a high-risk, high-reward type of aquatic food production. From China's current situation, offshore aquaculture has good economic benefits, and coupled with a series of incentive industrial policies implemented by the government, the enthusiasm of enterprises to invest in offshore aquaculture is still limited, and economic risks are the main reason [8]. The economic risks of offshore aquaculture mainly include the following aspects.

(1) with nearshore waters, the physical and biological environments of offshore waters are more complex. In addition, offshore aquaculture infrastructures and

target species are also more expensive and more vulnerable to external risks. The most critical issue is that the development of insurance systems specifically for offshore aquaculture is slow.

(2) offshore aquaculture infrastructures need to have the ability to resist wind and waves, the construction and operation costs are high, and it is necessary to ensure that their products have a certain competitive advantage in the market and that there are enough consumers willing to pay a higher price for them. This requires that the target organisms of offshore aquaculture be different from those of traditional aquaculture methods aquaculture to achieve staggered competition. At the current stage of development, these factors also constitute important economic risks faced by offshore aquaculture.

(3) recent years, with the rise of offshore aquaculture, marine engineering equipment enterprises have continuously increased their investment in the research and development of offshore aquaculture infrastructures. This high initial investment further exacerbates the economic risks of offshore aquaculture.

***Social risks***

The emerging offshore aquaculture industry is quite different from the characteristics of the entire aquatic food industry. It has high capital investment requirements, high technical requirements, and high product prices, and can only be carried out and operated by enterprises of a certain scale and strength. In this context, whether the offshore aquaculture industry can transform the benefits it produces into fair social well-being is a question worth careful scrutiny.

(1) current orientation of offshore aquaculture is a relatively high-end form of aquatic food production. In terms of capital investment, its aquaculture infrastructures are large-scale and expensive. In terms of technical input, its aquaculture equipment is structurally complex, and the target species are generally high trophic level fish that need to be fed, so operators need to have rich professional knowledge in engineering technology or aquaculture. This situation will have a negative impact on food and nutrition security in specific regions.

(2) Gender equality is also a major focus when examining the social risks of an industry. Offshore aquaculture is precisely a representative of high-entry-threshold, high-return aquatic food production forms, and the employment opportunities in it are highly scarce resources for the entire fishing community. Gender-based barriers further reduce employment opportunities for women within this industry.

### 3.4.3 Recommendations

Offshore aquaculture, with its potential for large-scale, sustainable development, addresses the critical need for high-quality aquatic food sources. It also offers a powerful solution to the challenges of oversaturation in coastal marine activities and enables efficient collaboration among various blue economy sectors. As a leading offshore aquaculture country, China, alongside other major producers, should implement robust industry governance policies. These policies should inclusively consider the demands of coastal communities and stakeholders across the industry value chain. Key principles of offshore aquaculture governance include: (1) Scientifically assessing the environmental impacts of aquaculture activities, and ensuring that the transition to offshore aquaculture aligns with marine ecological conservation as far as possible; (2) Minimizing carbon emissions "from aquaculture activities and upstream and downstream industries, and exploring the feasibility of negative-emission mariculture practices in offshore waters; (3) Ensuring equitable access to the benefits derived from the production outcomes of offshore aquaculture for women and other marginalized groups; (4) Reducing operational risks and entry barriers in offshore aquaculture to foster a more diversified market pattern.

## References

1. Sweetman, A. K., Smith, A. J., de Jonge, D. S. W., et al. (2024). Evidence of dark oxygen production at the abyssal seafloor. *Nature Geoscience, 17*, 737–739. https://doi.org/10.1038/s41561-024-01480-8
2. Jaeckel, A., Harden-Davies, H., Amon, D. J., et al. (2023). Deep seabed mining lacks social legitimacy. *NPJ Ocean Sustainability, 2*(1). https://doi.org/10.1038/s44183-023-00009-7
3. Zhuo, X. J., Wang, W. T., Xu, X. W., Zhang, T. et al. (2024) *Scientific and technological development report on the exploitation and utilization of deep-sea mineral resources* (2024).
4. The United Nations Environment Programme (UNEP) (2024). *Deep-sea mining.* Accessed at https://wedocs.unep.org/bitstream/handle/20.500.11822/45494/deep_sea_mining.pdf?sequence=3&isAllowed=y.
5. Cao, L., Halpern, B. S., Troell, M., et al. (2023). Vulnerability of blue foods to human-induced environmental change. *Nature Sustainability, 6*, 1186–1198. https://doi.org/10.1038/s41893-023-01156-y
6. Gephart, J. A., Henriksson, P. J., Parker, R. W., Shepon, A., Gorospe, K. D., Bergman, K., et al. (2021). Environmental performance of blue foods. *Nature, 597*(7876), 360–365.
7. Cao, L., Wang, W., Yang, Y., Yang, C., Yuan, Z., Xiong, S., & Diana, J. (2007). Environmental impact of aquaculture and countermeasures to aquaculture pollution in China. *Environmental Science and Pollution Research-International, 14*, 452–462.
8. Lin, M. (2022). Developing large-scale deep sea aquaculture: Problems, modes and realization ways. *Journal of Management World, 38*(12), 39–60. (in Chinese).

# Chapter 3
# Biodiversity Conservation and Implementation of the Kunming-Montreal Global Biodiversity Framework (GBF)

## 1 Report Objectives

The following report is an output of the 2023–2024 Biodiversity Conservation and Implementation of the Kunming-Montreal Framework policy research project launched by China on January 10, 2024. Additionally, there was a request for a particular emphasis on gender issues as they relate to the Kunming–Montreal Framework, which is addressed in Box 1 and in recommendations throughout.

> **Box 1 Gender equity** Understanding different roles between women and men is key to combating biodiversity loss and ensuring equitable access to limited and life-giving resources. Women act as primary caretakers and natural resource managers, procuring water and firewood. Their activities are directly impacted by biodiversity loss, and the related impacts reinforce unequal rights. Target 23 of the Kunming–Montreal Global Biodiversity Framework aims to "Ensure Gender Equality and a Gender-Responsive Approach for Biodiversity Action".

The content of this chapter is extracted from the main findings of the 2023-2024 CCICED Special Policy Study (SPS) on "Biodiversity Conservation and Implementation of the Kunming-Montreal Global Biodiversity Framework (GBF)".



*Adhering to Green and Open Cooperation to Advance Modernization*,
https://doi.org/10.1007/978-981-95-5547-5_3

## 2 The Proposed Context of Biodiversity Conservation and the Implementation of the Kunming-Montreal GBF

### *2.1 The Global Trend of Biodiversity Loss Has Not Been Reversed, and the Pressure on Biodiversity Is Intensifying*

According to the fifth edition of the *Global Biodiversity Outlook*, global biodiversity loss continues [1]. Without a concerted, international effort, the current rate of global species extinction will accelerate and is already tens if not hundreds of times faster than the average over the past ten million years [2–5].

### *2.2 An International Consensus Has Been Reached to Strengthen Biodiversity Conservation*

In 2022, the 15th Conference of the Parties to the CBD (COP 15) agreed on the Kunming–Montreal Global Biodiversity Framework (hereinafter the Kunming-Montreal Framework). The Kunming–Montreal Framework is designed to provide governments with a roadmap to achieve the three goals of the CBD: biodiversity conservation, sustainable use, and benefit sharing.

## 3 Area-Based Biodiversity Conservation Experiences and Strategies for Achieving the 30 × 30 Target

### *3.1 Importance and Challenges of the 30 × 30 Target*

The Kunming–Montreal Framework sets out four long-term goals (by 2050) and 23 action targets (by 2030) Fig. 1.

**Box 2 Target 3 of the Kunming-Montreal Framework** Ensure and enable that by 2030 at least 30% of terrestrial, inland water, and of coastal and marine areas, especially areas of particular importance for biodiversity and ecosystem functions and services, are effectively conserved and managed through ecologically representative, well-connected and equitably governed systems of protected areas and other effective area-based conservation measures, recognizing indigenous and traditional territories where applicable, and integrated into wider landscapes, seascapes and the ocean, while ensuring that any sustainable use, where appropriate in such areas, is fully consistent with conservation

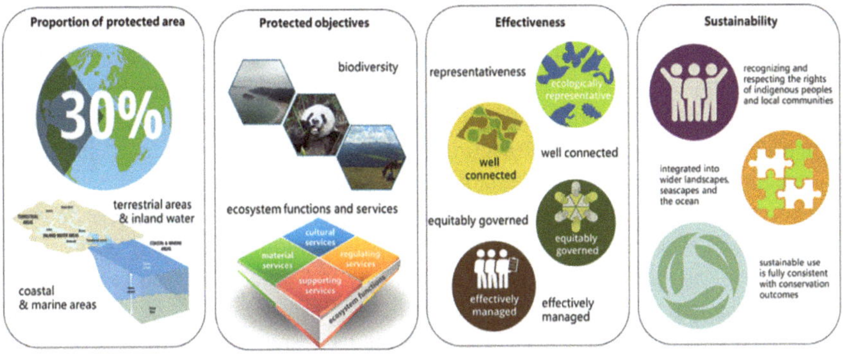


**Fig. 1** Core content of the 30 × 30 goal (refer to the reports of WWF and IUCN WCPA) [1]

outcomes, recognizing and respecting the rights of indigenous peoples and local communities, including over their traditional territories.

## *3.2 Progress of China's Protected Areas System and Analysis of its Effectiveness in Implementing the 30 × 30 Target*

***Progress of China's Protected Areas System***: Over the past 10 years, the development of China's nature protected areas system has included establishing a national park system in 2013, a major policy shift toward establishing a protected area system with national parks as the main body (first proposed in 2017), and the official establishment of the first batch of national parks in 2021 [6].

***Analysis of the Effectiveness of Construction of China's Protected Areas System to the Implementation of the 30 × 30 Target***: By the end of 2021, China had set up nearly 10,000 protected areas of various levels and types, accounting for about 18% of the country's land area [7, 8], with about 90% of terrestrial ecosystem types and 65% of higher plant communities currently represented [9, 10]

## *3.3 Progress in the Construction of an Ecological Conservation Redline in China and an Analysis of the Effectiveness in Implementing the 30 × 30 Target*

***Recent progress for China's ECR***: The ecological conservation redline (ECR) concept has gradually risen to the level of national strategy and legislation [11]

ECRs focus on three major areas: (i) protecting important ecological functions, maintaining ecosystem services, ensuring the supply of ecological products, and providing ecological support for sustainable development; (ii) protecting sensitive areas of the environment, mitigating and controlling ecological disasters, and buffering human settlements; and (iii) protecting key species and ecosystems, maintaining biodiversity, and promoting sustainable use of biological resources.

***Analysis of the Effectiveness of Implementing China's ECR to achieve 30 × 30 Target***: The ECR covers the key areas of the national ecological space, therefore effectively promoting biodiversity conservation in situ in China. China has established a rigid management system to ensure effectiveness. "Ensuring that the ecosystem function is not reduced, the designated area is not reduced, and the land use characteristic is not changed" is the ultimate goal of strictly managing ECR.

## 3.4 Progress in International Area-Based Biodiversity Conservation (Protected Areas and Other Effective Area-Based Conservation Measures) with Analysis of Implementation of the 30 × 30 Target

A draft outline of country reporting commitments for CBD COP is given in Fig. 2. [12]

***Status, Representativeness, and Effectiveness of Protected Area Systems and Conservation Measures in the International Community***: According to the World Database on Protected Areas, as of January 2024, the number of protected areas in the world is 295,579, with an area of 66,748,700 km2 [13]. Terrestrial and inland water protected areas cover 16.06% of the land surface, and marine protected areas cover 8.16% of coastal and ocean areas. The UNEP World Conservation Monitoring Centre lists 19,707 protected areas as having had an assessment, but it has been almost 15 years since a global assessment of results. The last such assessment found that around 40% showed major deficiencies in management [14].

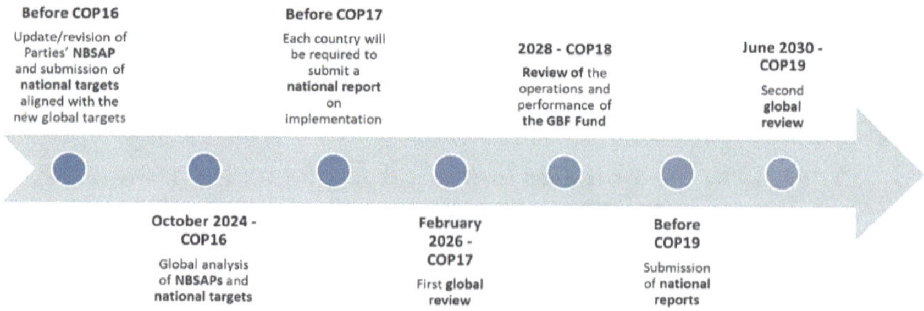


**Fig. 2** A draft outline of country reporting commitments. *Source* Aubert and Dudley, 2024 [briefing for the European Parliament]

***Challenges Defining OECMs in a National Context and Determining "What Counts" Toward the 30 x 30 Target***: OECMs aim to promote the in-situ conservation of global biodiversity by identifying and recognizing areas outside protected areas that are "effectively conserved". In 2019, IUCN published Recognizing and Reporting Other Effective Area-based Conservation Measures, which included OECM screening tools, case studies, and processes for monitoring and reporting OECMs. [15] In 2023, the IUCN World Commission on Protected Areas (WCPA) published the Site-Level Tool for Identifying Other Effective Area Conservation Measures. [16]

***Analysis of the Effectiveness of the International Community in Implementing the 30 x 30 Target and Addressing OECMs***: In some ways, the demands on OECMs are now stricter than for protected areas. Several challenges can be identified: (1) Disagreements about the extent to which livestock pasture, managed forests, and certain agricultural systems should or should not be recognized as OECMs. (2) Uncertainty about what kinds of fishing are acceptable within marine—and, to a lesser extent, freshwater—OECMs. (3) Whether "potential OECMs" might be recognized in areas that are currently undergoing restoration. (4) How monitoring can be organized in OECMs, many of which are being managed by people not skilled in or perhaps interested in biodiversity conservation. [17, 18]

Prospects for Achieving Target 3 Globally, Including Components Related to Management Effectiveness, Ecological Representation, and Equity: Today, there is a strong commitment to addressing the whole of Target 3. This includes monitoring: (1) Biodiversity. (2) Ecosystem services. (3) Management effectiveness, requiring at least basic headline indicators applied across the full range of protected areas and OECMs. (4) Integration of protected areas and OECMs into the wider landscape, represented by indicators of ecological connectivity. (5) Rights and equity, particularly for Indigenous Peoples and local communities. (6) Agreement on what kinds of sustainable use is acceptable in protected areas and in OECMs, and ways to measure that sustainability is being achieved.

***Challenges and Best Practices in Selecting Areas and Building the 'Social License to Protect,' Leveraging Support from and Generating Socio-Economic Benefits to Local Communities***: Protected areas and OECMs need to be supported by the majority of people in surrounding communities if they are to have the chance of being effective in the long term. For Indigenous Peoples, there is now a formal requirement of Free, Prior, and Informed Consent (FPIC) for any activity on their traditional land or water [19]

## *3.5 Policies and Recommendations*

This study puts forward the following suggestions; (1) **Funding**: In addition to central and local sources of finance, investigation of novel and diversified funding channels is recommended, with pilots and cost-benefit analyses carried out in some national parks and other nature protected areas. (2) **System planning and effectiveness**:

**Formulate and introduce a plan for China to implement the 30 × 30 target domestically, committing to exceed the 30% global benchmark if possible.** (3) **Management effectiveness**: Set a baseline and agree on methods to ***monitor trends*** in both (i) management and (ii) biodiversity and ecosystem services. (4) **OECMs**: Countries should consider determining the criteria for recognizing OECMs according to their national conditions, clarifying current challenges of implementing OECMs, and integrating OECMs into the path to the 30 × 30 Target. (5) **Sustainable use**: It is important to agree to guidance on what can be included within the term in protected areas and OECMs. (6) **Broader aspects of 30 × 30**: It is important to include all aspects, including greater stakeholder participation and influence, including Indigenous Peoples and local communities as appropriate, and a focus on both equity and gender equality. (7) **Ecological conservation redline**: Document, disseminate, and apply China's ECR model and establish a multi-functional ecological protection redline model that is in line with international guidance for protected areas and OECMs. (8) **Strengthen the connection between the IUCN protected area definition and management categories and OECMs**. (9) **Introduce the idea of "important carbon sink ecological function areas" to help the vision of carbon peak and carbon neutrality.** China can further improve the ECR results according to China's climate change goals by including carbon storage and sequestration in site-selection methodologies. (10) **Establish an integrated technical package for the delineation of ECRs**. Users will be able to use the *Ecological Redline Delineation Package* to identify ecological redlines at the regional, regional, national, and international levels. (11) **Assess the benefits of ecological conservation redlines.** A variety of approaches, including economic accounting, value assessment, public policy, and environmental impact assessment, are recommended to carry out a comprehensive assessment. (12) **Apply the results of the redline benefit assessment to the formulation of green development policies.** Local governments can introduce and formulate policies such as compensation for environmental protection, quantitative monitoring of green development performance, and preferential green development, as well as explore the establishment of an ecological security normative system that matches local conditions. (13) **China's policy role**: **Formulate and introduce a plan for China to implement 30 × 30 globally.**

## 4 Progress and Management Strategies in the Sustainable Management of Agriculture for Biodiversity

### *4.1 Importance and Challenges of Sustainable Utilization of Agricultural Biodiversity*

It is critical to cut across the "silos" of agriculture and environment, with environmentalists needing to consider change and progress from the standpoint of the agrifood sector and agriculture policymakers needing to prioritize shifting the sector from a source of negative environmental impacts to becoming an agent of change for sustainability.

## 4.2 Analysis of the Effectiveness of Modern Agrobiodiversity Conservation Measures and Their Implementation of Kunming–Montreal Framework Targets in China

*Agricultural Pollution Risk Prevention and Control*:

(1) **Fertilizer and Pesticide Reduction**: China's total pesticide application increased from 1.386 million tonnes in 2004 to a peak of 1.808 million tonnes in 2013 and then fell back to 1.2392 million tonnes in 2021, down 31.4% from 2013.
(2) **Decrease of plastic film residue in farmland and annual increase in recycling rates**: Starting in 2016, the annual usage of agricultural plastic films began to decline year by year, reaching 2.358 million tons by 2021, a decrease of 9.4% compared to 2015 [20].
(3) **Legal framework for pesticide packaging waste management and establishment of recycling and resource utilization system**: In February 2022, the Ministry of Agriculture and Rural Affairs led the formulation of the "14th Five-Year Plan for National Pesticide Industry Development", which outlines specific measures to promote the recycling and utilization of pesticide packaging waste.
(4) Advancement of livestock manure recycling and implementation of pollution prevention measures: The "Solid Waste Pollution Prevention and Control Law of the People's Republic of China", implemented in September 2020, stipulates that large-scale livestock operations must establish facilities for the harmless treatment and resource utilization of manure, ensuring their proper functioning.

*Sustainable Agricultural Management*

(1) **Continuous development and growth of organic agriculture, making China the fourth largest organic cultivation country in the world**: According to the "2023 Annual Report on Organic Product Certification and Industry Development", by 2022, China's organic crop production area had reached 4.752 million hectares, establishing the country as the fourth largest organic agriculture producer globally.
(2) **Diversification of agricultural waste utilization methods and steady improvement in resource utilization levels and capabilities**: As of now, the comprehensive utilization rate of livestock and poultry manure nationwide has reached 78%, and a new development pattern of integrated crop-livestock farming and agricultural-circular systems has begun to take shape.
(3) **Significant improvement in agricultural productivity assessment indicators and substantial increase in comprehensive farmland productivity**.

(4) **Significant achievements in agricultural resource protection, and the addition of 91 new national resource repositories**: Last year, the Chinese government approved and funded 72 national crop germplasm resource banks and 19 national microbe resource banks [21].
(5) **Diversification of dietary structure and the gradual formation of a "big food concept"**: The Central Document No. 1 of 2023 emphasizes the establishment of a "Big Food Concept," aimed at accelerating the construction of a diversified food supply system that integrates grain, economy, and feed while combining agriculture, forestry, animal husbandry, and aquaculture, alongside plant, animal, and microbial contributions.

## *4.3 Analysis of the Effectiveness of International Conservation Measures in the Agricultural Sector and Implementation of the Kunming–Montreal Framework*

The substantial overlap between regions rich in agricultural biodiversity and areas of high poverty indicates that the mere presence of biodiversity does not guarantee better yields or improved livelihoods for farmers. Therefore, a key challenge facing developing countries is how to enhance farmers' quality of life while simultaneously protecting agricultural biodiversity [22].

## *4.4 Research on Key Scientific Theories and Practices of Sustainable and Regenerative Agriculture in China*

China is promoting agrobiodiversity conservation by adopting sustainable agricultural technologies to increase the resilience, resilience, and productivity of agroecosystems. In addition, the ecological cycle model and the ecological model of reducing carbon and increasing efficiency have also become the focus of sustainable development of modern agriculture.

## *4.5 Policy Recommendations—Agriculture*

(1) **Improve policy frameworks to implement agricultural biodiversity conservation**: Enhance the policies, laws, and ecological compensation measures for agricultural biodiversity protection, implementing the "China Biodiversity Conservation Strategy and Action Plan (2023-2030)."

(2) **Enhance action measures to practice agricultural biodiversity conservation**: Promote the development of green substitutes for fertilizers and chemicals to reduce excessive reliance on chemical inputs. Implement techniques such as precision fertilization, conservation tillage, cover cropping, crop rotation, and intercropping. Actively promote the development of biodegradable or easily recyclable pesticide packaging and agricultural films. Utilize large-scale land use planning tools, such as the ecological red line system. Incorporate biodiversity indicators into the evaluation system for soil health and agricultural ecological development outcomes.

(3) **Strengthen capacity building to promote agricultural biodiversity conservation**: Strengthen training in green organic agriculture knowledge and skills, encouraging women to play a larger role in agricultural production management. Promote the use of smart agriculture and digital technologies. Increase the deployment of pesticide packaging waste and agricultural film recycling facilities in rural areas, establish recycling demonstration projects, accumulate experience, and promote widely. Encourage importers and other relevant companies to support the development of food and agricultural product traceability systems, ensuring transparency and sustainability in the food supply chain and preventing procurement practices that involve deforestation or habitat conversion detrimental to sustainable agriculture. Promote biodiversity—friendly agricultural demonstration zones, establish a monitoring network for agricultural biodiversity, regularly conduct surveys and assessments of agricultural biodiversity, summarize experiences, and gradually disseminate successful practices.

(4) **Enhance awareness and advocate for agricultural biodiversity conservation**: Promote the "Big Food concept", alleviating pressures on domestic and international food production ecosystems. Encourage dietary shifts toward whole grains, fruits, vegetables, and legumes, while avoiding ultra-processed foods, refined grains, and red meat. Ensure sustainable grazing lands and maintain healthy, high-yield wild-capture fisheries, as well as terrestrial and coastal aquaculture.

(5) **Strengthen safeguard measures to enhance agricultural biodiversity conservation**: Increase financial investment in agricultural biodiversity protection, ensuring funds are allocated for scientific research, monitoring, public education, and ecological restoration. Enhance policy support for organic and sustainable agriculture through financial subsidies and tax incentives to lower agricultural production costs, encourage farmers to adopt eco-friendly practices, promote biodiversity-friendly planting and breeding methods, reduce reliance on single crops, and enhance agricultural biodiversity. Ensure the effective use of financial resources by formulating clear funding usage plans, prioritizing projects and activities related to agricultural biodiversity, and strengthening the evaluation and supervision of fund usage to ensure effective implementation.

# 5 International Cooperation and Mobilization of Funds for Biodiversity Conservation

## *5.1 International Cooperation and Resource Mobilization for Biodiversity Conservation*

In recent years, the international community has put increasing emphasis on exploring alternative financing opportunities for biodiversity conservation. SCI Research estimates that as of the third quarter of 2023, the investment in 134 biodiversity-related funds is USD 59 billion. From 2013 to 2018, China's foreign aid amounted to CNY 270.2 billion (USD 37.87 billion),[1] including projects to support environmental protection, combat climate change, and promote sustainable development. According to incomplete statistics, from 2006 to 2020, China implemented more than 70 foreign aid projects related to biodiversity conservation.

## *5.2 Domestic Fiscal and Tax Policies and Incentives for Biodiversity Conservation*

In the decade from 2011 to 2020, China invested a total of CNY 2.16 trillion (USD 302.7 billion) in core funds for biodiversity conservation, CNY 4.14 trillion (USD 580.3 billion) in broad funds, and CNY 4.86 trillion (USD 681.2 billion) in comprehensive funds. Since 2016, China's total environmental public welfare donations have reached CNY 2.523 billion (USD 354 million), of which CNY 1.757 billion (USD 246 million) has been used for biodiversity-related funding, accounting for 69%. From 2016 to 2020, the growth rate of public welfare donations for biodiversity conservation also maintained a rapid growth rate, reaching 21%, 35%, 43%, and 29%, respectively [23].

## *5.3 Innovative Market-Based Financing Mechanisms for Biodiversity Conservation*

***Private Sector Transparency and Support for Nature-Positive Investment Both Domestically and Internationally***: The Kunming–Montreal Framework and the Strategy for Resource Mobilization established in Decision 15/7 rightly emphasized the need for a significant increase in the level of financial resources from all sources, including domestic, international, public, and private resources.

[1] Calculated based on real-time exchange rates, similarly hereinafter.

***Biodiversity Credits, Regulated Offsets, and Related Mechanisms***: China does not at present have any biodiversity credit markets as such, but it has considerable experience over the last two decades with the development of nature markets. Although still in its preliminary stages, this has begun to take shape through pilot projects. It is crucial to recognize the fundamental differences between biodiversity markets and carbon markets, necessitating innovative approaches in the biodiversity sphere. Establishing a benchmark standard to define the integrity of biodiversity credit projects and their outcomes, along with systems enabling credible claims from purchasers, is particularly important.

***Sustainability-Linked Sovereign Debt Instruments***: Sustainability-linked sovereign debt can contribute to solving sovereign debt issues in several ways. First, it directly rewards countries with lower costs of debt repayment when they achieve positive nature and climate outcomes. Second, it encourages investments that reduce risks and boost resilience and economic productivity, ultimately lowering debt repayment costs for the entire country. Third, it supports broader sustainable development by promoting economic growth and productivity and providing financial flexibility for increased public spending. Lastly, it reduces the need for complicated debt restructuring by promoting smarter risk-sharing between debtors and creditors.

***Alignment of Financial Flows in Line with the GBF's Approach to Mainstreaming***: Measuring the degree of misalignment of each of these types of institutions and then supporting their plans to improve the situation as part of wider nature and climate transition planning has been the subject of growing attention in the last two years or so. But strong policy and regulatory signals are also an indispensable part of action to achieve realignment of harmful financial flows and China has the opportunity both to strengthen how it achieves this aim domestically and to share its learnings with a wider international audience.

## 5.4 Synergistic Mechanisms for Biodiversity Finance and Climate Finance

***Prioritizing Solutions That Address Multiple Threats***: Actively promote the development and demonstration of biodiversity conservation and climate change response, promote nature-based solutions, play an effective role in the coordinated response to the biodiversity crisis and climate change crisis, and raise funds to support the coordinated promotion of biodiversity conservation and climate change response.

***Key Mechanisms for Financial Synergies and Safeguards to Ensure Biodiversity and Climate Benefits***: Approaches include promoting synergies between climate change and biodiversity, integrating the monetary policy tools of carbon emission reduction, optimizing the co-benefits of biodiversity and climate investment, and financing and identifying the synergies between biodiversity and other societal challenges.

## *5.5 Policy Recommendations*

At present, China's biodiversity conservation investment and financing mechanism is still in the stage of exploration and construction and is facing a large demand for funds.

***At a Domestic Level***: (1) Substantially increase domestic public spending related to biodiversity and, as far as possible, gradually align public finance with the goals and targets of the Kunming–Montreal Framework. (2) **Give preference and support to biodiversity conservation in financing and policies. Mobilize and guide all stakeholders to participate in biodiversity governance and build an institutional system and coordination mechanism for national action**. (3) Establish a special fund for biodiversity in China and integrate and coordinate scattered conservation resources through government guidance. (4) **Promote innovation in financial products related to biodiversity to improve financial access and reduce the financing gap**. Financial institutions can play a key role in financing biodiversity conservation projects through the issuance of biodiversity bonds. (5) **Eliminate harmful incentives**: China should strengthen biodiversity-related risk management, effectively identify and assess the biodiversity-related risks of investment and financing projects of financial institutions and carry out pilot projects on biodiversity risks in various pilot zones for green financial reform and innovation to assist financial institutions in effectively identifying and assessing the biodiversity-related risks of investment and financing projects. (6) **Promote transparency**: China should roll out private sector disclosure requirements rapidly and explore the establishment of harmonized and standardized biodiversity disclosure standards and frameworks and include within these reporting requirements for Chinese companies operating overseas, as well as domestically. (7) **Biodiversity credits**: China should share its experience of these new nature markets by participating actively in relevant international fora, including the International Advisory Panel on Biodiversity Credits (IAPB). (8) **Debt instruments**: China should explore the scope for the use of sustainability-linked sovereign debt instruments, and the opportunity this offers for China to demonstrate a major international leadership role in this area.

***At an International Level***: Application of the ECR principles and practices to infrastructure and other development projects financed by China overseas is worth exploring as a way to reduce the environmental and social impacts of such projects and enhance their contribution to biodiversity conservation. China also has an important opportunity to enhance its support to other developing countries, least developed countries, and Small Island Developing States in meeting their obligations under the Kunming–Montreal Framework, in particular Target 20.

# 6 Policy Recommendations

## 6.1 *On the Basis of Existing Protected Areas and China's ECRs, Establish an Area-Based Ecological Conservation Model in Line with International Standards, and Promote the Implementation of the 30 × 30 Target Globally, Demonstrating China's Leadership in this Sphere*

(1) Accelerate the introduction of China's plan for implementation of 30 × 30 and cooperate with stakeholders to set up an integrated technical package. (2) Commit to full achievement by 2030 combined with an aspirational commitment to exceed 30% and encourage other countries to exceed 30% in protected areas and other effective OECMs by 2030. (3) Strengthen research to clarify the similarities and differences between the concepts, spatial delineation, and identification criteria of protected areas as defined by the CBD and IUCN, ECR, and OECMs, and establish a multi-functional ecological reserve model that conforms to international standards. (4) Produce a series of technical support documents to support the implementation of Targets 1, 2, and 3 of the Kunming–Montreal Global Biodiversity Framework. (5) China can also play an important role in supporting other countries to achieve 30 × 30 alongside China's foreign policy goals, such as the Belt and Road Initiative.

## 6.2 *At a Global Level, Further Strengthen the Effectiveness and Representativeness of Protected Areas and Enhance High-Quality Protection of Key Ecosystems and Species Habitats*

(1) Systematically assess innovative measures and lessons learned in the existing global area-based conservation system; designate national parks; build and optimize an effective protected area system. (2) Focus particular attention on the representativeness of area-based conservation, protection of intact ecosystems, governance, and management. Improve institutional mechanisms and the effectiveness of protected area management. (3) Strengthen research on connectivity between protected areas to strengthen the construction and maintenance of ecological corridors. (4) Improve evaluation systems for the management effectiveness of protected areas, in consultation with all those affected. (5) Promote innovative techniques for biodiversity survey, monitoring, and assessment of key ecosystems and species habitats.

## 6.3 *In Consultation with all Affected Stakeholders, Accelerate the Development of a New, more Sustainable Approach to Agricultural Productivity and Improve Agricultural Ecological Functions and Biodiversity as the Basis for Ensuring Food Security*

(1) Deepen the implementation of Target 7 of the Kunming-Montreal Framework and continue to reduce agricultural pollution. Strengthen the research, development, and use of green agricultural inputs, prohibit the production and use of highly toxic and high-risk fertilizers and pesticides, and encourage the use of organic fertilizers and biopesticides. (2) Deepen the implementation of Target 10 of the Kunming-Montreal Framework and promote sustainable agricultural management. (3) Strengthen capacity building with the goal of developing new agricultural productivity. (4) Further promote and advocate the "Big Food" concept by encouraging a shift towards diversified dietary structures. Promote the transition from industrialized and intensive crop supply systems to diversified food supply sources. (5) Strengthen the effective implementation of fundamental agricultural support measures, increase financial investment in the protection of agricultural biodiversity, and ensure the funds are directed towards scientific research, monitoring, public education, and ecological restoration.

## 6.4 *Ensure China's Responsibility and Leadership in Resource Mobilization and Financial Investment for Biodiversity Conservation, Accelerate the Implementation of Market Incentive Policies, and Identify, Eliminate, Phase out, or Reform Incentives that Are Harmful to Biodiversity*

(1) Continue to play a leading role in actively participating in biodiversity conservation, promoting the implementation of the Global Biodiversity Framework Fund. (2) Ensure strong links are made between monetary policy tools for carbon emission reduction and the climate finance mechanism, actively promote the development and demonstration of biodiversity conservation and climate change response actions, strengthen the co-benefits and synergies of biodiversity and climate investment and financing, and share China's domestic experience with other countries and institutions working on nature and climate finance. (3) Coordinate the use of multiple funding sources and explore a combination of sustainability-related sovereign debt instruments.

# References


1. CBD (Convention on Biological Diversity). (2022). Kunming-Montreal Global Biodiversity Framework (CBD/COP/DEC/15/4). https://www.cbd.int/doc/decisions/cop-15/cop-15-dec-04-en.pdf
2. WWF and IUCN WCPA. A guide to inclusive, equitable and effective implementation of target 3 of the Kunming-Montreal global biodiversity framework. 2023. https://www.worldwildlife.org/publications/30×30-a-guide-to-inclusive-equitable-and-effective-implementation-of-target-3-of-the-kunming-montreal-global-biodiversity-framework
3. Jem, W., Venegas-Li, R., Grantham, H., et al. (2023). Priorities for protected area expansion so nations can meet their Kunming-Montreal global biodiversity framework commitments. *Integrative Conservation, 2*(3), 140–155.
4. Dudley, N. (2008). *Guidelines for applying protected area management categories*. IUCN.
5. Wang, W., Zhou, Y., Tian, Y., et al. (2022). Research progress on biodiversity conservation in nature reserves. *Biodiversity Science, 30*(10), 22459.
6. Gao, J., Mengjia, X., & Changxin, Z. (2019). The development and achievements of protected areas in China in the past 70 years. *China Environmental Management, 11*(4), 25–29.
7. Information Office of the State Council of the People's Republic of China (2021). *White paper on biodiversity conservation in China*. https://www.gov.cn/zhengce/2021-10/08/content_5641289.htm
8. Biodiversity Committee of the Chinese Academy of Sciences. (2024). Report on the State of biodiversity in China (2021–2022). *Biodiversity Science, 32*(S), 23286, 1–34.
9. Ministry of Ecology and Environment. (2024). *China's biodiversity conservation strategy and action plan (2023–2030)*. https://www.mee.gov.cn/ywdt/hjywnews/202401/W020240123333807288143.pdf
10. Fan, X., Xu, W., Zang, Z., et al. (2023). Representativeness of China's protected areas in conserving its diverse terrestrial ecosystems. *Ecosystem Health and Sustainability, 9*, 0029.
11. Xu, D., Zou, C., Xu, M., et al. (2015). Construction of ecological security pattern based on ecological protection redline. *Biodiversity Science, 6*, 740–746.
12. Aubert, G & Dudley, N. (2024). *Briefing to the European Parliament*. Institute for European Environmental Policy.
13. UNEP. World conservation monitoring Centre. Protected areas map of the world. Available at http://www.protectedplanet.net
14. Leverington, F., Lemos Costa, K., Pavese, H., et al. (2010). A global analysis of protected area management effectiveness. *Environmental Management, 46*, 685–698.
15. IUCN-WCPA Task Force on OECMs. (2019). *Recognizing and reporting other effective area-based conservation measures*. IUCN.
16. Jonas, H. D., Mackinnon, K., Marnewick, D., et al. (2023). *Site-level tool for identifying other effective area-based conservation measures (OECMs): First edition*. IUCN WCPA Technical Report Series 6, IUCN.
17. Sharma, M., Pasha, M. K. S., Nightingale, M., & Mackinnon, K. (2023). *Status of other effective area-based conservation measures (OECMs) in Asia*. IUCN Asia Regional Office.
18. Marnewick, D., Stevens, C., Jonas, H., et al. (2021). Assessing the extent and contribution of OECMs in South Africa. *Parks, 27*(1), 57–70. https://doi.org/10.2305/IUCN.CH.2021.PARKS-27-1DM.en
19. Oxfam Australia. Guide to Free, Prior and Informed Consent. Victoria.
20. NBS. (2023). *China statistical yearbook 2023*. China Statistics Press.
21. Zhu, Y. Y., Chen, H. R., Fan, J. H., et al. (2000). Genetic diversity and disease control in rice. *Nature, 406*(6797), 718–722.
22. Snapp, S. S., Blackie, M. J., Gilbert, R. A., et al. (2010). Biodiversity can support a greener revolution in Africa. *Proceedings of the National Academy of Sciences of the United States of America, 107*(48), 20840–20845.
23. Li, Y., Li, Y., Zhang, Y., et al. (2023). Recent progress and prospects of the funding of the Convention on biological diversity. *Biodiversity Science, 31*(04), 25–30.

# Chapter 4
# Collaborative Mechanism for Carbon Reduction, Pollution Reduction, Green Expansion and Growth

## 1 Green Development and Economic Growth: How to Promote Economic Growth by Emission Mitigation

### *1.1 The Potential and Obstacles of Emission Mitigation in Promoting Economic Growth*

#### 1.1.1 The Potential of Emission Mitigation in Promoting Economic Growth

**First, emission mitigation becomes a key factor in driving new green growth**. The process of reducing pollution and carbon emissions stimulates innovation in production technologies and transforms production methods, leading to leapfrogging development in green and low-carbon economies and industries. **Second, emission mitigation forces technological progress and efficiency improvements in traditional sectors**. Pollution and carbon reduction have made environmental factors important for economic development, promoting policies that balance economic incentives and environmental constraints, forcing traditional high-carbon industries to upgrade technologies and improve efficiencies, thus strengthening core economic growth capabilities. **Third, emission mitigation enhances the overall quality of economic growth**. Emission mitigation leads to ecological improvements, increasing the value of ecological resources, and multi-dimensionally enhancing the overall quality of regional economic growth. **Fourth, emission mitigation can contribute to economic resilience and efficiency. Reforming and adjusting the energy structure may decrease reliance on imported fossil fuels, reduce the economic**

The content of this chapter is extracted from the main findings of the 2023-2024 CCICED Special Policy Study (SPS) on "Collaborative Mechanism for Carbon Reduction, Pollution Reduction, Green Expansion and Growth".



*Adhering to Green and Open Cooperation to Advance Modernization*,
https://doi.org/10.1007/978-981-95-5547-5_4

**system's dependence on fossil fuels and associated impacts, lower operational risks, mitigate potential economic losses from international uncertainties, and enhance the domestic economy's resilience and adaptability.**

### 1.1.2 The Obstacles of Emission Mitigation in Promoting Economic Growth

**First, limitations in awareness and perspectives.** The importance of emission mitigation for sustainable economic development is systemic and long-term, leading some economic entities to not fully recognize the intrinsic relationship. **Second, the traditional development model and its inertia.** The traditional high-carbon development model has dominated economic growth which creates significant obstacles to transitioning to a low-carbon development model. **Third, short-term cost shocks and resulting inequalities.** These short-term cost and benefit shocks create significant economic pressures on specific industries, regions, and populations. **Fourth, objective obstacles in policy mechanism transmission.** On one hand, some local policies fail to take into account the specific conditions of different regions and industries, and exhibit a "one-size-fits-all" phenomenon, lacking effective mechanisms to achieve "dual carbon" goals. On the other hand, some sectors lack flexible guiding policies prevents the market mechanism from fully playing its role.

## *1.2 Four Major Topics of Policy Research*

### 1.2.1 How to Simultaneously Promote Economic Growth and Emission Mitigation Through Efficiency Improvements?

**First, investing in innovative technologies.** In 2023, China's new energy investments primarily focused on building a new energy system, with major technological investments centered on high-efficiency motors, energy-saving inverters, and smart grids, which have broad application prospects. **Second, adopting energy and resource efficiency measures.** Since the 13th Five-Year Plan, China's circular economy has made positive progress. In 2020, the output rate of major resources increased by approximately 26% compared to 2015, energy consumption per unit of GDP significantly decreased, and water consumption per unit of GDP cumulatively reduced by 28%. **Third, using clean production methods.** The Chinese government revised the "Cleaner Production Promotion Law" in 2012 to provide policy support and technical guidance for enterprises to achieve cleaner production, improve resource use efficiency, and reduce or avoid pollutant generation. **Fourth, increasing job opportunities and creating new markets.** According to the Ministry of Human Resources and Social Security of China, the green economic transition is expected to create tens of millions of new jobs in China over the next five years, enhancing market demand and social consumption capacity.

### 1.2.2 How to Effectively Internalize External Costs?

(1) **Carbon pricing provides benchmarks and references for the operation of market mechanisms.** As an important policy tool, carbon pricing sets a price on $CO_2$ emissions, incentivizing businesses to reduce emissions and invest in clean technologies. (2) **Carbon emissions trading systems can activate market mechanisms and avoid unnecessary transaction costs**. In 2023, the cumulative trading volume in China's national carbon market reached 212 million tons, with a transaction value of 14.44 billion yuan. (3) **Environmental taxes and other command-and-control policies accelerate market mechanism formation**. Since 2018, China has implemented an environmental protection tax to replace the previous pollution charge system. (4) **Green finance leverages financial tools to share costs and risks across society**. On one hand, large-scale high-tech venture financing is needed for carbon-neutral technology innovation; on the other hand, financial products designed to help high-carbon industries adapt to carbon-neutral policy environments support industrial structure transformation. (5) **Policy guidance and public advocacy encourage businesses and individuals to take environmental responsibility**.

### 1.2.3 How Does Emission Mitigation Drive Economic "Creative Destruction"?

**First, emission mitigation creates new job opportunities and markets, forming new economic growth poles**. Investments in green energy, technology innovation, and carbon mitigation can contribute to economic development, market prospects, and employment. **Second, emission mitigation changes traditional industrial structures, promoting the reorganization of production divisions**. Emission mitigation pressures force traditional or highly polluting industries to upgrade technologies and green transitions, reducing environmental impact and improving competitiveness. **Third, emission reduction has formed a new resource concept and accelerated the spatial reconstruction of industrial layout**. On the one hand, from the past simple mineral resources, etc., to a wider range of green resources, including ecological services, cultural resources, and even climate. On the other hand, the infrastructure that supports industrial spatial optimization also needs to be rearranged accordingly. **Fourth, emission mitigation generates new business models, products, and services, driving economic model transformations**. Emission mitigation policies and green technology development play a clear role in creating new business models and services.

### 1.2.4 How Does Emission Mitigation Promote the Dematerialization of the Economy?

**First, investing in technological innovation and intellectual property protection**. Promote the development of high-tech and digital economies, enrich productivity types, create new market opportunities, drive intangible economic growth, and enhance economic development. **Second, encouraging the development of cultural and creative industries**. Encouraging the development of cultural and creative industries in digital media, arts, and entertainment through policy support such as cultural funding, intellectual property protection, and market promotion can enrich the supply of spiritual products, stimulate the consumption of spiritual products, and promote steady industry growth. **Third, investing in sustainable agriculture and food industries**. Investing in sustainable agriculture and food industries encourages the development of organic agriculture and agricultural technology, actively changing traditional agricultural and food industry development methods, improving food quality and agricultural sustainability. **Fourth, encouraging the protection and sustainable use of ecological services and natural capital**. Ensure the sustainability of natural resources, help maintain ecosystems and provide sustainable ecological services such as water source protection and climate regulation. **Fifth, encouraging and supporting intangible economic growth**. Through innovative policies, education, and training, improving labor skills and creativity leverages emission mitigation to achieve more sustainable, creative, and intangible economic growth.

## *1.3 The Role of Women's Participation and Gender Mainstreaming in Promoting Economic Growth Through Emission Mitigation in the Context of "Dual Carbon" Goals*

**First, encouraging women's participation in emission mitigation policy formulation and implementation can enhance social equity**. Governments and organizations should establish gender equality policies to encourage women's involvement in emission mitigation decisions and projects, ensuring adequate representation and leadership positions for women in government, businesses, and NGOs. **Second, gender mainstreaming can enrich the decision-making process**. When formulating emission mitigation policies, attention should be paid to gender equality, and appropriate proportions of female experts should be invited to participate in policy formulation and evaluation. **Third, addressing employment gender discrimination and leveraging social capital**.

### 1.4 Policy Recommendations

**First, change the Traditional Perspective on Emission Mitigation Burdens and Establish a New Policy Narrative that Promotes Growth through Emission Mitigation. Second, emphasize Technological Innovation to Achieve Green Low-Carbon Development Driven by Technology. Third, uphold the Logic of Green Development and Vigorously Promote Green Consumption to Drive Green Production. Fourth, Promote Just Transition Under the Dual-Carbon Goals with a Systematic Approach.**

## 2 Air Quality PM Standards Upgrading Timeline, Original Control for Pollution Reduction and Decarbonization Potentials Estimation

### 2.1 Background

While significant strides have been made in improving China's air quality, we are clearly aware that the air pollution situation in China is still well below international standards. While as the world's largest energy consumer, China is also facing a pressing demand in reducing $CO_2$ emissions. In 2020, China pledged to peak its $CO_2$ emissions by 2030 and reach carbon neutrality by 2060.

### 2.2 The Progress of Ambient Air Quality Standards

#### 2.2.1 Air Quality Standards in Developed Countries, Regions, and Organizations

In 1987, the World Health Organization (WHO) released the first European Air Quality Guidelines based on health protection. In 1971, the United States Environmental Protection Agency (EPA) issued the first nationwide air quality standards. The European Union first issued air quality limit values in the 1980s and subsequently revised and improved them. The latest version of the EU Ambient Air Quality Directive (2008/50/EC) was issued in 2008.

### 2.2.2 The Development and Revision History of China's Ambient Air Quality Standards

China's ambient air quality standards were first implemented in 1982. In 1996, based on the need to manage coal smoke-type air pollution and considering the need for pollution control from urban motor vehicle emissions, China revised the Ambient Air Quality Standards for the first time, renaming them to the Ambient Air Quality Standards. In 2012, given the increasingly severe regional $PM_{2.5}$ air pollution caused by the complex coal smoke-motor vehicle pollution, China completed the second revision, further adding $PM_{2.5}$ and $O_3$-8h concentration limits, and tightening the concentration limits for pollutants like PM10 and $NO_2$. Compared to the standards (or guideline values) of the United States, the European Union, and WHO, China's concentration limits for the six basic pollutants are generally more lenient. With new revision proposals already put forward by the US and EU, it is necessary for China to conduct research on the assessment and revision of its ambient air quality standards.

### 2.2.3 Recommendations for the Evaluation Methods of Ambient Air Quality in China

The revision of air quality standards should refer to the latest research results on the impact of atmospheric pollution on human health, both domestically and internationally [1]. The revision of air quality standards should be accompanied by the optimization of air quality evaluation methods.

## *2.3 Pathways for Synergetic Control of Air Pollution and Carbon Emissions*

### 2.3.1 Progress in Structural Adjustment

China has made remarkable progress in optimizing its energy structure. By 2021, the nationwide installed capacity of non-fossil fuel power generation surged to 1.12 billion kilowatts, accounting for 47.0% of the national total, surpassing coal power (i.e., 46.7%) for the first time [2]. China has gained remarkable achievements in transforming and restructuring industries. In 2022, the share of high-tech manufacturing in the value added of large-scale industries increased to 15.1%. China has made substantial strides in establishing a low-carbon transportation system. In 2022, the penetration rate of New Energy Vehicle (NEV) sells further increased to 25.6%. The transportation structure has also been gradually optimized.

### 2.3.2 Progress and Achievements in Synergetic Control of Air Pollution and $CO_2$

China has made significant progress in coordinated governance of pollution reduction and carbon emission reduction. At the national scale, between 2015 and 2021, the industrial and residential sectors have achieved positive synergies between $CO_2$ emissions reduction and $PM_{2.5}$ pollution control. However, the implementation of some "coal-to-gas" policies has led to increases in $CO_2$ emissions (51–79%; 80–150 million tons) and $PM_{2.5}$ pollution (86–142%; ~0.01 μg/m$^3$) from natural gas combustion processes.

### 2.3.3 Pathway of Air Quality Under the Carbon Neutrality Pledge

According to evaluations by the Chinese Academy of Environmental Planning and Tsinghua University [3, 4], achieving the carbon peak and carbon neutrality goals plays a significant role in promoting the coordinated reduction of atmospheric pollutants and achieving air quality standards (Fig. 1). If carbon emission reduction policies are intensified before 2030 to achieve an earlier peak (2025), the early-time peak-net zero-clean air scenario as shown in Fig. 1, compared to the current carbon neutrality path, additional $PM_{2.5}$ concentration reduction benefits of 2.54 μg/m$^3$, 2.00 μg/m$^3$, and 0.98 μg/m$^3$ would be realized in 2030, 2040, and 2050, respectively.

However, the benefits of coordinated improvements are not static; the synergistic. Different carbon neutrality pathways will lead to profound transformations in China's energy system, resulting in differentiated emission source patterns for the energy industry by 2060. It is worth emphasizing that in 2021, the World Health Organization (WHO) tightened air quality guidelines, setting higher requirements for the

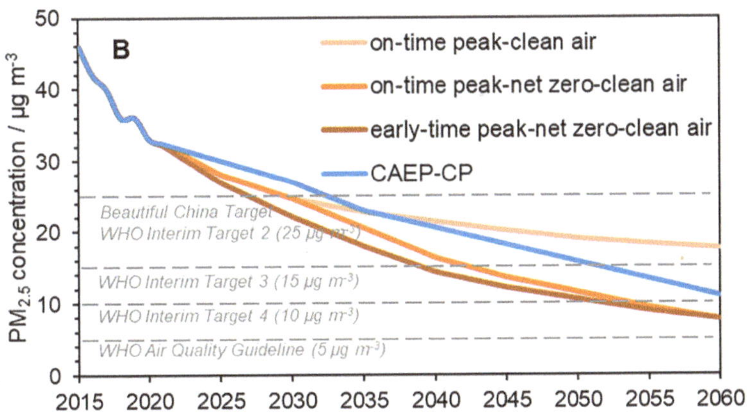


**Fig. 1** Schematic diagram of the medium- and long-term pathway for co-control air pollution and $CO_2$ emissions (A: $CO_2$ emission, B: $PM_{2.5}$ concentration)

synergistic improvement of air quality under China's dual carbon pathways from the perspective of protecting public health.

Due to the varying energy structures and industrial structures across different regions of China, the synergistic reduction effects of air pollutants under the carbon neutrality target vary significantly. Therefore, it is necessary to formulate region-specific synergistic pathways for carbon reduction and pollution control to maximize the synergistic effects of carbon reduction and pollution control while achieving the dual carbon goals. Similarly, the synergistic emission reduction potential varies greatly between different industries.

The electrification of terminal energy consumption is an important pathway for achieving synergistic reductions in carbon and pollutants in industries other than power, enabling varying degrees of positive synergistic reductions. The power industry must take on the transfers brought about by the electrification of other industries, creating a clean, low-carbon, safe, and efficient power system, which is also the only path to fully unleashing the synergistic carbon reduction potential of various industries.

## *2.4 Synergetic Control Policies and Tools*

### 2.4.1 Construction of a Coordinated Governance System for Pollution Reduction and Carbon Emission Reduction

The "Outline of the 14th Five-Year Plan for National Economic and Social Development of the People's Republic of China and the Long-Range Objectives Through the Year 2035", released in March 2021, the "Implementation Plan for Synergizing Pollution Reduction and Carbon Reduction", issued in June 2022 by seven departments, including the Ministry of Ecology and Environment.

### 2.4.2 Evaluation Indicator System for Synergetic Control of Pollution Reduction and Carbon Emission Reduction

The evaluation of synergistic effects of pollution reduction and carbon reduction mainly aims to achieve three objectives: (1) Identify Weaknesses and Issues: By assessing the progress of various regions, the evaluation aims to identify weak areas and major problems in the work process. (2) Facilitate Horizontal Comparison: Through evaluation based on a unified baseline, the work of different regions can be compared horizontally. (3) Enhance Systematic Understanding: Based on comprehensive evaluations, the evaluation aims to enhance the systematic understanding of the synergistic effects of pollution reduction and carbon reduction.

To achieve the aforementioned three objectives, the design of the evaluation system needs to be closely aligned with the goals and tasks of coordinated pollution reduction and carbon emission reduction: (1) Quantitative Evaluation: A persuasive weighting system should be designed. The final evaluation scores should enable horizontal and vertical comparisons. (2) Scalable Evaluation: The system should ensure extensibility to facilitate evaluations from the provincial level down to the city level, and even to districts and counties. (3) Focused Evaluation: The design of the evaluation system should revolve around the key areas of pollution reduction and carbon emission reduction. (4) Flexible Evaluation: The system should offer flexibility in selecting evaluation indicators based on local resources, development stage characteristics, and work priorities.

To achieve the goal of evaluating the synergy between pollution reduction and carbon emission reduction, Chinese Academy of Environmental Planning has developed the Index for Synergies in Environmental Protection and Carbon Reduction (ISEC Index). Its connotations are mainly reflected in the following three aspects: synergistic effect, synergistic pathways and synergistic management. Based on the aforementioned three aspects, an evaluation system has been constructed encompassing six primary indicators: environmental quality, carbon emission levels, synergy degree of coupling, synergy degree of structural adjustment measures, synergy degree of governance pathways, and synergy degree of ecological and environmental management.

## 2.5 *Policy Recommendations*

(1) Timely Revision of Air Quality Standards: It is suggested that the annual $PM_{2.5}$ concentration standard be set below 25 micrograms per cubic meter, and an $O_3$ peak concentration indicator for the warm season be added. It is recommended to conduct research on concentration limit standards and pollution control measures for air pollutants with short-term greenhouse effects. (2) Design Coordinated Emission Reduction Pathways Meeting Multiple Goals: Incorporate air quality co-benefits into the decision-making for carbon neutrality technology pathways, based on the dual constraints of health benefits and economic costs. (3) Strengthen Coordinated Governance Paths for Key Regions and Sectors: Identify key industries and develop localized coordinated pathways for carbon and pollution reduction.

# 3 Electricity Sector Carbon Reduction and Pollution Control

## *3.1 International Progress and Lessons on Carbon Reduction and Pollution Reduction in Power Systems*

Climate policy design takes three main forms: performance standards, economic signals, and other supporting policies including research and development, which are mutually reinforcing and have helped usher in this era of low-cost wind, solar, and now grid-scale battery storage [5]. Clean energy standards provide certainty to utilities and market participants that allow for integrated least-cost planning of generation, transmission, and distributed energy resources, supported by competitive markets.

Clean energy and emissions targets take six main forms: (1) **Markets**: To meet goals at a low cost, utilities and customers must structure procurement to reduce the risk for clean energy and set market rules to fairly reward the attributes and value that renewable energy provides the marketplace. (2) **Electricity prices**: Pricing policy such as time-of-use rates and virtual power plants can help align customer incentives better with system costs and support efficient integration of renewable energy. (3) **Price support**: Long-term contracts or direct subsidies can help promote revenue stability for clean energy resources. Special subsidies can also help newer technologies compete and scale as needed. (4) **Carbon pricing**: In organized energy markets, carbon prices raise market prices to make low-carbon energy sources more profitable. (5) **Land-use policy**: Policymakers can provide accessible maps to developers to increase transparency to help steer development toward low-conflict areas. (6) **Reliability**: System operators should move beyond a single-resource approach to reliability, instead evaluating portfolios using modern modeling techniques that include detailed weather datasets and probabilistic risk assessments.

## *3.2 Analysis of China's Power Supply and Demand Development Trend and Driving Factors*

### 3.2.1 Electricity Demand Growth Exceeds Expectations

Electricity has grown more than 20% faster than GDP for four consecutive years from 2020 to 2023 (Fig. 2). These trends provide evidence that China is on a development path that relies heavily on electricity growth.

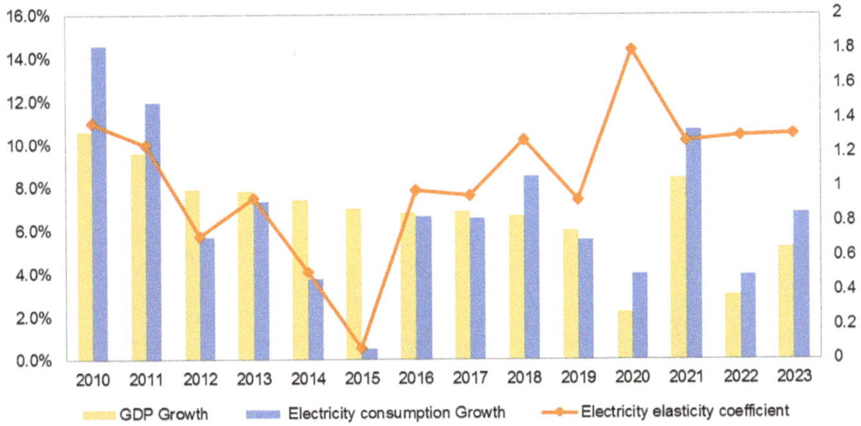


Data Source: China Statistics Bureau

**Fig. 2** Growth of China's total social electricity consumption, 2010–2023. *Data Source* China Statistics Bureau

### 3.2.2 Driving Force Analysis

Electricity has steadily increased as a share of final energy consumption from 18.6% in 2011 to 27% in 2021. Since 2018, electricity has steadily replaced coal and oil, to the tune of more than 150 TWh, causing a 2.3% increase in electricity consumption since 2018. According to the China Academy of Information and Communications Technology, from 2020 to 2022, the electricity consumption of big-scale data centers in China's information and communication industry will continue to maintain an average annual growth rate of 15%, and reach 270 TWh in 2022, accounting for 3% of the total electricity consumption in society. The internal structural adjustment of the manufacturing industry has also brought about changes in the electricity consumption pattern, driving rapid growth in electricity consumption.

## *3.3 China's Power System Development Outlook*

### 3.3.1 Economic Growth and End-Use Electrification Drive Continued Growth in Electricity Demand

The growth of China's economy has driven the growth of electricity consumption. The electricity demand will increase from 9.2 PWh in 2023 to 13 ~ 14 PWh in 2030 and to 16 ~ 18 PWh in 2040 (Fig. 3). The increase in electrification of end-uses across sectors is driving sustained growth in electricity demand. By 2030 and 2040, China's electrification rate of end-use sectors burgeons from the current 26.8% to 33% ~ 35% and 40% ~ 45%, respectively.

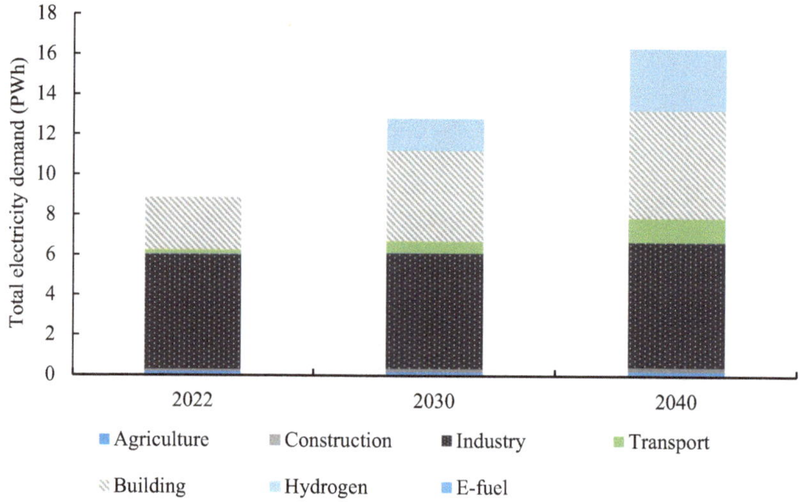


**Fig. 3** Total electricity demand by sector

### 3.3.2 Non-fossil Energy Support the New Electricity System Development

The proportion of non-fossil energy installed capacity will increase from 53 to 60 ~ 69% in 2030 and 75 ~ 91% in 2040, becoming the mainstay of the power system and supporting the development of new power systems. (Fig. 4).

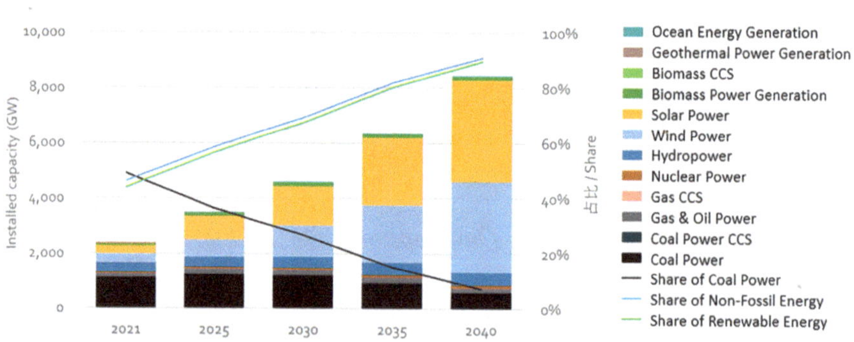


**Fig. 4** Outlook for the future installed base of different power sources

### 3.3.3 Energy Storage as an Important Support for the New Electricity System

After 2030, coal-fired power will gradually withdraw as a flexible regulatory resource, and demand side response, new energy storage, pumped storage, and other flexible resources will play a more important role in grid regulation. Clean peaking and flexibility resources will reach 8.9 ~ 9.5 TW by 2030 and 23 ~ 26 TW by 2040, with demand-side response providing up to half of the regulatory capacity.

### 3.3.4 Cross-Provincial, Cross-Regional Energy Transmission Will Continue to Grow

By 2030, the cross-regional power transmission will exceed 6.1 PWh, and by 2040, the cross-regional power transmission will exceed 8.6 PWh (divided into six regions in China: Northeast, North China, Northwest, Central, Southern, and East China) (Fig. 5). By 2040, the proportion of centralized photovoltaic installed capacity in the "Three North" regions will reach 68%, with more than half located in the northwest region, and both centralized and distributed development will be emphasized in the central-eastern southern regions. Offshore wind power is mainly deployed in the eastern and southern regions of China.

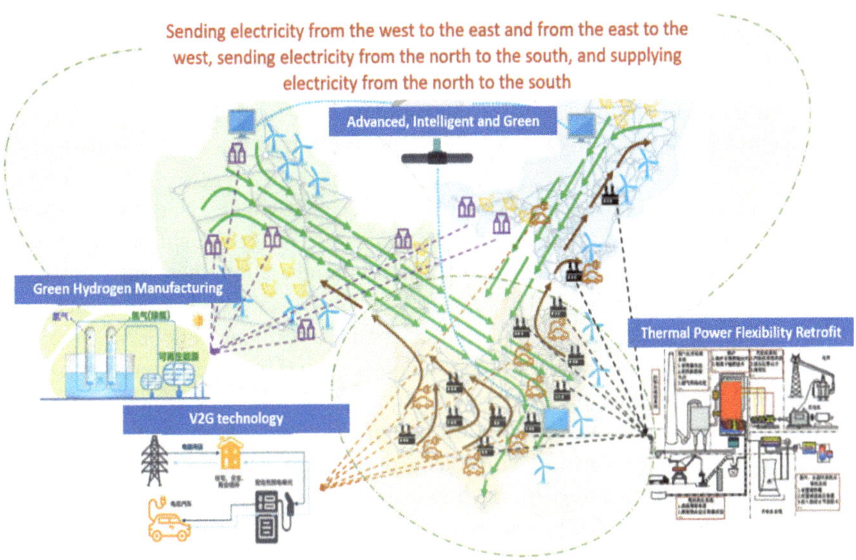


**Fig. 5** Future cross-province and cross-region electricity transfers

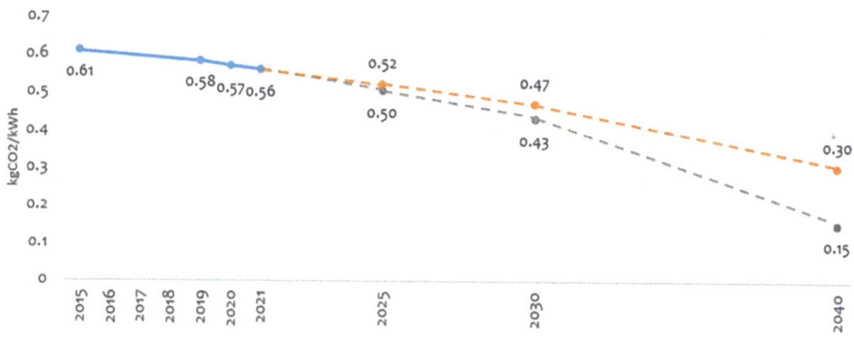


**Fig. 6** Average carbon emission factors for China's power grid (dashed line is predicted)

### 3.3.5 Electricity Sector Low-Carbon Transition Accelerates

The carbon emissions intensity of China's electricity industry will accelerate during the 15th Five Year Plan, with carbon emissions per kilowatt hour decreasing to 0.43 ~ 0.47 kg $CO_2$ by 2030. After 2030, the rate of decrease in carbon emissions intensity from electricity will further accelerate, reaching 0.15 ~ 0.3 kg $CO_2$/kWh by 2040 (Fig. 6).

## *3.4 Policy Recommendations*

### 3.4.1 Actively Promoting the Development of Non-fossil Fuels and Promoting Low-Carbon Transformation of the Power Sector

First, develop non-fossil energy, mainly wind, photovoltaic, and peak shaving resources. Second, ensure the connection between the development plan of the renewable energy industry and the national spatial planning. Strengthen the assessment of renewable energy resource development reserves, establish a national database of exploitable resources, timely incorporate spatial information such as exploitable utilization scope into the same level national land spatial basic information platform and national land spatial planning "one map", and reasonably arrange the scale, layout, and development and construction timing of new land for renewable energy projects. Third, on the basis of vigorously developing green electricity, we need to accelerate the demonstration and application of hydrogen production through electrolysis of water and the production of liquid fuels through electricity.

### 3.4.2 Upgrading and Promoting the Cleaner Transformation of Coal-Fired Power, Enhancing the Efficiency, Cleanliness, Low-Carbon, Flexible, and Intelligent Level of Coal-Fired Power

First, deepen the clean and low-carbon transformation of coal-fired power. Increase the proportion of large capacity, high parameters, and low pollution clean and efficient coal-fired power units. Explore the coupling of "coal-fired power + renewable energy + energy storage", deeply expand the regulation capacity of coal-fired power, increase the demonstration application of CCUS, and promote coal-fired power units to achieve ultra clean emissions based on accelerating ultra-low emission transformation.

Second, promote the transformation of coal-fired power towards basic and system-regulated power sources. Continuously increasing technological investment in the flexibility transformation technology of coal-fired power generation will improve coal's ability to provide flexibility and reliability services or serve as emergency backup while running less. Continuously optimize and establish an intelligent scheduling operation mode based on local conditions, improve the meteorological observation, resource evaluation, and forecasting technology capabilities, enhance the accuracy of wind and photovoltaic output prediction, achieving coupled operation of coal-fired power and new energy, and optimizing the overall power generation efficiency of coal-fired power.

Third, optimize supporting policies and support. Increase the support for special funds/loans for the transformation and upgrading of coal-fired units, and introduce financial and tax reduction and refund policies for the transformation and upgrading of coal-fired units by region, unit, and content. Accelerate the improvement of spot markets and auxiliary service mechanisms, increase paid peak shaving compensation efforts, and appropriately integrate coal-fired power into these new market mechanisms. Improve the coordination between the retirement, standby, and life extension mechanisms of coal-fired power plants and the requirements for renovation and upgrading.

### 3.4.3 Build an Energy Storage System Adapted to the New Power System and Improve the Stability of the Power Grid

First, actively guide the coordinated development of new energy and energy storage. Conduct scientific research on the scale and proportion of new energy distribution and storage, innovate the integrated development model of hybrid "new energy + energy storage" projects. Strengthen the application and promotion of smart centralized control technology, and achieve coordinated complementarity between wind, solar and storage.

Second, promote the large-scale development of new energy storage in various links such as generation, network, and load. Promote the development of various short-, medium-, and long-term energy storage technologies such as electrochemical energy storage, compressed air energy storage, gravity energy storage, and hydrogen energy. Coordinate distributed new energy, adjustable industrial load, electric vehicles, new energy storage and other resources, promote the demonstration and application of integrated source network load storage projects.

Third, establish a new energy supply and demand model of "source load interaction". Use prices and fund technology demonstrations to promote the potential of industrial flexible loads such as electrolytic tanks, as well as demand side resource regulation in building air conditioning, new energy storage, electric vehicles, etc., to provide demand response that can enhance system flexibility and peak shaving. Increase the difference between peak and off-peak electricity prices for electric vehicles, and develop programs to manage charging of electric vehicle fleets in ways that reduce customer costs and provide grid benefits. Invest in demonstrations of bidirectional vehicle charging and discharging, fully tapping into the energy storage and regulation potential of widely distributed electric vehicles.

### 3.4.4 Accelerate the Development of Major Cutting-Edge Energy Technology Innovation, and Consolidate the Foundation of Low-Carbon Energy Transformation

First, strengthen research on cutting-edge technologies in electricity and focus on solving key core technical problems in electricity. Encourage specialized and innovative enterprises to carry out breakthroughs in key technologies. Increase support for intelligent manufacturing and digital upgrading of the new energy industry, and achieve deep integration of the new energy industry and information technology.

Second, actively cultivate disruptive technologies and future industries in the energy sector. Increase basic research on new generation energy "disruptive technologies", and focus on breaking through new perspectives and principles of technologies. Improve the policy support system for technology industries, and explore effective commercialization paths for industrial development.

### 3.4.5 Build a Unified Modern Energy and Power Market System to Promote Optimized Allocation of Electricity Resources Nationwide

First, accelerate the construction of a unified national electricity market system and guide the coordinated operation of electricity markets at all levels. Accelerate the construction of an efficient and collaborative electricity market system.

Second, strengthen national power interconnection, promote the optimization of clean energy allocation on a larger scale, and break the long-standing limitations of power transmission within the scope of power grid operation zones. Strengthen the interconnection between the western Mongolian power grid and the State Grid, and promote more transmission of Inner Mongolia's wind and solar resources to the eastern regions. Optimize and improve policy mechanisms and achieve cross-regional power transmission and mutual assistance between the State Grid, Southern Power Grid, and West Mongolia Power Grid.

Third, bring consistency to environmental attribute markets. Promote the organic connection between the green electricity market, green certificate market, and carbon trading market. Achieve the coupled development of the "electricity certificate carbon" market, and avoid duplicate environmental value payments and incentives. Strengthen the data exchange and standardize green electricity, green certificates, and carbon markets to establish a unified supervision platform for the entire process of green electricity environmental value transfer.

# 4 Electrification of Industrial Heating in China

## *4.1 Background*

To meet China's climate targets and attain a zero-carbon industrial sector, decarbonizing these industrial heating processes is necessity. Nearly 100% of that industrial heating energy in China is supplied by fossil fuel sources. Ferrous metals and non-metallic minerals are two industries that account for much of China's needed high-temperature (over 1100 °C) process heat (Fig. 7a); over a third of China's industrial heating needs fall in this temperature bracket (Fig. 8b). Low temperature industrial heat (below 150 °C) and low-to-medium temperature heat (150–300 °C) each also accounted for a little more than a quarter of the industrial sector's process heating demands (Fig. 7b).

## *4.2 Electrical Technologies for Industrial Heat*

The Techno-Economic Comparison section of this report (below) focuses on three specific electrical technologies: electric resistance boilers, industrial heat pumps, and thermal batteries, because these technologies have broad applicability across many industrial sub-sectors and can be readily compared to fossil fueled alternatives.

Heat pumps and thermal batteries can reduce electricity costs for industrial firms relative to other electrical technologies. Heat pumps reduce costs by achieving extremely high energy efficiencies, while thermal batteries enable firms to purchase particularly inexpensive electricity (i.e., during hours when the grid is over-supplied or from cheap off-grid wind and solar generation).

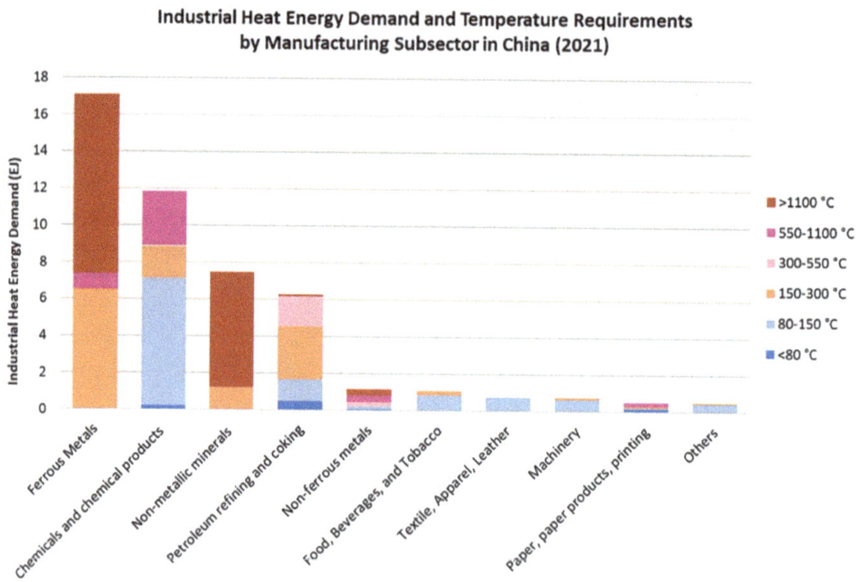


a

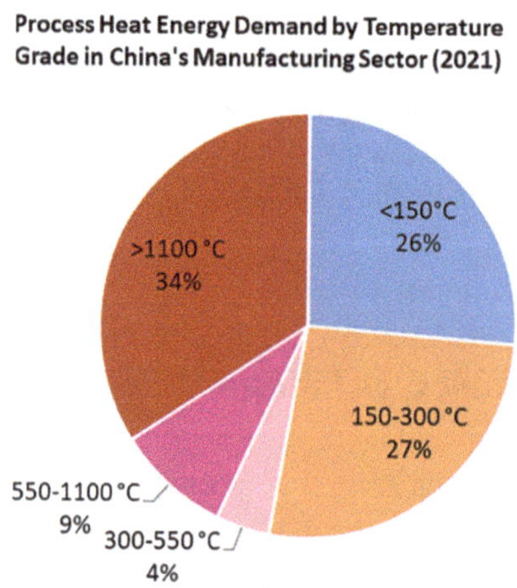


b

**Fig. 7** **a** Process heat energy demand and temperature requirements by manufacturing subsector in China (2021), and **b** process heat energy demand by temperature grade in China's manufacturing sector (2021). *Sources* [6, 7]

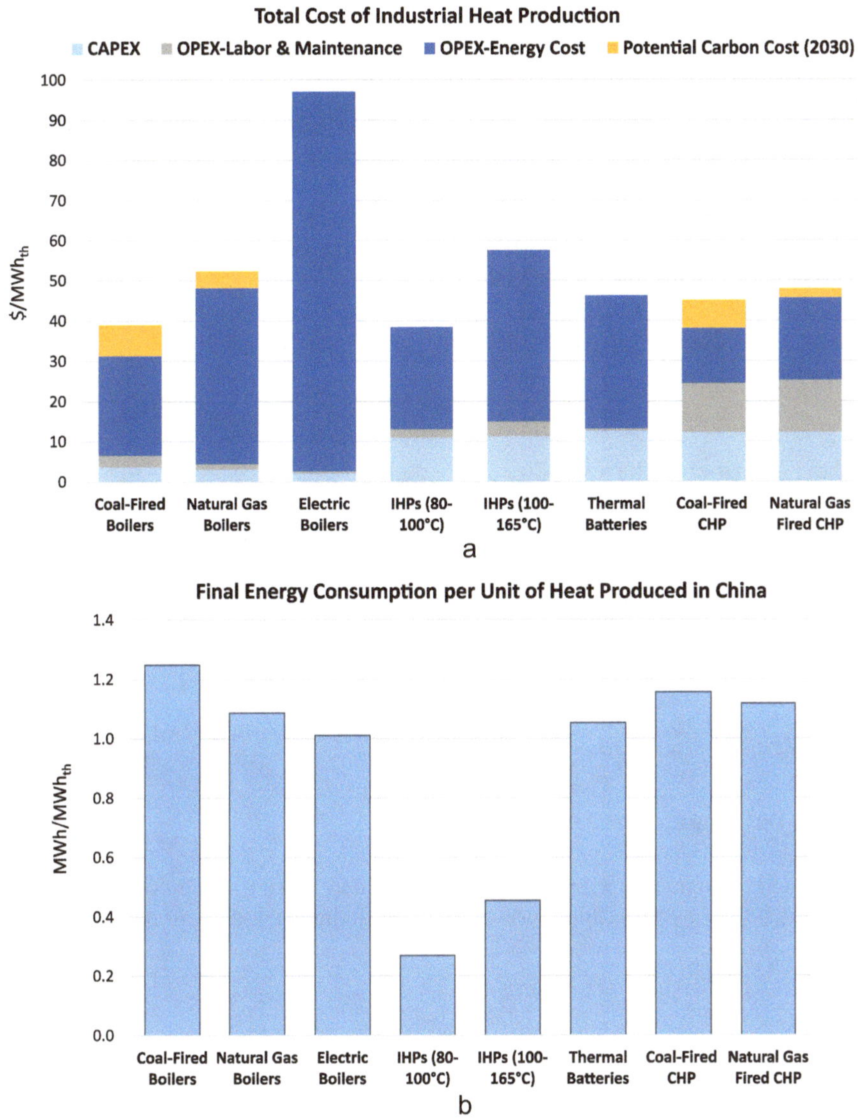


**Fig. 8** **a** Total levelized cost incorporating capital expenditure, energy and non-energy operational expenditure, and forecast 2030 carbon pricing costs of various industrial heat production technologies. **b** We used MWh/MWhth as the unit and indicator to assess final energy intensity per unit of heat production

### 4.2.1 Industrial Heat Pumps

Industrial heat pumps (IHPs) can be the most efficient and cost-effective method to supply low-temperature heat for industrial processes, which can achieve efficiencies several times higher than other electrical technologies.

The uniqueness of these efficiency gains cannot be overstated: no other electrical heating technology or fuel combustion method can generate heat at an efficiency over 100%. That said, upgrading industrial facilities to use heat pumps can be technically challenging. The potential for industrial heat pumps to address the needs of China's industries is large. Approximately 7000 PJ in 2021, or 13% of China's industrial non-electricity energy demand falls within the feasible temperature ranges for heat pumps [8].

### 4.2.2 Thermal Batteries

Most industrial heat demand is at temperatures too high to be provided by heat pumps. For these processes, thermal batteries are often a viable option. While other batteries, such as lithium-ion, also have the ability to store energy, thermal batteries have much lower capital costs per unit capacity.

## *4.3 Techno-Economic Comparison*

### 4.3.1 Data Inputs and Assumptions

A detailed comparative analysis of three electrified heating technologies and their alternatives, focused on China, was conducted; findings are discussed below.

### 4.3.2 Findings

***Cost Comparison***

A comparison of the levelized cost of various industrial heat technologies (Fig. 8a) indicates that for temperatures under 100 °C, industrial heat pumps are highly competitive with CHP variants and considerably cheaper than natural gas or electric boilers in China. When a 2030 estimated carbon cost is considered, industrial heat pumps will be the lowest-cost option to provide industrial heating for temperatures under 100 °C.

Heat produced from high-temperature industrial heat pumps is only 10 and 20% more costly than heat from natural gas boilers and natural gas-fired CHP respectively, so heat pumps could be competitive with natural gas at these temperature ranges. The International Energy Agency and Tsinghua University found that an industrial heat pump with a COP of 3 had a cost of \$40/$MWh_{th}$ [9], in between the estimated costs of the two heat pump variants in Fig. 8a. Industrial thermal batteries can meet heat needs at temperatures far higher than heat pumps at a cost of \$46/$MWh_{th}$ (¥314/$MWh_{th}$), in between the two heat pump variants. Thermal batteries are competitive with the two natural gas-fired technologies today and will be similarly competitive with coal-fired technologies after considering a 2030 carbon price.

***Energy Consumption***

Industrial heat pumps are much more energy efficient than all other heat-producing technologies included in this analysis. The efficiency differences between non-heat pump technologies are small but generally favor electrical technologies due to the better efficiency of electrical resistance versus combustion.

***$CO_2$ Emissions***

In terms of $CO_2$ emissions, industrial heat pumps are one of the least carbon-intensive options for heat production in China. As shown in Fig. 9a, $CO_2$ emissions from lower-temperature (80–100 °C) industrial heat pumps are 26%, 60%, and 73% lower than natural gas, coal, and electric boilers, respectively, when emissions from the generation of electricity to power the heat pump are included.

To achieve China's carbon neutrality goals, China's power sector needs to reach zero or negative emissions before 2050 [10]. Assuming China achieves its target, China's grid emissions factor in 2050 will be zero. As a result, industrial heat pumps, thermal batteries, and electric boilers will have no $CO_2$ emissions (Fig. 9b). Simultaneously, CHP units will no longer offset $CO_2$ emissions from the electric grid, so their emissions intensities will be higher than in 2021. This highlights that decarbonizing the power sector in China is necessary to decarbonize industrial heat production via electrification.

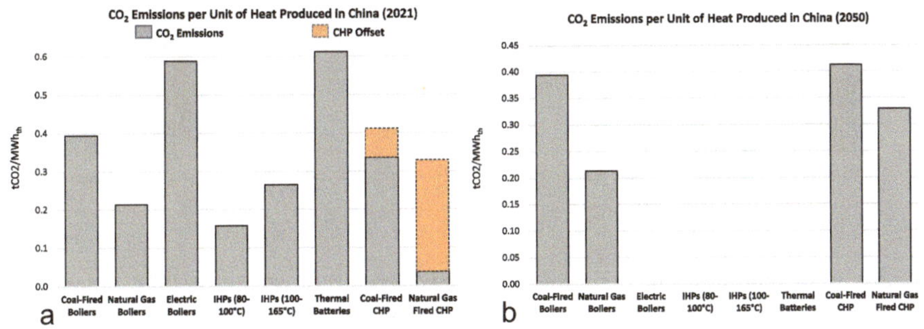


**Fig. 9** **a** $CO_2$ emissions per unit of heat produced by various technologies in China (2021). *Notes* (1) based on China's grid emission factor of 2021; (2) considers offsetting grid $CO_2$ emissions due to electricity generation onsite in CHP units. **b** $CO_2$ emissions per unit of heat produced by various technologies in China (2050), assuming a fully decarbonized electric grid

### *Conventional Pollutant Emissions*

In 2050, assuming China meets its targets, the grid will be decarbonized, and electrical heating technology will therefore not cause conventional pollutant emissions (Fig. 10, electrical technologies boxed in green).

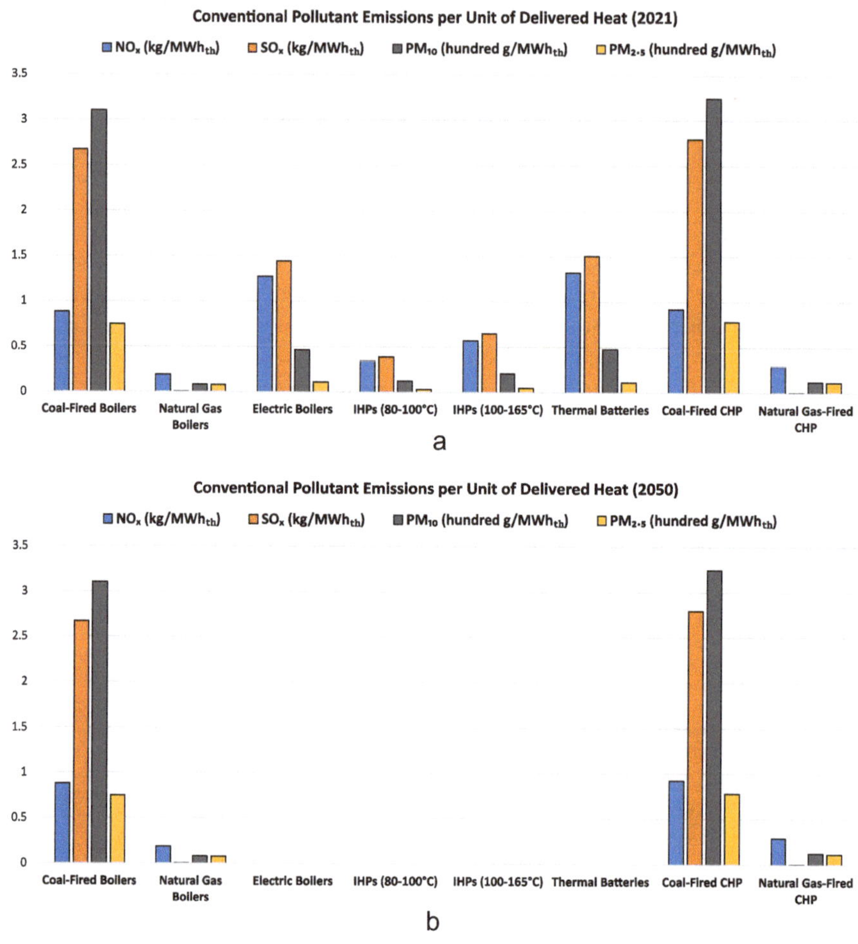


**Fig. 10** Conventional pollutant emissions per unit of delivered heat in China in 2021. Technologies boxed in green would have no pollutant emissions by 2050, assuming a fully decarbonized electric grid

### 4.4 *Barriers and Policy Options*

There are several barriers to this energy transition. Principal among them is the price difference between renewable electricity and fossil fuels and the broader challenges with shifting the electricity sector entirely to zero-carbon sources.

## 5 Development and Implementation of Carbon and Pollution Reduction Legislation

### 5.1 *Introduction*

The existing environmental laws and regulations and their enforcement still fall short of the requirements of the concept of synergistic carbon and pollution reduction. First, there is no legislation at the legal level for the "Climate Change Response Law". Second, the existing climate change legislation is scattered and not of high rank. Third, the relationship between climate legislation and related environmental, resource, and energy regulations is unclear. Fourth, the enforcement of climate regulations is ineffective. Current legislative and judicial practices provide opportunities for implementing synergistic carbon and pollution reduction. At the legislative level, the codification of the ecological and environmental code offers an opportunity to enhance the coordination of carbon and pollution reduction rules at a higher overall level.

### 5.2 *Implementing the Concept of Synergistic Carbon and Pollution Reduction Through the Compilation of the Ecological and Environmental Code*

#### 5.2.1 Basic Approach to Implementing the Concept of Synergistic Carbon and Pollution Reduction in the Compilation of the Ecological and Environmental Code

***Progress, Basic Structure, and Main Content of the Compilation of the Ecological and Environmental Code***

The compilation of the ecological and environmental code aims to sort out, classify, and integrate the current environmental legislative system, which mainly comprises 34 environmental laws, and to comprehensively and systematically update the existing environmental legal system in accordance with the concept of ecological civilization. Since 2019, the Environmental Resources Law Research Association of the China Law Society has organized environmental law scholars nationwide to draft the "Expert Proposal of the Ecological and Environmental Code (Draft)", laying the foundation for the Legislative Affairs Commission of the NPC to carry out the compilation work of the code.

***Basic Approach to Implementing the Concept of Synergistic Carbon and Pollution Reduction***

The "Guiding Opinions on Coordinating and Strengthening the Work Related to Climate Change and Ecological Environmental Protection" (Huan Zong He [2021] No. 4) issued by the Ministry of Ecology and Environment in 2021 established "highlighting synergistic efficiency" as one of the basic principles, requiring the "coordinated control of greenhouse gases and pollutants emissions, coordinated promotion of climate change adaptation and ecological protection and restoration, and supporting the in-depth battle against pollution and the peaking of carbon dioxide emissions." In the five-part structure of the ecological and environmental code, the general provisions, the section on pollution control, and the section on green and low-carbon development are the key sections for implementing the concept of synergistic carbon and pollution reduction.

### 5.2.2 Implementation of the Synergistic Concept of Carbon and Pollution Reduction in the Pollution Control Section

***Core Concept, Basic Structure, and Main Content of the Pollution Control Section***

The pollution control section is compiled based on current pollution prevention legislation, with the goal of scientifically organizing existing pollution prevention laws to achieve systematization and incorporating the latest achievements in ecological civilization system and mechanism reform.

***Reflecting the Synergistic Concept of Carbon and Pollution Reduction in the Section***

The pollution control section is compiled based on existing pollution prevention legislation, focusing on creating a system for synergistic governance of pollution and carbon reduction, mainly reflected in the following three aspects: (1) The legislative concept of pollution control is updated. (2) The general systems of pollution control are improved. (3) Specialized systems are created in the chapter on air pollution control.

### 5.2.3 Core Concept, Basic Structure, and Main Content of the Green and Low-Carbon Development Section

The Third Plenary Session of the Twentieth Central Committee of the Communist Party of China (CPC) explicitly called for "accelerating the comprehensive green transformation of economic and social development" and promoting "green and low-carbon development". The compilation strategy for the green and low-carbon development section follows the logic of "general to specific, domestic to international, and established to emerging." It forms an economic closed loop of "clean production (upstream)—green circulation (midstream)—green consumption (downstream)," making selective adjustments to existing legal norms while maintaining an appropriate level of openness and adaptability.

## 5.3 *Implementing the Synergistic Concept of Carbon and Pollution Reduction Through Improved Environmental Public Interest Litigation*

### 5.3.1 The Basic Approach to Implementing the Synergistic Concept of Carbon and Pollution Reduction

Air pollution public interest litigation serves as a crucial tool for implementing the synergistic governance concept of carbon and pollution reduction. On one hand, the large-scale emission of greenhouse gases causing global warming can exacerbate the health and ecological impacts of air pollution. On the other hand, air pollution can intensify the effects of global warming. Due to the commonality of sources between air pollution and climate change, controlling atmospheric pollutant emissions and reducing greenhouse gas emissions have synergistic effects. Therefore, current air pollution public interest litigation can directly serve the purposes of climate change litigation, using judicial pressure to force adjustments in industrial and energy structures, thereby reducing greenhouse gas emissions.

### 5.3.2 Civil Public Interest Litigation Pathways for Implementing the Synergistic Concept of Carbon and Pollution Reduction

In the dual-carbon field, the Supreme People's Court issued judicial service guidance for the "dual carbon" initiative in 2023, mobilizing judicial actions for "dual carbon" independent initiatives. Besides the "abandoning wind and solar energy" civil public interest litigation, air pollution civil public interest litigation also serves as a primary pathway for implementing the synergistic concept of pollution and carbon reduction.

Successful judicial experiences in air pollution civil public interest litigation can provide important insights for China's climate change litigation. Firstly, at the level of defendants and specific polluting behaviors involved, the defendants in air pollution civil public interest litigation are highly consistent with those in climate change litigation. Secondly, concerning the consequences of damage, in air pollution civil public interest litigation, the existence of damage is self-evident as long as there is excessive or unauthorized emission, despite the degree of damage still needing to be proven by the plaintiff. Lastly, the choice of remedies by the court significantly influences whether air pollution civil public interest litigation can also reduce climate pollutant emissions..

## 5.4 Policy Recommendations

### 5.4.1 Expanding the Carbon Reduction Function of the Pollution Control System in the Pollution Control Section of the Ecological and Environmental Code

First, the legislative concept of pollution control should be updated to introduce the principle of synergistic governance for pollution and carbon reduction. Second, the general system of pollution control should be improved to meet the needs of coordinated management of pollution reduction and carbon reduction. Third, a special system should be created in the section on air pollution control to build a coordinated control system for air pollutants and greenhouse gases, addressing the specific needs of key regions and industries.

### 5.4.2 Providing Specialized Legal Regimes for Carbon Reduction in the Green and Low-Carbon Development Section of the Ecological and Environmental Code

First, the section on comprehensive utilization of resources should promote energy conservation and low-carbon practices across the entire chain of production, logistics, and consumption. Second, the section on green energy should establish institutional norms for increasing renewable energy and energy conservation, achieving the synergy of carbon and pollution reduction at the source through the transformation and upgrading of the energy structure. Third, the section on addressing climate change should focus on mitigation and adaptation, establishing systems for controlling greenhouse gas emissions, statistical accounting, allocation of carbon emission quotas, carbon trading, regulation, voluntary emission reductions, climatic risk evaluation, and enhancement of carbon sink capacity.

### 5.4.3 Implementing the Synergistic Concept of Carbon and Pollution Reduction Through Multi-stakeholder Participation in Environmental Public Interest Litigation

First, procuratorial organs should explore public interest litigation in the areas of air pollution control and carbon sink protection to ensure that the approvals of high-emission projects are carried out in strict accordance with the law. Second, the government should initiate ecological and environmental damage compensation litigation related to the synergy of pollution reduction and carbon reduction. Third, social organizations should have clear pathways to file civil public interest litigation related to pollution reduction and carbon reduction. Fourth, courts should address complex issues in climate justice, such as causality and remedies, in public interest litigation cases.

# 6 Policy Recommendations

1. **Establish a new policy narrative that promotes emission reduction alongside economic growth by integrating policies with market dynamics, enhancing the availability of financial resources for emission reduction technologies, improving the institutional framework, and driving green development towards a just transition.**

   (1) Increase awareness of green development concepts among governments, industries, and the public. Promote and publicize the opportunities for green growth created by emission reduction. (2) Strengthen policy incentives to energize green finance and the carbon trading system, and attract social capital. Create a financial policy system that is market-led with government support, and refine performance assessments to focus on carbon reduction, pollution reduction, green expansion and growth. By implementing stringent environmental standards, encourage traditional industries to upgrade and accelerate the development of low-carbon sectors and expand investments in renewable energy to ensure economic sustainability. (3) Advocate for green development, promote the dematerialization of the economy, and prioritize investments in sustainable agriculture, food, and cultural creative industries. Support ecological services and the preservation of natural capital, considering the interests of all parties to pursue a fair and green transition, and strengthen international cooperation in emission reduction. Ensure that large-scale investments in new infrastructure or the rehabilitation of existing facilities meet the highest environmental and greenhouse gas emission standards.

2. **Revise air quality standards in a timely manner, design pathways for coordinated multi-objective emission reductions, and strengthen collaborative governance in key regions and critical sectors.**

   (1) Drawing on current local air pollution data, health impact studies, and benchmarks, and inspired by research, guidelines, and methods from the World Health Organization and other developed countries, promptly revise environmental air quality standards. Enforce stricter limits on particulate matter concentrations and enhance the urban air quality management system with a focus on compliance. (2) Begin by considering health benefits and economic costs to integrate air quality synergistic benefits into the decision-making factors for choosing pathways to carbon neutrality, and conduct research on multi-objective control strategies. Choose pathways to carbon

neutrality that primarily rely on renewable energy, establish a zero-carbon energy industry system, and promote substantial long-term improvements in air quality. (3) Identify key industries, develop localized paths for synergistic carbon reduction and pollution control, fully unleash the synergistic potential for carbon reduction across various industries, and efficiently drive the achievement of the dual carbon goals, thereby continuously improving air quality and public health.

3. **Actively promote the development of non-fossil energy sources, continuously advance the upgrading and transformation of coal-powered plants, build and enhance energy storage systems, improve grid stability, and drive the power sector's low-carbon transition.**

   (1) Set clear targets for renewable energy development. Integrate renewable energy development planning with national land use planning, and foster seamless integration between the green electricity market, green certificate market, and carbon trading market. (2) Be sure that the transition is designed so that it increases energy security along the way. Further decarbonize remaining coal energy through technologies like "coal power + energy storage" and carbon capture, utilization, and storage (CCUS). (3) Coordinate distributed renewable energy, adjustable industrial loads, electric vehicles, and innovative energy storage. Strengthen national power interconnectivity and promote the optimized distribution of clean energy.

4. **Promote the electrification of industrial heating by advancing the use of technologies such as industrial heat pumps and thermal batteries, and leverage fiscal and tax policies to support technology research and development and achieve cost parity with coal power. These measures will help accelerate the process of industrial electrification.**

   (1) Incentivizing the transition using equipment rebates, retooling grants, and access to low- interest financing mechanisms to offset the capital expenditures related to adopting these technologies. (2) Enhancing existing energy-efficiency standards, emissions standards, and green public procurement programs can likewise encourage the transition to direct electrification. (3) China's research laboratories can collaborate with private industry on research and development (R&D) programs to move these early-stage technologies forward. (4) Create a competitive landscape between coal and electricity by carbon pricing or by subsidizing the cost of clean electricity and the cost of upgrades to support electrification. (5) Inter-provincial electricity trading and optimization of China's Green Electricity Certificate system can help facilitate access to clean electricity.

5. **The General Provisions and Pollution Control sections of the Environmental Code enhance the carbon reduction function of pollution control systems, while the Green and Low-Carbon Development sections provide specialized legal frameworks for carbon reduction, encouraging diverse entities to**

**participate in environmental public interest litigation and implementing the concept of synergistic enhancement through carbon reduction and pollution control.**

(1) Update legislative principles, clarifying that governments at all levels and their environmental departments are legally responsible for supervising and managing pollution prevention and greenhouse gas control. (2) Refine the general pollution control system and pollution control and create specialized systems within the section on air pollution control. (3) The prosecutor's office further explores the development of civil and administrative public interest litigation in areas while the government attempts to initiate litigation related to ecological environmental damage compensation that is tied to synergistic carbon reduction and pollution control.

## References


1. Xue, T., Wang, R., Wang, M., et al. (2024). Health benefits from the rapid reduction in ambient exposure to air pollutants after China's clean air actions: Progress in efficacy and geographic equality. *National Science Review*. https://doi.org/10.1093/nsr/nwad263
2. China NBS. (2022). *China energy statistical yearbook* 2011. http://www.stats.gov.cn/english/
3. Cheng, J., Tong, D., Liu, Y., et al. (2023). A synergistic approach to air pollution control and carbon neutrality in China can avoid millions of premature deaths annually by 2060. *One Earth*. https://doi.org/10.1016/j.oneear.2023.07.007
4. Shi, X., Zheng, Y., Lei, Y., et al. (2021). Air quality benefits of achieving carbon neutrality in China. *Science of the Total Environment*. https://doi.org/10.1016/j.scitotenv.2021.148784
5. Harvey, H., Orvis, R., & Rissman, J. (2018). *Designing Climate Solutions A Policy Guide for Low-Carbon Energy 1 版*. Island Press Washington, DC. https://doi.org/10.5822/978-1-61091-957-9
6. NBS. (2023). *China energy statistical yearbook* 2022. China Statistics Press.
7. Rissman J. (2022). *Decarbonizing low-temperature industrial heat in the U.S.* San Francisco.
8. Sawe, N., Yu, H., Rissman, J., Tian, Z., & Zhou, N. (2024). *Clean industry in China: A techno-economic comparison of electrified heat technologies, barriers, and policy options*. Lawrence Berkeley National Laboratory. https://escholarship.org/uc/item/7tg4x3n3
9. University. IEA & T. (2024). *The future of heat pumps in China*. https://www.iea.org/reports/the-future-of-heat-pumps-in-china
10. Yu, S., Fu, S., Behrendt, J., Chai, Q., Chen, L., Chen, W., Cheng, X., Clarke, L., Du, X., Guo, F., Hultman, N., Khanna, N., Krey, V., Li, M., Liu, J., Lu, H., Lou, J., Mei, C., Qin, X., ... Zhou, N. (2022). *Synthesis report 2022 on China's carbon neutrality: electrification in China's carbon neutrality pathways. energy foundation China*. https://www.efchina.org/Attachments/Report/report-lceg-20221104/Synthesis-Report-2022-Electrification-in-Chinas-Carbon-Neutrality-Pathway.pdf

# Chapter 5
# Green Development and Climate Adaptation for Urban and Rural Areas

## 1 Introduction

Cities face increasing climate risks due to high concentrations of population and economic activity. The countryside is also facing a changing climate, affecting the rural population and the harvest yields they depend on. The progressive nature of climate change, combined with already densely populated areas, increases difficulty of action. At the same time, due to their social and economic strength and importance, densely populated urban and rural areas can become transformational centers for climate adaptation.

## 2 Conceptual Assessment Framework for Climate Adaptation

### *2.1 Climate Adaptation Resilience Assessment Methods*

In recent years, in order to promote the effective implementation of climate adaptation measures and accelerate the transition, international organizations such as the United Nations, national governments such as USA and EU, 100 Resilient Cities [1], academia and consultants have launched a wide variety of tools to assess (a) the need for adaptation, (b) the benefits of climate adaptation measures and (c) the enabling conditions and quality of the tools to accelerate implementation.

Despite this fragmentation of climate adaptation assessment frameworks, three phases of climate adaptation and resilience can be identified: (1) Preparedness and

---

The content of this chapter is extracted from the main findings of the 2023–2024 CCICED Special Policy Study (SPS) on "Green Development and Climate Adaptation for Urban and Rural Areas"

 
*Adhering to Green and Open Cooperation to Advance Modernization*,
https://doi.org/10.1007/978-981-95-5547-5_5

precaution, (2) Relief and (3) Recovery/rebuilding. The framework should not only be focusing on disaster risk reduction, for 'sudden' disasters, but also include slow changes like sea level rise and the consequences of increasing average temperatures.

### 2.2 *Five Capacities and Their Enabling Conditions for Creating Climate Resilience*

A brief outline of the five capacities is given below: (1) Threshold capacity: Refers to the capacity to cope with disaster risk thresholds established in urban and rural areas to avoid disaster losses due to extreme climate events. (2) Coping capacity: Refers to the ability of an urban or rural area to reduce disaster losses following a disaster caused by extreme weather. (3) Recovery capacity: Refers to the capacity of urban and rural areas to recover from a disaster to a state consistent with or better than the state before the disaster. (4) Adaptive capacity: Refers to the ability to cope with uncertainty about future climate change, as well as uncertainty about demographic, economic, technological and other developments. (5) Transformative capacity: Refers to the ability of the current social, economic and governance system to transform and pro-actively change itself in the face of expected disaster risks.

Enabling conditions are needed to be able to take action in strengthening the five capacities. It includes [2–5]: (1) Sense of urgency; (2) Systems approach; (3) Capacity development; (4) Trusted knowledge, data and information; (5) Financing; (6) Citizen's engagement; (7) Policy, legal & regulatory framework; (8) Innovation; (9) Public-private system collaboration; (10) (Fair) sharing of risks and rewards.

## 3 Major Case Studies: Examples of Climate Adaptation Practices

### 3.1 *Case 1: Climate Adaptation in the Chengdu-Chongqing Area*

#### 3.1.1 Trends in Climate Change

Over the past 50 years, the region has experienced a fluctuating upward trend in temperature, with an average increase rate of approximately 0.14 °C per decade, while precipitation has shown a linear decrease, with a decline rate of about 17.5 mm per decade. The frequency of extreme weather events, such as heatwaves and heavy rainfall, has increased, particularly in the southwestern areas.

Summer rainstorms are a frequent extreme weather event in Chengdu and Chongqing, exacerbated by global climate change and the urban "rain island effect." Recent years have seen an increase in extreme meteorological and hydrological events, particularly affecting the mountainous city of Chongqing at the confluence of two rivers. This has led to the dual threat of localized torrential floods and river flooding. In the Chengdu-Chongqing area in recent years, high temperatures, drought and other extreme weather continue to appear, forest fires show more frequent occurrences. The Chengdu-Chongqing region's energy structure relies heavily on hydropower, coal power, and natural gas, with hydropower comprising nearly 80% of Sichuan Province's total energy production. Seasonal climate variations significantly impact power generation, with a notable decrease in capacity during the dry season, leading to power shortages. The impact of climate change has gradually weakened overall grain production, with extreme weather conditions, particularly drought, leading to periodic reductions in yield.

### 3.1.2 Major Disasters and Issues Arising from Climate Change

It has continued to promote the construction of river embankments to meet the standards and enhance the ability to defend against floods; it has organized the excavation of fire protection zones and optimized the layout and design of infrastructure, and has upgraded the level of operation and management of the infrastructure, and so on. It has strengthened risk research and identification, organized geo-engineers and grid clerks to carry out flood and rain "triple checks", refined the granularity of heavy precipitation warnings, and enhanced climate change monitoring, early warning and risk management capabilities. It includes the implementation of a daily dispatch system for disaster situations at the municipal level, the formulation of technical guidelines such as the Technical Measures for Responding to Grain and Oil Crop Production in Hot and Drought Weather, making full use of water conservancy facilities to increase the capacity for emergency irrigation, organizing late-autumn production to make up for the poor, organizing drought-resistant agricultural service teams to go deep into the countryside, and strengthening the capacity for regional coordination and the mobilization of personnel and materials, among other things.

### 3.1.3 Current Gaps in Climate Adaptation

***Threshold capacity***: The Chengdu-Chongqing region has insufficient capacity to deal with various climate disasters. In terms of flood disaster response, the flood responding pattern of "people retreat when water advances while people advance when water recedes" has not been effectively promoted. In terms of forest fire response, the adaptability of forest structure is insufficient, with a lack of fire-resistant and drought-resistant tree species, and the construction of firebreak systems needs to be improved.

***Coping capacity***: Firstly, the capacity for climate change risk prediction is not strong, and the capacity for precise identification, assessment and planning response to risks is relatively weak. Secondly, there is room for improvement in public awareness of disaster prevention and relief as well as climate change response. Thirdly, the basic functions of urban lifeline systems.

***Adaptive capacity***: Firstly, there is a deficiency in the adaptive capacity of key facilities. Secondly, the impact of climate on power supply is significant.

***Transformative capacity—the capacity to change***: There is a lack of inter-departmental collaboration and smooth information sharing in the Chengdu-Chongqing region. The joint dispatching mechanism for water resources and energy across regions is yet to be improved. e have not yet been widely promoted.

## 3.2 Case 2: Climate Adaptation in the Taihu Lake Basin

### 3.2.1 Trends in Climate Change

In recent years, the Taihu Lake basin has experienced a significant increase in rainfall during the flood season, along with a rise in extreme weather events, leading to an elevated risk of low-probability natural disasters. At the same time, there is a clear warming trend across the basin, with the impact of heatwaves in Shanghai intensifying, setting multiple high-temperature records in 2022.

### 3.2.2 Major Disasters and Issues Arising from Climate Change

Strong convective rainstorms are prone to cause localized pluvial flooding in the city. Data from Shanghai between 2004 and 2015 show that thermal convective precipitation occurs on average every 11 days during summer under the influence of subtropical high pressure. Typhoons are frequent, and "wind, storm, tide and flood" are the most serious disasters. The persistent high temperature and low rainfall will easily lead to "flood season reverse dryness" in the basin. The intrusion of salty tide affects the safe and stable supply of raw water from the Yangtze River in Shanghai. Monitoring data shows that in 2022, the incursion of the salty tide in the Yangtze River Estuary will cause the number of consecutive days of non-take of water in the three water sources along the river in Shanghai to exceed 27 days. Continued high temperatures exacerbate the threat of sea level rise. According to the observation data of Wusong tide gauge station, the rate of rise during 1912–1960 was 0.92 mm/a, and has accelerated significantly from 1960 to the present, reaching 2.03 mm/a.

### 3.2.3 Overview of Priority Climate Adaptation Measures in Place

***Taking the large polder as a unit, continuously expanding the scope of urban flood protection and defense standards***

Through the small dikes into large dikes, joint dikes and dikes, etc., the flood control standard of important cities in the basin basically reaches the standard of once in 50 years and above.

***Gradual expansion of urban ecological space and increased ability to adapt to high temperatures***

Municipalities are actively carrying out green space expansion projects, both to effectively curb urban sprawl and to mitigate the urban heat island effect. In 2022, the greening rate of the built-up areas in the basin will reach 41%, an increase of 12% compared with 2010, exceeding the national average (39%).

### 3.2.4 Current Gaps in Climate Adaptation

***Threshold capacity***

First, the local flood control and drainage facilities in response to catastrophes are short of capacity. Secondly, the reduction of natural regulatory spaces and the intensification of aggressive construction have exacerbated the risk of urban waterlogging.

***Coping capacity***

Firstly, there is room for improvement in the level of early warning and prediction of sudden meteorological disasters and extreme disasters. Secondly, although the emergency plan system framework has been basically established, the linkage and precision are not high, and the comprehensive guidance role has not been fully exerted. Thirdly, the guidance role of planning related to resilience in disaster prevention for particular spaces is insufficient. Fourthly, the safety and disaster prevention supporting facilities are insufficient at the community and grassroots levels, and the public's awareness of disaster prevention is not strong.

***Adaptive capacity***

Firstly, the flood prevention standards between upstream and downstream cities in the basin are inconsistent, making it difficult to manage intercity rainstorm and flood in a coordinated manner. Secondly, the linkage capacity among water sources is relatively weak, and the impact of continuous high temperatures combined with rising sea levels poses risks to water supply safety.

***Transformative capacity—the capacity to change***

Firstly, there is insufficient understanding of the various new environmental changes brought about by climate change. Secondly, the overall coordination function of the emergency management department has not been fully exerted.

## 3.3 Case 3: Climate Adaptation in the Pearl River Estuary Region

### 3.3.1 Trends in Climate Change

Climate change in the Pearl River Estuary has led to an increase in both the frequency and intensity of heavy rainfall, with storm centers shifting toward highly urbanized areas. At the same time, sea level rise, in combination with storm surges, threatens low-lying coastal areas.

### 3.3.2 Major Disasters and Issues Arising from Climate Change

Climate warming has increased the frequency and intensity of extreme rainfall in urban areas, leading to an increase in the frequency and extent of flood disasters in urban areas. Due to the higher water levels along the coast, the natural drainage capacity of the city decreases, and the urban sewage discharge is difficult or even backed up, increasing the difficulty of flood discharge and drainage, and aggravating flood disasters. The salty tide extends upstream to the water intake of water supply plants, affecting the normal public water supply, which is hazardous to human health, and can also affect the normal production of industrial enterprises. Since the end of 2021, rainfall in the Pearl River Basin has been persistently low, and the Pearl River Basin has suffered the most severe drought in 60 years (Fig. 1).

### 3.3.3 Overview of Priority Climate Adaptation Measures in Place

***Pearl River Basin Sectors Link Up to Cope with Salt Tide Intrusion Disasters***

During the 2022 Pearl River Delta salty tide intrusion disaster, the Ministry of Water Resources instructed a basin-based unit to release water from a cluster of reservoirs in the Xijiang and Dongjiang river basins in order to replenish the downstream river part and suppress salty water intrusion.

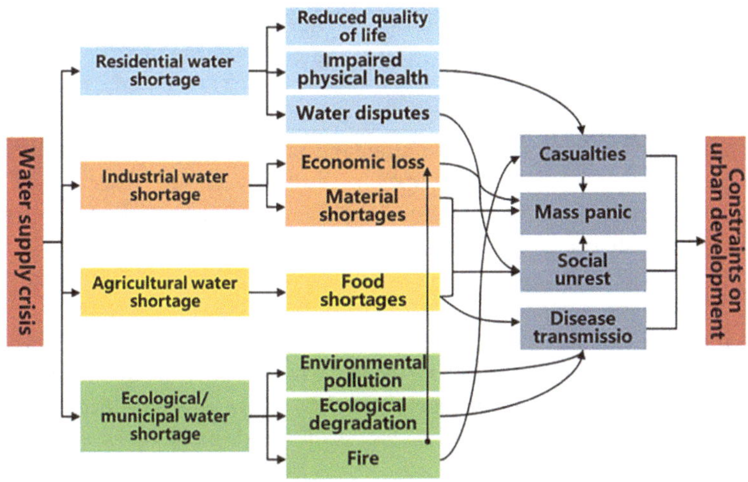


**Fig. 1** Schematic dissemination of the disaster chain of water supply crisis in the Pearl river delta

***Guangdong, Hong Kong and Macao cooperate closely to carry out climate response and maritime emergency rescue***

In 2024, The Guangdong Provincial Government signed the "Framework Agreement on Guangdong-Hong Kong-Macao Co-operation in Emergency Management and Greater Bay Area Co-operation in Emergency Rescue Operations" with Hong Kong and Macao, to strengthen the co-operation and exchanges between Guangdong, Hong Kong and Macao in emergency management and rescue.

***Hong Kong and Shenzhen Joint Social Forces to Address Climate Change—A Case Study of Mangrove Protection and Restoration***

Learning from international experience and strengthening cooperation with Hong Kong, cities such as Shenzhen have begun to actively introduce NGOs that have also begun to join in the protection and management of mangrove forests.

### 3.3.4 Current Gaps in Climate Adaptation

*Threshold capacity*

Firstly, the resilience of urban spatial structure is insufficient. Secondly, the disaster prevention capacities of urban vulnerable nodes such as underground spaces are not strong.

*Coping capacity*

Firstly, there is a shortage in the monitoring and early warning capacities for meteorological disasters. Secondly, the disaster resilience of major infrastructures such as urban lifeline projects needs to be enhanced. Thirdly, there is a need to enhance the disaster prevention awareness among communities and residents.

***Recovery capacity***

The quantity and pumping capacity of drainage facilities allocated internally within the underground spaces in some cities is insufficient. The reserve and inter-departmental dispatch mechanisms for flood control equipment like water pumps are not well-established.

***Adaptive capacity***

Firstly, the risk of annual storm surges and flood disasters in coastal cities is exacerbated by rising sea levels, leading to a decrease in the actual flood and storm surge disaster prevention capacities of cities. Secondly, there is an insufficient capacity to adapt to salt tide intrusion and coastal erosion, making coastal ecosystems more vulnerable to damage.

***Transformative capacity—the capacity to change***

The systematicness of disaster prevention work and inter-departmental collaboration still need to be further improved. The coordination of flood prevention within regions and river basins remains to be perfected.

## 3.4 Case 4: Climate Adaptation in the Rhine-Meuse-Scheldt Delta

The Rhine-Meuse-Scheldt delta is flat and low-lying, but densely populated, highly urbanized, industrialized and agriculturally intensive. This makes their vulnerability or damage sensitivity so high that climate hazards are economically and socially unacceptable. Extreme weather events and near-misses have been on the rise over the past few decades, and exposure to extreme weather and sea-level rise will be severely exacerbated.

### 3.4.1 Threshold Capacity; Damage Prevention

Significant funds have been invested in creating more retention capacity along the Rhine tributaries, the Maas, the Scheldt and their tributaries to withstand flooding.

#### 3.4.2 Coping Capacity

Planning and design goals should include minimizing damage from extreme events that exceed the threshold or design capacity of the water supply system, such as the "water bomb of 2021" and the drought of 2018.

#### 3.4.3 Recovery Capacity

Throughout the Delta region, preparations for disaster recovery are limited. In some cases, such as in Flanders, the financing of recovery is partly covered by private insurance; usually, the national government has to cover the losses. And, as seen in Germany and Belgium, "rebuilding better" is not always allowed.

#### 3.4.4 Adaptive Capacity

Rigid, grey adaptation measures are robust and have a long service life. However, an excessively long lifespan may be counterproductive when new conditions and/or emerging social needs are taken into account.

#### 3.4.5 Transformative Capacity—the Capacity to Change

Climate adaptation is a continuous and collaborative learning process that requires research into new solutions, pathways and methods. Research-Based Design (RBD) has proven to be a valuable approach to this complex challenge, creating innovative solutions that are flexible in the long term. Sustained funding for adaptation measures is crucial. Belgium, the Dutch Delta Fund provides continuous financial support. An implicit obligation to consider climate adaptation as an essential element in such urban renewal and rural redevelopment project is found in the European flood directive and in the water assessment ("watertoets"), obligatory part of the environmental permitting system in the Netherlands and Belgium.

### *3.5 Concluding Discussion*

**These four regions have revealed very similar challenges and similar deficiencies in their ability to cope with the climate crisis, including**: (1) insufficient resilience of the urban spatial pattern in adapting to climate change, and major infrastructures are aged and not designed for future climate conditions; (2) regulations, protocols and standards for addressing climate change still need to be upgraded; (3) the overall coordination function of the emergency management department has not been brought into full play, and the level of early warning of sudden meteorological disasters and prediction of extreme disasters needs to be upgraded; (4) Insufficient risk awareness among community residents, their ability to assess climate risks, and their options for risk-reducing interventions (5) inter-city, regional and river basin wide climate crisis response plans have yet to be connected; (6) The risks faced by small and medium-sized cities, small towns and villages are intensifying.

## 4 Lessons from Disasters: How to Deliver Precaution, Relief and Rebuilding

Cases and findings are grouped according to three typical phases and terminology commonly used in disaster risk reduction: (1) Precaution and Preparedness. (2) Relief—aid, compounding risks, vulnerabilities. (3) Rebuilding—adaptation, resilience, innovation (Fig. 2).

For this part the four major cases portrayed in the former context have been considered, as well as sixteen other cases.

**Major Cases (Chap. 3)**: Case 1: Climate adaptation in the Chengdu-Chongqing area; Case 2: Climate adaptation in the Taihu Lake basin; Case 3: Climate adaptation in the Pearl River Estuary region; Case 4: Climate adaptation in the Rhine-Meuse-Scheldt Delta.

**Precaution & Preparedness**: Case 5: The "July 20" extremely heavy rainstorm Disaster in Zhengzhou, Henan Province in 2021; Case 6: Early Warning

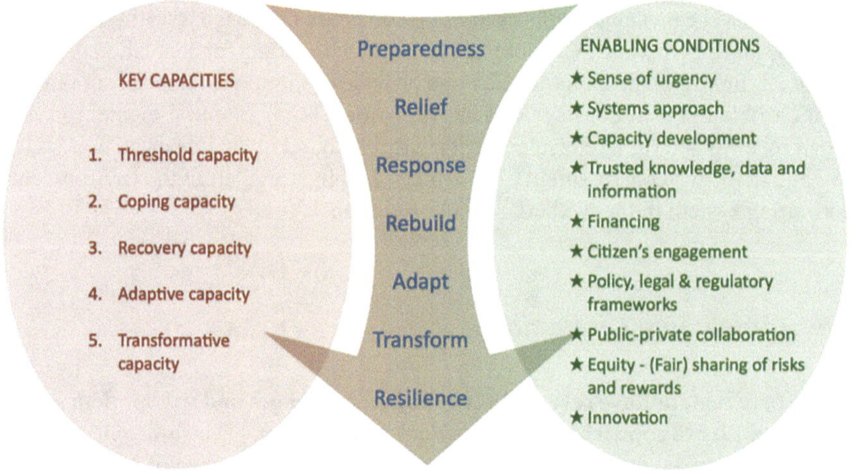


**Fig. 2** Phases of disaster risk reduction and its common terminology related to the five key capacities to develop climate resilience and their enabling conditions

For All(United Nations); Case 7: Bangladesh—Early Warning Systems; Case 8: Singapore—Water Wise.

**Relief—aid, compounding risks, vulnerabilities**: Case 9: Rare and catastrophic flood in Chongqing on August 20th, 2020; Case 10: California wildfires; Case 11: Repeated flooding in Nairobi; Case 12: Flooding and Rebuilding in the Aachen region 2022.

**Rebuilding—adaptation, resilience, innovation**: Case 13: Extremely heavy rainstorm disaster in Beijing Tianjin Hebei in 2023; Case 14: New York—Hurricane Sandy Response; Case 15: Flanders—Post Floods, Preparedness/Resilient Waterland.

**Other cases**: Case 16: High temperature and drought disasters in the Sichuan Chongqing region; Case 17: Mozambique—Partially Prepared; Case 18: Peru—Flood Resilience Task Force; Case 19: Water At The Heart Of Climate Action; Case 20: Water As Leverage—Innovative Preparedness.

More analyses of flooding-related cases: https: //www.zurich.com/knowledge/topics/flood- resilience/zurichs-flood-resilience-program.

## 4.1 *Precaution & Preparedness*

### 4.1.1 Case 5: The "July 20" Extremely Heavy Rainstorm Disaster in Zhengzhou, Henan Province in 2021

From July 17 to 23, 2021, Henan Province experienced a rare extremely heavy rainstorm. Insufficient preparedness in the prevention of, preparation for and response to flooding has been an important cause of the seriousness of these impacts. The problems included insufficient ability to issue timely and unambiguous warnings; attention to the extreme meteorological information was inadequate. In addition, the emergency response was late, and not soundly linked to the flood warnings.

### 4.1.2 Cases6 and 7: Early Warnings for All and Bangladesh—Early Warning Systems

In March 2022, United Nations Secretary-General António Guterres [6] initiated the global Early Warnings for All Initiative, to ensure within five years universal protection from hazardous weather, water, or climate events through life-saving early warning systems. WMO, UNDRR, ITU and IFRC with partners developed a people-centered Multi-Hazard Early Warning System (MHEWS), as an integrated information system for upcoming weather or climate events, while supporting institutions and communities in preparedness, relief and response actions.

Bangladesh leapfrogged from a nation suffering from disasters, to one being fully prepared, capable of moving millions of people out of harm's way, in the lead up to disasters. Thanks to Bangladesh's successful Cyclone Preparedness Program (CPP)—an early warning system where more than 76,000 volunteers, half of whom are women, with the best available data on forecasting go around to even the most remote communities and guide everyone across the country to multifunctional shelter-infrastructure.

### 4.1.3 Case 8: Singapore—Water Wise

Singapore from its independence saw itself confronted with freshwater dependency and embarked on a bold water resilience approach. Singapore implemented a three-pronged water management system to meet the growing water demand: collect every drop of water; reuse water endlessly; and desalinate more seawater [7].

## *4.2 Relief—Aid, Compounding Risks, Vulnerabilities*

### 4.2.1 Case 9: Rare and Catastrophic Flood in Chongqing on August 20th, 2020

On August 18–20, 2020, the Yangtze River area in Chongqing suffered floods. Subsequent measures taken to cope with disasters include a joint approach to dispatching excess water in the basin and ensuring ample retention capacity. In addition, emergency management and evacuation systems at various levels were linked: city, district, street, and community levels. Over and above this, the prevention concept has shifted from engineering resistance to coexisting with floods.

### 4.2.2 Case 10: California Wildfires

The Rim Fire in California started on August 17, 2013 and burned over 1041 $km^2$. It resulted in significant impacts to the local economy, threatened access to clean water, disrupted the lives of thousands of people, and damaged air quality. With funding from a federal program (National Disaster Resilience Competition, coming from Rebuild by Design), a combination of local, state and federal agencies collaborated to address unmet recovery needs and build long-term resilience (Fig. 3).

### 4.2.3 Case 11: Repeated Flooding in Nairobi

In spring 2024, rains triggered flash floods in Kenya that killed at least 228 people, left 72 people missing, and displaced more than 212,000 people. This widespread

**Fig. 3** Community and watershed resilience program, California [8] The diagram shows the program's system-wide perspective considering forest recovery, economic recovery, and community building

destruction by the floods has revealed the urgent need for comprehensive flood management and mitigation strategies with government agencies, non-government agencies, and community groups working towards enhancing early warning systems, improving infrastructure and drainage systems, promoting sustainable land use practices, strengthening disaster preparedness and response, enhancing risk communication and public awareness as well as fostering cross-sectoral collaborations and partnerships.

### 4.2.4 Case 12: Flooding and Rebuilding in the Aachen Region 2022

The floods of mid-2021 caused widespread damage and loss of life in West Germany. Its recommendations point to an obvious need for smarter engineering, including a rethink of the region's many small bridges. The critical role of land use on hillsides and in narrow valleys is underlined once more, as is the need for good general access to risk information and for really fast warning in emergency situations; active preparedness and training at local level, even if these events are deemed to happen once in a lifetime.

## *4.3 Rebuilding—Adaptation, Resilience, Innovation*

### 4.3.1 Case 13: Extremely Heavy Rainstorm Disaster in Beijing Tianjin Hebei in 2023

After the Beijing Tianjin Hebei extremely heavy rainstorm disaster in 2023, the Chinese government acted along multiple lines. One line is planning guidance, led by the National Development and Reform Commission to organize the compilation of the "Plan for Enhancing Disaster Prevention and Reduction Capacity in North China Region with a Focus on Beijing Tianjin Hebei". Next to the reparation of damage, including the reconstruction of water conservancy facilities in river basins, strengthening the meteorological monitoring network in key areas, and improving urban drainage and flood prevention facilities.

### 4.3.2 Case 14: New York—Hurricane Sandy Response

Hurricane Sandy battered the Northeast coast of the United States in 2012, wreaking havoc across the region. President Obama's Hurricane Sandy Rebuilding Tasks Force launched Rebuild by Design, part policy process, part rebuilding program, and part design competition, bringing together all levels of government, stakeholders and residents to innovatively design and develop new standards of regional resilience.

### 4.3.3 Case 15: Flanders—Post Floods, Preparedness/Resilient Waterland

After extreme flooding in July 2021 across the rivers Maas and Demer, the Flemish Government appointed a multidisciplinary Expert Panel on flood protection. Formulating a strategy with 10 coherent actions. These include: water security goals; water, soil and climate security as lead factors in spatial planning; four cross-sectoral regional programs; a knowledge and innovation program. The Expert Panel's advice was the start of a recalibrated systemic approach for water safety and water security in Flanders [9].

## 4.4 *Summary of Lessons and Gaps*

### 4.4.1 Zooming out before Zooming in

Firstly, review major disaster events in history. A great help in this is modern scenario analysis, enabling to assess the potential problems that cities may encounter in extreme situations, in order to determine the scope of impact, the vulnerability of various systems, and possible measures to be taken. Secondly, conduct climate adaptation assessments at different spatial scales. Taking joint consideration of environmental protection and eco-system restoration requirements is required for high quality spatial development in addition to keeping safe, and finding overall balanced solutions.

### 4.4.2 Immediate Lessons and Injunctions

(1) This part highlights the **important opportunity—post disaster—to create an exception** to rules and habits—and the methodologies for creating these exceptional processes—that 'normally' stand in the way of progressive action, collaboration and unorthodox thinking, all needed to leapfrog for resilience. (2) Essential while rebuilding, and essential to maintain, is **a strong orientation on the future,** not the past. (3) Risk-informed planning and delivery of infrastructure and services is key in addressing vulnerabilities. Thus, information on climate risks should be **mainstreamed into spatial planning** and the delivery of urban infrastructure and services, while strengthening local capacity and celebrating successes. (4) Learning requires pausing and thinking, after the event. **Understanding what went wrong**, in engineering terms and otherwise, is essential to build back better [10, 11]. (5) Learning from disasters is costly and painful. Learning from what happened elsewhere, and **joining forces** is of course much preferred. For All and Water as Leverage [12]. (6) Cases also showed that **innovative technologies** at all levels of the information flow from monitoring to decision making are essential to adopt to risks caused by either climate change or other factors. (7) All cases illuminate the importance of **governance and process**: strong institutions; political will and leadership, trust and ambition.

# 5 Equity and Gender Aspects; Due Consideration for Vulnerable People and Areas

## *5.1 Social Inequities in Climate Change*

### 5.1.1 Urban-Rural and Regional Disparities

Worldwide urban-rural and regional disparities are evident and sharp disparities exist even between urban areas. Less developed areas face bigger challenges in coping with climate change due to relatively poor infrastructure, limited economic resources and inadequate public services. At the same time, the vulnerability of rural and less developed areas to climate change involves demographic issues (young people moving to cities for career reasons). Rural areas usually have a higher proportion of elderly groups that can be seen as more vulnerable. In addition, the degradation of agricultural land, such as soil salinization, and the instability of water supply sources threaten agricultural production and exacerbate the vulnerability of these areas.

### 5.1.2 Gender Disparities: Global Background Related to Vulnerabilities

Research about gender disparities in *adaptation* to climate change is extremely limited [13]. But multifaceted data from around the world show that women are more vulnerable to climate change and suffer disproportionately from the multiple impacts of climate change, including disaster casualties, displacement, gender-based violence, and violations of livelihoods and multiple rights. Moreover, considering that other vulnerable groups also face high vulnerability similar to women, this report focuses on vulnerable groups in a broader sense, including, for example, women, children, elderly and unrepresented groups. Such differences have to be factored into the design and implementation of adaptation policies.

## *5.2 Climate Adaptation Actions May Further Widen Social Inequities*

Climate adaptation actions are often seen as the responsibility of the government rather than the public interest of society. It resulted in limited public participation in the decision-making and implementation process and vulnerable groups. Overall, government bias in the formulation and implementation of climate adaptation policies is an important cause of increased social inequity. The unbalanced distribution of resources not only exacerbates the gap between urban and rural development, but also weakens the ability of rural areas to the realization of an effective response to climate change—thus increasing the climate risks for urban areas as well.

## 5.3 *Actionable Improvements for Social Equity in Climate Adaptation*

Social equity is the responsibility of all. Climate change adaptation policies should prevent further social inequities. On the one hand the planning process must be inclusive and involve everyone (IISD); on the other hand, interventions should be steered towards equitable policy outcomes, i.e., there is a need to identify the beneficiaries and disadvantaged of adaptation policy decisions, and to assess how it affects vulnerable groups [14] Climate adaptation in rural situations often has a larger-scale regional impact. Urban and rural problems must be tackled in an integrated manner and indeed. This can start with small-scale local elements, where possible with as much input from residents and those directly involved. At the same time an overview of larger interconnections is needed, thus an integrated approach to interventions in the physical and social system is indispensable.

## 5.4 *Common Prosperity Policy as a Tool*

Common prosperity as a policy is part of China's Ecological civilization paradigm [15]. It represents a strategic shift from a focus on (urban) growth and wealth accumulation to a more balanced and inclusive development model by improving rural infrastructure (public services, digital connectivity, etc.), and reducing economic disparities.

There are currently still significant differences between urban and rural areas in their approach to flood risks and adaptation measures [16]. Urban adaptation measures even sometimes have a big spatial impact on rural areas, a form of environmental injustice. Governments need to focus on the requirements of social equity and development and the promotion of people's well-being, carefully designing climate adaptation and investment planning, rationally arranging the implementation of actions and the allocation of resources and encouraging the participation of residents and social organizations in climate adaptation actions.

# 6 Experiences Show: Action, Knowledge and Investment Required at Multiple Scales

Overall insights include: (1) **A need to move fast**. This is in view of the existing adaptation gap in many situations, lead times to prepare interventions, and worrying climate projections. (2). **A need to move beyond fragmented and reactive approaches**, which lead to restoring the status quo or, at best, incremental changes. (3) **A need to be location-specific and build collective intelligence**. National governments and, for some aspects, intergovernmental organizations like the European Union and the United Nations, have to lead especially in setting ambition levels, expanding and spreading scientific understanding and creative design, and requiring assessment and mandating collaboration at scales that match the challenges. (4) **Strong coordination among different governmental bodies is needed**. Land use planning, development planning, infrastructure construction, stormwater management, flood control, greenspace management and community governance are all sharing parts of the responsibilities of adaptation work, but coordination mechanisms to make synergies are often missing at various levels. (5) **Adaptation goals setting for multiple purposes would promote alignments and facilitate trade-off for implementation.** Issues of equity should be considered when designing and implementing adaptation measures.

## *6.1 Emerging Insights Related to Spatial Scale*

### 6.1.1 Urban Area

A persistent challenge remains moving beyond reactive responses to past events in specific locations and time-constrained circumstances. Room for improvement seems to exist in: (1) Developing an ambitious and broadly supported mission-based approach [2]. (2) Learning from what went wrong, and went well. (3) Complementing city-based assessments with assessment of national or regional approaches. (4) The increasing frequency of extreme events poses a serious challenge to urban water management. (5) The accelerated urbanization has outpaced the development of essential water supply infrastructure. (6) Traditional water detention and management structures are dismantled to make way for construction in the process of urban expansion, further increasing flood hazards and the vulnerability of water supply capacity. (7) Aging of the infrastructure for drainage, water supply and water quality management is a rapidly increasing challenge for the near future.

To cope with the negative impact of climate change, urban expansion and urban renewal should consider medium- and long-term climate change impact. Measures or strategies adopted include: (1) Protecting the underground space in cities as well as using it as a potential resource to build a more flexible and sustainable water management system. (2) Investigating the risk of nature-based solutions failing under extreme conditions through stress tests and considering the combination of nature-based, 'green' solutions with 'grey' engineering infrastructure. Combining nature-based, 'green' solutions with 'grey' municipal engineering infrastructure forms a multi-level, 'green-grey' flood prevention and floodwater utilization system (*also see details from Shanghai case in Chap.* 3). (3) Minimize the damage of extreme events that overload the capacity of the drainage system.

### 6.1.2 Rural Areas

Rural areas in China face unique climate change challenges. The climate-related challenges include health risks, agricultural production reduction; water-electricity cuts require more measures in rural areas: (1) Water resource management is required to become more efficient through improving traditional facilities and expansion of rainfall collecting facilities. (2) Drought requires more public awareness and participation. (3) Social capital and community culture play vital roles in recovering from mental and social damage. (4) Increasing the financial investment in rural areas to improve old infrastructure, agricultural production capacity, and water resource management, etc., to promote the fairness between rural areas and urban areas (*also see details in the part on < Equity Aspects>*).

### 6.1.3 Delta Regions

Specific issues emerging from the cases include: (1) Increasing salt-water intrusion in coastal urban agglomerations. Countermeasures to salt-water intrusion include potentially costly engineering such as managed aquifer recharge, cross-river connectivity and off-shore freshwater lakes. (2) Soil subsidence, other than in permafrost situations is a large complicating factor in flood risk management. Lasting solutions would always deal with groundwater regimes, and thus with use of the land, for urban, agricultural or other economic purposes, or for nature.

### 6.1.4 Mountainous and Hilly Areas

Mountainous and hilly areas stand out as a category of their own, in particular in terms of flash flood and mudslide risks caused by extreme rainstorms. In China, since 1949, more than 190,000 people have died by flash floods in mountainous areas.

### 6.1.5 Integrated River Basin Management

Integrated river basin management considers the river basin as a whole to formulate collaborative management and development plans. At the planning stage, coherence of the interventions at all spatial scales is essential for achieving the management goals and objectives set in the plan. At operation stage, both engineering and non-engineering measures of the basin management are essential for the coping capacity, i.e. to minimize the damage of extreme events.

## *6.2 Emerging Insights Related to Time Horizon, Investment and Finance*

### 6.2.1 Strategies at Different Time Horizons

**A planning horizon of 2050 and an exploratory horizon of 2100** is recommended for planning and upgrading infrastructure for urban expansion and urban and rural renewal projects. **At the medium-term** scale, measures are needed to adapt to the challenges of climate change in the medium term, tuned with developments in technology, economic development and societal demands. **At the short-term** scale, contingency strategies such as establishing emergency response mechanisms, optimizing water resource dispatch systems and raising public awareness can be taken to improve the short-term adaptation capacity. **Keeping** the threshold, coping and recovery capacities **up to date** requires attention every 5–10 years. Assessments and stress-tests are to be repeated to find new gaps in the resilience of areas, systems and populations.

### 6.2.2 Investment and Financing Strategies

A new perspective to the revenue stream of climate adaptation: (1) The economics and financing of investments in adaptation to climate change differ significantly from the economics and financing of mitigation of climate change. Adaptation to climate change can avoid enormous damage and painful losses. Blue-green and grey interventions hardly ever generate direct financial revenue stream. (2) Mixed public private financing approaches requires careful management of expectations, clear accounting of the various kinds of costs and benefits and clear legal agreements between the public and private partners in the project. (3) Globally, there is an increasing focus on infrastructure investments to mitigate the economic impact of climate-related disasters. However, there remains an overemphasis on grey infrastructure. This highlights an urgent need to demonstrate the economic viability of investing in sustainable Blue-Green Infrastructure (BGI) and to create an enabling environment for scaling up these initiatives. (4) Adaptability to new, currently unknown climate conditions and other requirements is key to smart investments.

Overhaul of infrastructure and building stock: (1) The coming years will see a surge in overhaul of buildings and infrastructure, urban renewal projects and, sometimes, spatial layout and land use adaptation. (2) In many Western countries, buildings and important infrastructure often date from shortly after the second world war. They, too, are now up for major maintenance.(3) A useful strategic notion by the China-EU Water Platform reminds of the need to upgrade, or change, existing infrastructure and design and install new infrastructure and minimize the cost and maximize benefits, society-wide, across these two kinds of operation.

Current investment portfolios: The key challenge at present is to achieve an ambitious acceleration of adaptation to climate change within the current decade. Typically, good opportunities to improve climate resilience can be found in large-scale long-gestation projects that are initiated for non-climate reasons. Therefore, in many urban and peri-urban situations, it will be necessary to launch an acceleration programme, typically covering 2025–2035.

The financial industry and adaptation to climate change: Ill-orchestrated insurance practice like in Germany and The Netherlands can be a real impediment to building back better after disasters. Recovery from a disaster provides an opportunity for improving resilience that ought to be seized. At the level of general oversight of the financial industry and its portfolios of lending and investment, national regulators are becoming more aware of climate risks, and are requesting lenders to assess and limit climate and biodiversity risks of their lending portfolios.

## 6.3 *Emerging Insights Related to Institutional Geography*

### 6.3.1 Collaboration at National, Regional and Sectoral Levels

A key challenge is to rise above fragmentation between departments in government bodies, between government bodies, between government and private sector agendas, and between world views. Collaboration across jurisdictions, relevant to adaptation to climate change, comes in many forms and shapes.

### 6.3.2 Knowledge Management

Because of the large funding involved, it is unavoidable that setting up and maintaining **scientific and advisory institutions** for adaptation to climate change is taken care of by national governments. Two critical conditions apply: stable funding, and a mandate to provide independent and unsolicited advice to governments and societal actors. Five tools, or approaches, are lauded in the cases analyzed and work visits made for this SPS. They are:

(1) Stress-testing: Core question is how, considering the five capacities, systems and organisations in a given area are capable of dealing with extreme scenarios. For a useful and realistic result, stress-testing should also be applied to the area of the physical water system, the river basin, which normally is larger than a city, or comprise more than one jurisdiction. (2) Broad-based scenario approaches: There is much to be gained from applying the full width of modern scenario-based analysis. In particular, spatially differentiated scenarios for the key undercurrents in society enable to

adaptation strategies with an integrated view of a territory, with a sufficiently long time horizon. (3) Advances in strategically exploring and planning climate adaptation interventions. (4) Advances in effective early warning: Effective early warning protects life from the impact of natural disasters. (5) For river basins, digital twinning technology has emerged as an excellent tool in increasing the adaptation to climate change. During extreme events, based on the real-time monitoring, digital twin river basin systems enable managers to forecast warnings, previews and preplans, risk assessment and decision support systems.

Periodic assessment: Periodic, systematic and authoritative assessment of climate risks and of adaptation capacities is essential to make things move, build alliances, engage citizens, check on equity and fairness, consider new information and technology.

In the context of China, two concluding comments pertain: (1) the city level. The opportunity exists in the form of the "City Health Examination" system of the Ministry of Housing and Urban-Rural Development (MOHURD), which currently provides the basis for the establishment of a climate resilience assessment system. (2) the national level. This level of assessment should address the strategic, long-haul issues.

# 7 Operational Assessment Framework for Adaptation to Climate Change

## *7.1 Role of the Operational Assessment Framework*

In the conceptual assessment framework, the five capacities and ten enabling conditions for resilience cover a wide range of potential weaknesses, including recovery capacity, long-term adaptability and the transformative capacity of society. Therefore, it is recommended to integrate these five capacities in the assessment framework, to position these in the assessment process and in its contents, creating a solid basis for an adaptation plan and its successful implementation in practice.

## *7.2 Contents and Steps of the Assessment*

### 7.2.1 Context and Climate Adaptation Background Assessment

The climate change trends and scenarios are to be studied and presented for all relevant spatial scales and major impacts faced by cities, regions and watersheds. Also, the basic characteristics and development trends of the project area are to be mapped and analysed.

### 7.2.2 Climate Change Risk and Opportunity Assessment

Assess the climate change risks faced in cities, regions and river basins in terms of disaster hazards, exposure and vulnerability (terminology from IPCC, 2012 [17]) by performing "climate stress tests" to study the impact of extreme climatic events. On this basis, the vulnerability of the community at the local, regional and river basin scale is identified, as well as the main challenges and weaknesses in the resilience of the physical and governance systems to cope with the risks.

### 7.2.3 Assessment of a Climate Adaptation Action Plan

Starting from the goal of enhancing the climate resilience of the area, adaptation interventions can now be studied from the perspective of the gaps found in the five capacities to creating resilience (threshold capacity, coping capacity, recovery capacity, adaptive capacity, and transformative capacity) and the weaknesses in the enabling conditions required to achieve these goals. This step results in a climate adaptation plan that is coherent with the adaptation strategy and plans at the other spatial scales.

### 7.2.4 Performance Evaluation of Climate Adaptation Implementation

To be able to implement this plan the enabling conditions should be met. Substantial efforts and funding are often needed for this. Monitoring of the progress and ex-post evaluation of the performance are standard elements of good governance; these provide input for a new cycle of climate adaptation planning.

## *7.3 Assessment Challenges*

### 7.3.1 Strengthen Synergies to Jointly Conduct Integrated Assessments of Climate Resilience

Synergies between different regions and government departments is required in climate adaptation assessments. This includes defining a unified vision and goal for climate adaptation, integrating the risks and interests of different regions and sectors; establishing a unified framework and data and methodological basis for climate adaptation assessment through information-sharing; and establishing synergistic mechanisms across regions and sectors, as well as between different levels of governance and departments within cities, regions and ministries. It is necessary to shift from the sub-assessment of individual hazards and the "cost-benefit" evaluation

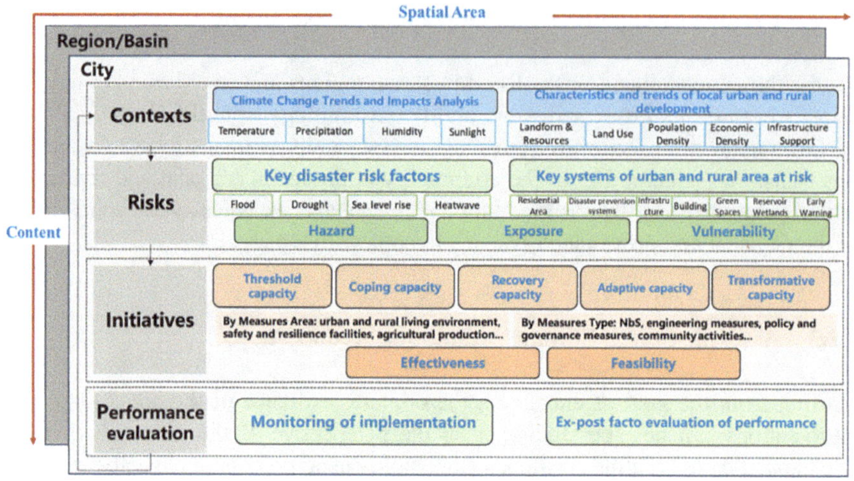


**Fig. 4** Proposed operational procedure to assess climate resilience and to plan adaptation interventions in urban and rural areas. Based on an analysis of the local situation, now and in the future, and on a vulnerability analysis to find the gaps in the five capacities required for a climate resilient environment, coherent adaptation plans can be made at the local, city, regional and the river basin scale

of individual measures to a comprehensive, multi-objective decision-making assessment based on integrated risk & opportunity management, taking decisions despite the many uncertainties for the long-term future (Fig. 4).

### 7.3.2 Identify Key Risks of Area, Adaptive Capacity of Facilities and Transformative Capacity of Organisations

Different regions and areas should carry out climate change risk assessments based on their own geographical characteristics, focusing on identifying the key disaster risks and outstanding problems faced locally. First, different regions and cities should carry out climate change risk assessments based on their own geographical characteristics, focusing on identifying the key disaster risks and problems faced locally. Second, the adaptive capacity and the longevity of the existing infrastructural facilities in urban and rural systems is to be quantified in the light of the changing exposure to extreme conditions and the changing vulnerability of their function.

### 7.3.3 Analyse the Correlation Between Different Disaster Risks; Develop Scenarios for Adaptation

To conduct a comprehensive assessment, it is necessary to consider the short-term extreme weather risks as well as the risks resulting from long-term climate change, and. Attention should be paid to the correlation and interactions between different hazards as well as the interactions between different facilities and systems in urban areas and the river basin they are part of.

### 7.3.4 Differentiated Assessments Should Be Carried out Based on Spatial Perspective

A differentiated assessment should be carried out, taking into full consideration the great variability that exists between different regions in a river basin and between cities, city districts, villages and countryside, as well as the differences in climate adaptation capacity at the city level and at the community level. Priorities should be clarified for urban and rural settlements in different spatial regions, taking into account their own characteristics, such as estuarine delta areas, densely populated small to medium-sized watersheds, and ecologically fragile areas in the upper reaches of watersheds.

### 7.3.5 Assess the Implementation Feasibility and Performance of Climate Adaptation Measures

The current cost-benefit evaluation systems have difficulties in estimating the benefits of interventions. Long-term monitoring and assessment of the social, economic and ecological effects of climate adaptation measures and policies would help to improve these economic and financial evaluation models.

### 7.3.6 Equity in Adaptation and Public Participation in Climate Adaptation Assessments

Climate adaptation assessments should strengthen attention to less developed regions and vulnerable groups, and assess whether climate adaptation measures embody the principles of social equity and gender equality. Public participation should be strengthened in climate adaptation assessments and planning. The public, enterprises and other stakeholders should be encouraged to work together with government departments to carry out climate adaptation assessments, plan and implement adaptation measures.

## 8 Policy Recommendations

### *8.1 Climate Risks Are Increasing and Disaster Losses Are Huge; Action on Adaptation Is Urgent*

National, regional and local governments therefore should urgently elevate the political and governance priority of adaptation to climate change across all relevant ministries and departments, accelerate climate adaptation in all policy domains, and enhance capacity-building in the five key dimensions of climate resilience from national to local level. Climate adaptation strategies should address both the challenges of the increasing weather extremes and the slow-onset changes. Urban and rural authorities and communities should be supported throughout the entire adaptation process.

### *8.2 The National Government Should Develop a Systematic and Comprehensive Policy Framework for Short- and Long-Term Action, from the Central to the Local Level, across Regions/Basins and Sectors, and with Active Participation of the Society, Industry and Local Communities*

Priority actions are: (1) Develop and implement systematic and comprehensive climate adaptation programs and near-term action plans from national to local governments and incorporate them into government development strategies at all levels and through all ministries and sectors. (2) Establish and commit to pro-active future oriented adaptation goals and targets across all policy domains, and establish a foundational system of regulations, administrative systems, financial and fiscal systems, policy systems and knowledge and data systems that effectively support climate adaptation strategies. (3) Create long-term stable adaptation funds as required by the long lead-time of adaptation strategies and implementation of measures, and coordinate government investment in different areas such as ecological restoration, disaster prevention and relief, and urban infrastructure construction. (4) Promote and support regional collaboration of climate adaptation strategies across administrative borders in key areas with high population and economic densities.

## 8.3 The National Government Should Establish a Climate Adaptation Assessment Framework and Mechanism to Identify Gaps in Climate Resilience and Priorities for Climate Adaptation in Key Areas and Facilities, and Provide a Solid Foundation for Urban and Rural Areas to Monitor and Improve Their Climate Adaptation Policy and Implementation Capacity

A localized framework for climate adaptation assessment in urban and rural areas is required, taking into account natural resources and conditions, climate change characteristics and the level of socio-economic development and the potential interaction with other institutional policy programs and assessments. (1) Address the actual risks and problems in climate adaptation assessments. Identify how resilient urban and rural communities are now and what are the resilience gaps. (2) Evaluate existing laws, regulations, educational training programmes and funding mechanisms, based on climate adaptation requirements, and adjust these if needed. (3) Provide local and regional governments periodically with the latest insights.

## 8.4 The National Government Should Pay Attention to Social Equity and Gender Issues in Climate Adaptation and Should Accelerate the Enhancement of the Climate Adaptation Capacity of Less Developed Regions and Especially Vulnerable Groups

Priority actions are: (1) Raise the standards for disaster prevention infrastructure in vulnerable and underdeveloped areas as soon as possible, with increased financial investment to accelerate facility and capacity building. (2) Improve disaster relief services for vulnerable groups, such as disabled, elderly and low-income people. (3) Widely carry out publicity and popularization activities on climate adaptation, disaster prevention and mitigation to enhance public understanding of and support for climate adaptation and improve public participation.

## 8.5 Strengthen International Cooperation and Conduct Joint Multidisciplinary Research to Lay an Important Scientific Foundation for Climate Adaptation

For CCICED priorities are: (1) Strengthen sustained and in-depth international cooperation, to support learning from experiences around the globe and help to make leap-frog steps in climate adaptation policy development. (2) Conduct integrated, long-term and task-oriented policy research across disciplines and policy areas; and develop highly adaptive climate adaptation measures to avoid maladaptation in response to future development uncertainties. (3) Explore if the economy and financing of climate adaptation should be a new topic for CCICED as there are still major gaps in knowledge and experience hampering the implementation of climate adaptation.

## References

1. 100 Resilient Cities. (2015). *City resilience framework*. The Rockefeller Foundation/ARUP.
2. Mazzucato, M. (2019). *Governing missions in the European Union*. European Commission, DG for Research and Innovation.
3. Sommer, K. H. (2019). Study and portfolio review of the projects on industrial symbiosis in DG research and innovation: Findings and recommendations. *European Union*. https://doi.org/10.2777/381211
4. Brown, R. R., & Clarke, J. (2007). *Transitioning to water sensitive Urban Design: The story of Melbourne, Australia* (*report no.* 07/1). FAWB, Monash University.
5. ARB. (1997). *Resultaatgericht Beleid. Advies bij Beleidsontwikkeling, Communicatie en Samenwerking*. Ministerie van Verkeer en Waterstaat.
6. United Nations. (2022). *Secretary-general's message for World Meteorological Day*.
7. Public Utilities Board (PUB). (2016). *Our water, our future*.
8. Sierra Nevada Conservancy. (2018). *The community and watershed resilience program*.
9. Vlaamse Milieumaatschappij (VMM). (2022). Weerbaar waterland. *Advies van het expertspanel hoogwaterbeveiliging aan de Vlaase Regering*.
10. Zurich. (2014). *Zurich's flood resilience program*.
11. Ovink, H., & Boeijenga, J. (2018). *Too Big. Rebuild by Design: A transformative Approach to Climate Change*. ISBN: 978-94-6208-315-8.
12. World Water Atlas. (n.d.). *Water as leverage Programma*.
13. Roy, J., Prakash, A., Some, S. et al (2022). Synergies and trade-offs between climate change adaptation options and gender equality: A review of the global literature. *Humanities and Social Sciences Communications 9*, 251.
14. Global Commission on Adaptation (GCA) & World Resources Institute (WRI). (2019). *Unlocking the potential for transformative climate adaptation in cities*.
15. Pan, J. (2020). *Ecological civilization: A new development paradigm*. The Integrated Assessment Society (TIAS).
16. Palmer, J. (2023). Record flooding highlights rural-urban divide in China. *Foreign Policy*.
17. Intergovernmental Panel on Climate Change (IPCC). (2012). *Managing the Risks of Extreme Events and Disasters to Advance Climate Change Adaptation* (p. 582). [Field, C.B., V. Barros, T.F. Stocker, D. Qin, D.J. Dokken, K.L. Ebi, M.D. Mastrandrea, K.J. Mach, G.-K. Plattner, S.K. Allen, M. Tignor, and P.M. Midgley (eds.)]. Available from Cambridge University Press.

# Chapter 6
# Technology Innovation for Green and Low-Carbon Transition

## 1 The Importance of Technological Innovation for Green and Low-Carbon Transition

Green and low-carbon transition has become a crucial direction for sustainable global development. This transition is not only necessary for addressing climate change and solving environmental pollution but also serves as a fundamental pathway for achieving high-quality economic development. Technological innovation plays a pivotal role as the driving force behind this transition. Governments and enterprises worldwide are increasing investments in research and development (R&D) of green and low-carbon technologies and implementing a series of policy measures to support the development and application of green technologies. This innovation-driven development model is leading the world into a new phase of development. Green and low-carbon technological innovation helps reduce energy consumption, decrease greenhouse gas emissions, minimize pollutant discharge, and protect the environment and ecosystems. Besides, green technological innovation can foster the emergence of new industries and business models, such as those in the fields of renewable energy, energy efficiency, environmental protection technologies, and circular economy. These emerging industries not only stimulate economic growth but also create substantial employment opportunities and optimize economic structures.

In September 2023, during an inspection in Heilongjiang province, President Xi Jinping first mentioned the term "new quality productive forces". This was followed by the Central Economic Work Conference at the end of 2023, which called for the development of new quality productive forces. At the first collective study session of the Political Bureau of the Central Committee of the CPC in 2024, the focus was on "new quality productive forces", and the report on the work of the government (2024) listed accelerating the development of new quality productive forces as one

---

The content of this chapter is extracted from the main findings of the 2023-2024 CCICED Scoping Study on "Technology Innovation for Green and Low-Carbon Transition".

 
*Adhering to Green and Open Cooperation to Advance Modernization*,
https://doi.org/10.1007/978-981-95-5547-5_6

of the top ten tasks for the year. These significant arrangements clearly signal the importance and urgency of developing new quality productive forces.

President Xi Jinping stated, "Green development is the underlying theme of high-quality development, and new quality productive forces are essentially green productive forces". This important assertion highlights the intrinsic link between new quality productive forces and green development, providing theoretical guidance for accelerating the development of new quality productive forces and promoting the comprehensive green transition of economic and social development. It holds profound theoretical and practical significance.

This study focuses on green and low-carbon innovation from a policy research perspective, analyzing how policy guidance can accelerate green and low-carbon technological innovation to empower the rise of new quality productive forces.

## *1.1 Achieving Carbon Neutrality Depends on the Maturity and Wide Application of Green and Low-Carbon Technologies*

Achieving carbon neutrality requires global net greenhouse gas emissions to reach zero, which involves not only a fundamental transition of the energy structure but also deep emission reductions across industries such as transportation, construction, and agriculture. This goal is heavily reliant on the maturity and widespread application of green and low-carbon technologies. Firstly, the development of renewable energy technologies, such as solar, wind, and hydropower, forms the foundation for achieving carbon neutrality. The widespread adoption of these clean energy sources can gradually replace fossil fuels and reduce carbon emissions. Moreover, the development of hydrogen energy, energy storage, and smart grids ensures the stability and reliability of energy systems, facilitating large-scale integration of clean energy. Secondly, in the industrial and transportation sectors, breakthroughs and widespread adoption of technologies like green hydrogen, electrification, and carbon dioxide capture, utilization and storage (CCUS) are critical for addressing deep decarbonization challenges. Without the maturity and large-scale application of these technologies, achieving carbon neutrality will be difficult.

Green and low-carbon technologies are playing an increasingly prominent role in addressing climate change. However, many key technologies remain in the research and testing stages and have not yet reached the level of maturity required for large-scale commercial application. They still face technological bottlenecks and market acceptance challenges. According to the latest data from the International Energy Agency (IEA) [1], as shown in Fig. 1, technologies currently in the demonstration or prototype phase are expected to contribute 35% of the emission reductions needed in the 2050 net-zero emissions (NZE) scenario in energy system. Some technologies still need to be promoted and applied to support emission reduction targets. To achieve net-zero carbon emissions in the energy system by 2050, low-emission hydrogen

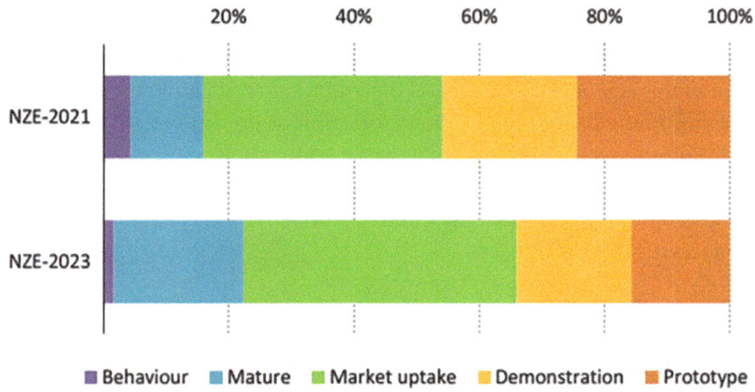


**Fig. 1** Global energy sector $CO_2$ emissions reductions by technology maturity in 2050 compared to the base year (IEA 2021 and 2023 net-zero scenarios). *Source* International Energy Agency (IEA), 2023

and hydrogen-based fuels, sustainable aviation fuels (SAF), and direct air capture and storage (DACS) must see significant growth in the coming years. By 2030, low-emission hydrogen production is expected to increase from less than one million tons in 2022 to 70 million tons. By 2030, SAF's share of aviation final energy consumption is expected to rise from nearly zero to about 11%. By 2030, DACS is expected to contribute an annual CO reduction of nearly 7 billion tons, up from almost zero today [2].

## 1.2 *Technological Innovation for Green and Low-Carbon Transition Will Empower the Rise of New Quality Productive Forces*

Technological innovation for green and low-carbon transition will drive profound changes in production methods. Traditional productive forces depend on high energy consumption and high-emission production models, while green and low-carbon technologies introduce low-energy consumption, low-emission, and low-pollution production methods through new technologies (such as renewable energy and clean production technologies). This not only improves production efficiency but also reduces negative environmental impacts, thereby reshaping production methods. This innovation provides strong driving power for sustainable economic growth and aligns with the global trend toward green economic transition. Green total factor productivity (GTFP) is an important indicator of an economy's capacity for efficient and sustainable development under resource and environmental constraints. Research has shown that green innovation can significantly enhance GTFP [3, 4].

Technological innovation for green and low-carbon transition will also strengthen the role of knowledge and technology in productive forces. With the rapid acceleration of technological advancement and innovation, green and low-carbon technologies have become essential components of modern productive forces. Enterprises and nations are focusing on R&D, innovative applications, and talent development, concentrating high-end technological resources to improve the competitiveness of green technologies. This knowledge-driven productivity has become a vital source of economic growth.

## *1.3 Technological Innovation for Green and Low-Carbon Transition Will Create Numerous Emerging Industries and Market Demands*

The green and low-carbon transition is driving the rise of many emerging industries globally and creating widespread market demand. This transition benefits from the combined forces of policy support, technological advancement, and market trends. First, the introduction of carbon neutrality goals and government incentive policies has created vast opportunities for the development of new fields, such as renewable energy, energy storage technologies, smart grids, and electric transportation. These policies direct capital toward green industries and stimulate market demand for low-carbon products and services.

Second, accelerated technological innovation is transforming traditional industries while creating new industrial chains and ecosystems. For instance, the widespread adoption of electric vehicles) has driven rapid growth in related industries, such as battery technology, charging infrastructure, and recycled materials, shaping a new market landscape. According to data from the International Energy Agency (IEA) [5], global sales of new energy vehicles reached approximately 14 million in 2023, accounting for 20% of total global car sales. This growth has not only accelerated the expansion of the new energy vehicle industry but also spurred the development of associated industries, such as battery production, charging infrastructure, and smart transportation systems. Notably, global battery manufacturing capacity is expanding rapidly, with 2023's capacity reaching 1500 GWh, a 50% increase from 2022. This figure is expected to further grow to 7500 GWh by 2030, almost five times the 2022 level. This explosive growth reflects the substantial demand for battery technologies, which is driving rapid development in materials, equipment, and recycling industries.

Additionally, the increasing environmental consciousness of consumers has made green consumption a significant market trend, driving the continued growth of green products and services. Finally, the global trend toward carbon neutrality has driven growing international demand for green technologies and products, making the green economy a crucial component of global trade. Thus, the green and low-carbon transition is not only an environmental imperative but also a new engine for economic growth, creating new industries and markets while fundamentally transforming global economic structures.

## 1.4 Green and Low-Carbon Transition Is Not Merely a Technological Upgrade but Also a Profound Transformation of Industrial and Social Structures

The essence of green and low-carbon transition is not simply technological iteration and upgrading. While innovative technologies are the means, the resulting deep changes in industrial and social structures are the ultimate goals. First, green transition requires systematic adjustments to traditional high-carbon industries, not just within the energy sector but across industries such as industrial manufacturing, transportation, and construction. Traditional high-carbon industries will gradually be replaced by new industries such as clean energy and green manufacturing, leading to significant changes in industrial structures. According to IEA statistics [6], in 2023, it is expected that 96% of newly installed utility-scale PV and onshore wind power generation will be cheaper than new coal and gas-fired power plants. Moreover, three-quarters of new wind and PV power plants will provide electricity at a lower cost than existing fossil fuel facilities. The cost competitiveness of wind and PV systems will continue to increase, leading to the gradual replacement of traditional high-carbon industries by clean energy.

Second, green and low-carbon transition will restructure industrial chains and upgrade value chains, promoting efficient resource allocation and circular utilization, forming a more ecological and intensive economic model. This transition will not only change production methods but will also impact employment structures and the division of labor within society, as demand for new green jobs and professions increases, while traditional occupations face pressure for transformation and upgrading. According to the IEA's 2050 Net-Zero Emissions (NZE) scenario, the energy supply sector is expected to create 14 million new jobs by 2030. During the same period, fossil fuel production may see a reduction of five million jobs [7].

Additionally, technological innovation for green and low-carbon transition will profoundly affect social consumption habits, lifestyles, and urban development models, promoting the development of low-carbon cities, green transportation, and sustainable communities, and shaping a new social form.

## 2 Key Fields and Directions in International Technological Innovation for Green and Low-Carbon Transition

This study uses text mining methods to identify key fields and technologies in green and low-carbon technological innovation. As shown in Fig. 2, critical areas of green and low-carbon technological innovation include energy transition, low-carbon industries, circular economy, green transportation and buildings, agriculture and land use, and digital technology, all of which exhibit strong cross-disciplinary features. From the deployment strategies of major developed countries, there are focused technological R&D efforts in these key areas, although the priority of specific breakthrough directions varies. Frontier technologies such as controlled nuclear fusion, green hydrogen technology, CCUS, next-generation energy storage batteries, long-duration energy storage, and industrial decarbonization are the focus of major developed countries.

**The United States focuses on fostering forward-looking and disruptive technologies in the green and low-carbon sectors.** Since 2021, the U.S. Department of Energy, through the Advanced Research Projects Agency-Energy (ARPA-E), has consistently supported transformative low-carbon energy technologies. These include key decarbonization technologies in electricity, renewable energy, nuclear energy, hydrogen, industry, and transportation, with a broad scope and an emphasis on disruptive technological innovations. Besides, the U.S. has deployed long-term scientific research programs, such as the "Energy Earthshots", which focuses on mobilizing national strengths in key areas of energy science and technology over the next

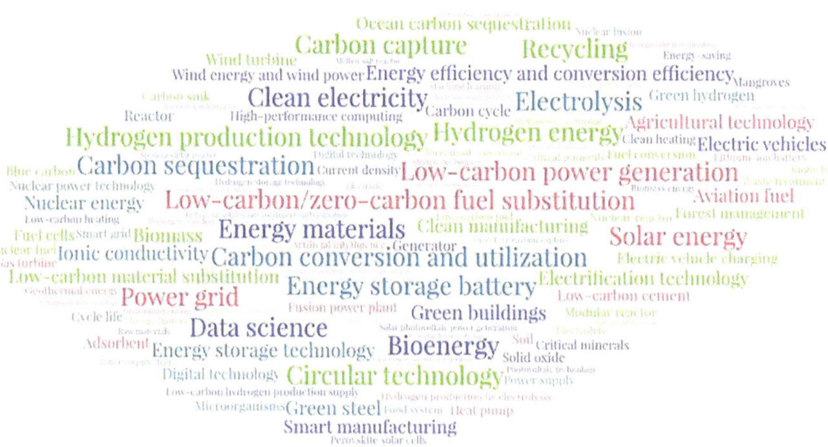


**Fig. 2** Key fields and technologies in green and low-carbon technological innovation. *Note* Based on the monthly monitoring product "Intelligence Dynamics for Dual Carbon" of the The National Science Library, Chinese Academy of Sciences, more than 1600 monitoring item data of this product from 2021 to date were collected, and a word cloud map was generated after data cleaning, word segmentation, and clustering

10–15 years to reduce costs, improve performance, and achieve large-scale deployment of clean energy. This plan has continuously funded projects in hydrogen energy, energy storage, negative carbon technologies, enhanced geothermal energy, clean fuels and products, industrial heating, and building decarbonization. Furthermore, in the industrial sector, the U.S. passed the Technologies for Industrial Emissions Reduction Development Program in 2023, outlining decarbonization technologies for key industries such as steel, cement, and chemicals, while advancing technologies with commercial potential through industrial demonstrations.

**The European Union emphasizes sustainable, inclusive, and secure green transition solutions.** The EU funds scientific research and technological development related to carbon neutrality through multiple programs, including the EU Research Framework Programmes, Innovation Fund, Modernisation Fund, Sustainable Europe Investment Plan, Connecting Europe Facility, and LIFE Programme. Horizon Europe, the EU's ninth research framework program (2021–2027), is the largest research program in Europe with a budget of €15.123 billion dedicated to climate, energy, and transportation research. It supports six major themes: (1) climate science and responses to climate neutrality; (2) cross-sector solutions for climate transition; (3) sustainable, secure, and competitive energy supply; (4) efficient, sustainable, and inclusive energy use; (5) clean and competitive solutions for all modes of transport; (6) secure, resilient transportation and smart transportation services.

Despite the EU's significant efforts to promote scientific research and technological development related to carbon neutrality through various funding programs, it still faces numerous challenges in the implementation process. Particularly in the clean technology sector, the EU's technological advantages have not yet fully translated into production capacity. For instance, while the EU has a technological edge in areas such as heat pumps, batteries, electrolyzers, and carbon capture technologies, it continues to struggle with limited production capacity and a reliance on imported key components [8]. The EU's lithium-ion battery production capacity remains low, electrolyzers have not yet reached large-scale manufacturing, and the expansion of carbon capture technology is hindered by inadequate infrastructure.

As a result, the EU has taken more proactive steps toward localizing production and scaling up these critical technologies to enhance its global competitiveness and reduce dependency on external supply chains. In 2024, the EU passed the *Net-Zero Industry Act*, identifying eight strategic net-zero technologies that are either commercially viable or soon to enter the market. These include solar photovoltaic and solar thermal technologies, onshore wind and offshore renewable energy technologies, batteries and storage technologies, heat pumps and geothermal energy technologies, electrolyzers and fuel cells, sustainable biogas/biomethane technologies, carbon capture and storage technologies, and grid technologies, all of which are supported for localized deployment.

**Germany has incorporated climate change technologies into its future research and innovation strategy.** In 2023, Germany released its latest top-level strategic plan for scientific and technological innovation—the *Future Research and Innovation Strategy*, aimed at more robustly promoting innovations related to sustainable greenhouse gas emissions reduction. Key areas of focus include developing new technologies and processes for zero-emission raw materials, promoting lightweight construction, creating innovation-friendly frameworks to tap into the potential of the bioeconomy, leveraging digital technologies to reduce resource consumption, establishing interoperable standards for a safe circular economy, leading the development of high-performance, secure, and energy-efficient communication systems such as 6G, rapidly advancing the green hydrogen economy to become a global leader in hydrogen technology, building a global battery research, production, and recycling center, continuing research on nuclear fusion, and developing novel concepts and technological pathways. Germany's goal is to create a climate-friendly, efficient, and resilient future energy system through innovation.

**The United Kingdom has focused on low-carbon industrial technologies for its "Green Industrial Revolution" initiative.** By setting comprehensive green innovation strategies aimed at achieving net-zero emissions, the UK has emphasized applied research on key industrial technologies, forward-looking planning, and large-scale application. The UK's primary goal in green innovation is to reduce the transition cost to net zero, cultivate better products and new business models, and prioritize key future technologies. In March 2021, the UK government launched a £1 billion Net Zero Innovation Portfolio, based on the Ten Point Plan for a Green Industrial Revolution. This portfolio focuses on ten priority areas, including floating offshore wind energy, bioenergy, hydrogen, flexible energy storage, advanced nuclear energy, and carbon capture, utilization, and storage (CCUS).

**Japan has broadly deployed decarbonization technologies in energy, industry, and environmental sectors to drive a societal green transition.** In its early 2020 *Environment Innovation Strategic*, Japan planned the deployment of key decarbonization technologies across five major areas—Non-fossil energy, industry, transportation, construction, and agriculture/forestry/fisheries—totaling 39 key technologies in 16 categories. By the end of 2020, under the *Green Growth Strategy*, Japan established a ¥2 trillion Green Innovation Fund and launched several long-term major funding programs targeting key industries such as offshore wind energy, ammonia and hydrogen fuels, next-generation batteries, novel geothermal energy, bio-based materials, and circular economy technologies. **Japan** also highly values the integration of digital technologies, such as artificial intelligence and the Internet of Things, into various sectors to achieve energy conservation and emission reduction. Moreover, Japan actively promotes the use of digital technologies in transportation, construction, energy, and other fields.

**South Korea has formulated a list of emerging and core technologies to achieve carbon neutrality.** In June 2022, the Korea Institute of Science and Technology Evaluation and Planning (KISTEP) identified 10 emerging technologies to assist in meeting South Korea's 2030 nationally determined contribution (NDC) targets. These include (1) carbon capture and utilization (CCU); (2) bio-based raw materials/products manufacturing technologies; (3) low-carbon steel production; (4) high-capacity and long-life secondary batteries; (5) clean hydrogen production; (6) ammonia-fueled power generation; (7) grid integration systems; (8) high-efficiency crystalline silicon solar cells; (9) large offshore wind power systems; and (10) rare earth element recycling. In October 2022, the South Korean government formally established the 2050 Carbon Neutrality Green Growth Commission, which, in addition to releasing *The Strategy For Promoting Carbon Neutrality And Green Growth*, also published the *Carbon Neutrality And Green Growth Technology Innovation Strategy*, involving 100 core technologies related to carbon neutrality. These technologies span energy, key industries, building efficiency, and transportation (Table 1).

**Table 1** Green and low-carbon key technology layout of major developed countries

| Country/ region | Key technology |
|---|---|
| US | Small modular reactors, nuclear fusion, green hydrogen, CCUS, battery energy storage, next generation low-carbon buildings, renewable energy, advanced nuclear energy, sustainable aviation fuel, industrial electrification, artificial intelligence technology in the energy field, electric vehicles, climate-smart agriculture, etc. |
| EU | Renewable energy, hydrogen energy, integrated energy systems, smart grids, energy storage, CCUS, industrial decarbonization and digital transformation, green buildings, sustainable and smart transportation, precision agriculture, organic agricultural ecosystems, bioeconomy, synthetic low-carbon fuels, etc. |
| Germany | Green hydrogen, energy storage, electric vehicles, smart grids, electrification of transportation networks, biofuels, fuel cells, low-emission industrial production technologies, climate and environmentally friendly buildings, digital energy systems, modernization of cogeneration, ecological agriculture, etc. |
| UK | Energy storage, hydrogen energy, offshore wind power, advanced nuclear energy, electric vehicles, electrification of transportation networks, zero-emission aircraft, sustainable transportation fuels, clean shipping, green buildings, industrial fuel transition, bioenergy, direct air capture and advanced CCUS, artificial intelligence in the energy field, etc. |
| Japan | Renewable energy, hydrogen and ammonia fuels, new geothermal energy, advanced nuclear energy, nuclear fusion, electric vehicles, energy storage, zero-emission ships, smart agriculture, forestry and fisheries, low-carbon semiconductors, aviation electrification, carbon resource utilization, net zero emission buildings, resource recycling and reuse, etc. |
| South Korea | Offshore wind power, crystalline silicon batteries, zero-energy buildings, smart grids, electric vehicles, hydrogen energy and fuel cells, resource recycling, hydrogen reduction ironmaking, low-carbon fuels, low-carbon semiconductors, bioenergy, CCUS, smart energy management systems, smart agriculture and fisheries, carbon sinks, etc. |

## 3 Main Policies and International Experiences in Technological Innovation for Green and Low-Carbon Transition

In order to promote the development of green and low-carbon technological innovation, countries must not only comprehensively consider the balance between ecological environment and economic development in policy design but also establish systematic policy support frameworks. The creation of green and low-carbon technological innovation often relies on a series of interconnected policy tools that can guide and promote innovation activities by regulating behavior, providing economic incentives, and creating supportive environments. By implementing effective regulatory tools, economic and financial tools, and "soft" tools, countries can stimulate the innovative motivation of market players while safeguarding environmental quality, gradually forming a complete pathway from policy formulation to innovation realization.

Innovation policy refers to all actions taken by public organizations to support the generation and dissemination of innovation, with the aim of influencing the innovation process to achieve predetermined political objectives through innovation. Innovation is defined as new products, services, processes, or business models used for commercial or non-commercial purposes [9].

Innovation policy tools are categorized into three main types: regulatory tools, economic and financial tools, and soft tools. These three categories are commonly referred to as "sticks", "carrots", and "sermons" [10].

**Regulatory tools play a crucial role in innovation policies.** They govern the interaction between society and markets through legal and institutional mechanisms. Specific measures include intellectual property protection, the management of researchers and higher education organizations, competition policies, and industry-specific regulations. These regulations are characterized by their mandatory and restrictive nature, ensuring that all stakeholders operate within a clear legal framework, thereby providing structured and institutional support for knowledge and innovation processes. This framework directly drives innovation and encourages companies to adapt to new regulations, enhancing their innovative capacity.

**Economic and financial tools are key in regulating and incentivizing innovation activities.** These tools mainly operate through offering economic incentives or deterrents to support specific social and economic activities. They can be categorized into positive incentives (such as cash transfers, grants, subsidies, low-interest loans, and loan guarantees) or negative incentives (such as taxes and fees). These tools are widely used to support public research organizations and corporate R&D activities. Traditionally, most of the existing financial instruments impact innovation from the supply side, but policymakers have increasingly recognized the importance of demand-side tools, such as driving innovation through public procurement.

**Soft tools typically complement regulatory and economic tools.** They are characterized by voluntary and non-coercive measures, achieved through providing advice, making normative appeals, or reaching voluntary or contractual agreements

**Table 2** Classification of policy tools for technological innovation for green and low-carbon transition

| Classification | Policy tools |
|---|---|
| Regulatory tools | Mandatory environmental regulations |
| | Green and low-carbon standards |
| | Emission trading systems |
| | Regulations and voluntary agreements |
| Economic and financial tools | Fiscal grants |
| | Tax incentives |
| | Low-interest loans, green bonds |
| | Venture capital, seed funds |
| Soft tools | Strategic planning |
| | Provision of public research facilities |
| | Technology incubator support |
| | Intellectual property protection |

to achieve governance objectives. Specific soft tools include support for research infrastructure, networking and collaboration platforms, information provision, data access, and innovation promotion and public outreach. Notably, the use of soft tools reflects the shifting role of governments, moving from being providers and regulators to facilitators and coordinators, with increasingly widespread application in the field of innovation policy.

Green and low-carbon technological innovation policies have both the common characteristics of traditional innovation policies and some special policy tools. Table 2 lists the commonly used policy tools of technological innovation for green and low-carbon transition [11–14]. This study focuses on the role of five types of policy tools, namely fiscal policy, financial policy, green and low-carbon standards, carbon market and strategic planning, in promoting green and low-carbon transition technological innovation.

### *3.1 Fiscal Policies*

**Major countries have increased fiscal investment in green and low-carbon technological innovation and green infrastructure to provide financial support for low-carbon industries and a fair transition.** The United States recently passed three pieces of legislation that can help reduce greenhouse gas emissions: the *Infrastructure Investment and Jobs Act* (IIJA, also known as the *Bipartisan Infrastructure Law*), the *Inflation Reduction Act of 2022* (IRA), and the *CHIPS and Science Act*. Compared to the average level of spending over the past decade, these laws have more than tripled U.S. government spending on energy transition, making the

U.S. a global leader in climate change and clean energy investment [15]. In 2019, the European Union established the world's largest low-carbon technology demonstration program—Innovation Fund—with plans to allocate €10 billion between 2020 and 2030 to support the commercialization of innovative low-carbon technologies. In 2021, Japan's government established a ¥2 trillion "Green Innovation Fund" to support decarbonization technology R&D and deployment in key industrial sectors. That same year, the United Kingdom announced a £1 billion Net Zero Innovation Portfolio over the next decade to leverage £3.5 billion from industrial and academic institutions, accelerating low-carbon technology innovation and creating new business models.

Many countries have utilized subsidies and tax incentives to create a favorable policy environment for green and low-carbon technological innovation. Since 2021, the European Union has revised the *Guidelines on State aid for environmental protection and energy*, the *Trans-European Networks for Energy* (TEN-E), the *Energy Taxation Directive*, the *Alternative Fuels Infrastructure*, and the *Renewable Energy Directive*. These revisions support low-carbon technology R&D and demonstrations, address unfavorable factors hindering the development of low-carbon technologies, reduce investment risks for low-carbon projects, and ensure the sustainability of project financing. Japan's 2021 tax reforms established the "Carbon neutral investment incentive", allowing companies to obtain a certain percentage of tax deductions for purchasing equipment with decarbonization effects as stipulated by law.

Japan's *Green Growth Strategy* serves as a typical example of a green economy that promotes the coordinated development of the environment and the economy, aiming to facilitate low-carbon transformation through government spending, tax incentives, and national subsidies to achieve the goal of carbon neutrality by 2050.

First, Japan has established a ¥20 trillion *Green Innovation Fund* to support the development of key green technologies, such as hydrogen supply chains, large-scale electrolyzers, and the research and development of next-generation ships and aircraft. This funding will continue to support technology development, demonstration, and commercialization over the next decade. Additionally, Japan provides special funding support for certain industries, particularly projects utilizing hydrogen technology in steel production and initiatives to reduce offshore wind energy costs.

Secondly, regarding tax incentives, Japan offers tax benefits for investments in carbon-neutral facilities, allowing companies to enjoy up to a 10% tax reduction or 50% special depreciation for qualifying investments, applicable to facilities such as power semiconductors, lithium-ion batteries, and fuel cells. Moreover, the research and development tax credit has been extended to companies that increased their R&D investments during the pandemic, with credits reaching up to 30%.

Finally, Japan directly supports low-carbon technologies and renewable energy projects through national subsidies, especially in high-emission industries, assisting companies in transitioning to low-carbon production methods through financial support. These measures not only advance the development of low-carbon technologies like hydrogen and wind energy but also lay a solid foundation for achieving carbon neutrality goals.

## 3.2 Financial Policies

Diverse financial support policies have provided funding for green and low-carbon technological and industrial innovation. Developed countries are using a variety of financing mechanisms to strongly support the development and market application of related technologies. These mechanisms include traditional financial instruments—such as credit, bonds, stocks, trusts, and financial leasing—which provide enterprises with abundant financial support. They also include venture capital and hybrid financing methods, which offer necessary financial backing for innovation projects at different stages of development.

For traditional financial tools, they provide diverse and ample funding support for companies, ensuring that green and low-carbon projects receive the necessary financial backing. For example, the European Union (EU) supports SMEs through the European Investment Fund (EIF) and the European Investment Bank (EIB) [16]. The EIF designs innovative financial products and cooperates with banks, guarantee institutions, leasing and microfinance institutions, and private equity and venture capital funds to meet SMEs' financing needs. In terms of equity tools, the EIF invests in venture capital and growth funds, supporting technology transfer and business incubators, and increasing the availability of venture capital for high-growth and innovative SMEs. Regarding debt tools, the EIF enhances the lending capacity of financial intermediaries by providing guarantees and credit enhancements through securitization. Additionally, EIF also has an established track record of supporting both established and first-time entrepreneurial teams, promoting best market practices and corporate governance. Its equity activities are primarily supported by resources from the EIB and the European Commission. By cooperating with numerous financial intermediaries, EIF provides two main guarantee tools: credit enhancement/securitization and guarantees/counter-guarantees for microcredit, SME loans or leasing portfolios, providing strong protection for SME financing. Japan's financial measures focus on enhancing sustainability disclosure, supporting public-private partnerships among financial institutions, and revitalizing the green bond market [17]. First, through the revised 2021 Corporate Governance Code, Japan has promoted companies to improve the quality of sustainability disclosures based on international frameworks such as the Task Force on Climate-Related Financial Disclosures (TCFD), thereby attracting more green investments. Second, regulatory authorities encourage financial institutions and local entities to develop business models that address climate change issues by providing support and knowledge sharing, which not only promotes local financial innovation but also strengthens the overall flexibility of the financial system. Moreover, Japan has revised its Green Bond Principles and developed a climate transition finance roadmap for high-emission sectors to ensure capital supply and stimulate the green bond market's prosperity. Additionally, Japan is committed to establishing an international green financial center, offering convenient information platforms and certification mechanisms, driving the development of the ESG bond market, and further incentivizing financial innovation and green investments.

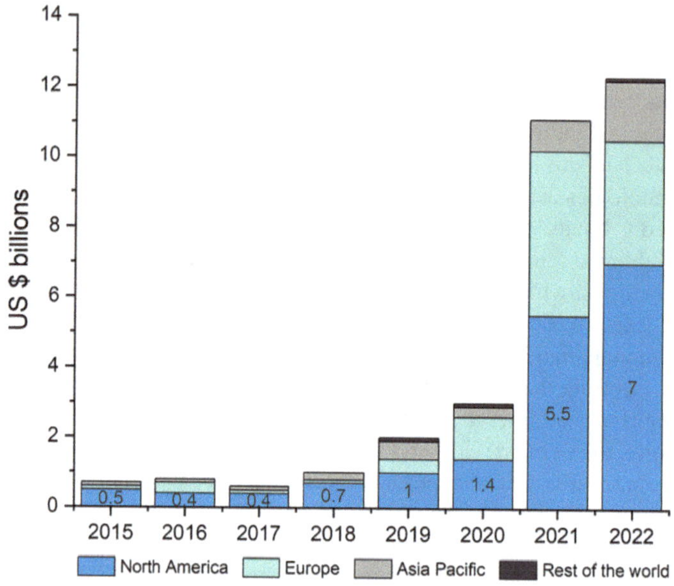


**Fig. 3** Venture capital investment in clean energy startups by region. *Source* Oliver Wyman, 2024

For venture capital and hybrid financing, these innovative financing methods are more focused on helping startups bridge the gap between R&D and market entry. Venture capital can provide pilot and demonstration funds for innovative ideas and future technologies, especially for small companies aiming to scale up in the market. Supported by comprehensive financial regulations and a mature venture capital system, as shown in Fig. 3, North American investors in 2022 invested $7 billion in clean energy startups, accounting for 57% of global venture capital in clean energy [18].

Clean energy technologies are not always well-suited to venture capital's typical investment model. Most venture capital funds sign buyback agreements, committing to return capital to investors within approximately five years. However, this short-term investment horizon often mismatches with the longer time frames and substantial upfront investments required by energy technology developers, especially those in the hardware domain. Additionally, energy technology developers typically need to overcome numerous technical, commercial, and regulatory hurdles before becoming sustainable businesses.

**Hybrid financing,** as an innovative financial tool, combines public and private funds through debt and equity financing and risk mitigation tools. It helps companies overcome the "valley of death" between technology development and large-scale deployment and commercialization. For example, the Breakthrough Energy Ventures-Europe fund, a collaboration between the European Commission and the European Investment Bank, uses equity investment to support energy technology development, leveraging private investment in R&D and innovation [19] (see case

1). This financing approach helps address the challenges of sustainable development, especially supporting the R&D of "public goods".

## 3.3 Green and Low-Carbon Standards

**Standards play a critical role in shaping the path of green and low-carbon technological innovation by regulating market behavior and creating demand for low-carbon innovations.** Different types of standards, such as technical standards and emissions-based standards, not only set targets for controlling emissions but also specify methods and equipment requirements, promoting the development of clean technologies. For instance, Japan's Green Transformation Technology Inventory (GXTI) provides a comprehensive tool for identifying green technology trends through patent searches [20]. In the U.S., the voluntary carbon market (VCM) integrates protocols and methodologies for sectors like energy, while the Department of Energy's carbon dioxide removal (CDR) procurement initiative supports the application of standards [21].

The European Union has implemented a series of forward-looking green and low-carbon standards, such as the *Ecodesign for Sustainable Products Regulation*, which significantly expands the scope of the Ecodesign Directive to cover nearly all physical products. This regulation requires products to consider their entire lifecycle environmental impact, including durability, reparability, recyclability, and resource efficiency. The regulation also introduces a digital product passport, allowing consumers and regulators to make more informed environmental decisions based on product sustainability data [22].

## 3.4 Carbon Markets

**Carbon markets are an effective policy tool for accelerating emission reductions and promoting investment in low-carbon technologies.** By controlling financial flows and establishing constraint mechanisms, they expedite the global transition to a low-carbon economy. The European Union launched its carbon market EU-ETS in 2005, which is now in its fourth phase. The EU-ETS initially controlled the total issuance of allowances in the primary market and then allowed regulated entities in industries subject to carbon emission regulations to trade allowances in the secondary market. This market mechanism drives companies to fulfill their emission reduction obligations. To prevent carbon leakage, the EU also introduced the Carbon Border Adjustment Mechanism (CBAM) and has proposed the scope of products and methods for calculation under this mechanism. After Brexit, the UK launched its own carbon trading scheme, UK-ETS, on January 1, 2021. The UK-ETS is a "cap-and-trade" system, a market-based pricing mechanism designed to incentivize and regulate emissions reductions cost-effectively. Japan has been developing its carbon trading system through multi-sector pilots for some time. In April 2023, Japan officially launched its GX-ETS trial carbon market, marking significant progress toward establishing a national carbon trading platform. In contrast, the United States primarily operates regional carbon trading markets, using carbon trading to push companies to adopt innovative low-carbon technologies and achieve emissions reductions.

In 2023, the European Parliament passed the EU-ETS reform plan, aiming to cut emissions in the ETS sectors by 62% by 2030 (compared to 2005 levels), an increase from the previous target of 43%. The reform achieved three key agreements: First, expansion of ETS coverage. ETS will be extended to the maritime sector for the first time, and the EU will continue to assess the feasibility of including waste incineration in the ETS. Additionally, a separate ETS II is planned to be established by 2027, covering road transport and commercial building heating fuels, with potential expansion to manufacturing sectors. Second, clarifying the transition period and the free allocation reduction schedule for the CBAM. From October 2023 to December 2025, the CBAM will be in a transition period where reporting is required but no taxes will be levied. From 2026 onwards, CBAM will be fully operational, with free emission allowances for EU industries gradually phased out by 2034. Third, improving green transition financing. All revenues from ETS auctions will be used for climate-related activities, including increasing funding for innovation and modernization. Starting in 2026, a Social Climate Fund will be established to support vulnerable groups affected by the climate transition, with funding primarily sourced from ETS II revenues, estimated to reach up to €65 billion ($68.9 trillion).

## 3.5 *Strategic Planning*

Strategic planning is a crucial tool for supporting the implementation of technological innovation outcomes for green and low-carbon transition, enhancing technological competitiveness, and establishing leadership in future green technologies. In the United States, the Biden administration has formulated a comprehensive green and low-carbon innovation strategy. This includes [23]:First, strengthening top-level design for green and low-carbon technological innovation. Through presidential executive orders, the U.S. government has outlined the ten critical green and low-carbon technology fields that need breakthroughs. The administration has also laid out the basic pathways for achieving carbon neutrality through cutting-edge science and technology innovation. Second, enhancing governmental leadership and coordination in climate technological innovation. Under the framework of the National Climate Task Force, the Climate Innovation Working Group has been established. This group, co-chaired by the White House Office of Domestic Climate Policy, the Office of Science and Technology Policy, and the Office of Management and Budget, is responsible for coordinating and organizing technological forces across federal government agencies to define and implement a unified innovation agenda.

Third, building a multi-departmental collaborative R&D and innovation organizational model. The U.S. Department of Energy (DOE) leads in designing and organizing the core tasks of the innovation chain, from basic research to technological breakthroughs, early-stage demonstrations, and deployment. The DOE coordinates with agencies such as the National Science Foundation, Department of Transportation, Department of Agriculture, and Department of the Interior in joint R&D efforts. Four, optimizing the mechanism for the transformation and implementation of scientific and technological achievements. The DOE has been reformed and reorganized to address weak links in the lifecycle of innovative technologies, from R&D to commercialization. The department has established the Office of Clean Energy Demonstrations to create an innovation chain that links technology development with commercialization. The office also deploys major scientific and technological tasks with the collaboration of various stakeholders, including research institutions, industries, and universities, leveraging the strengths of these diverse entities to form new research institutions (see Fig. 4).

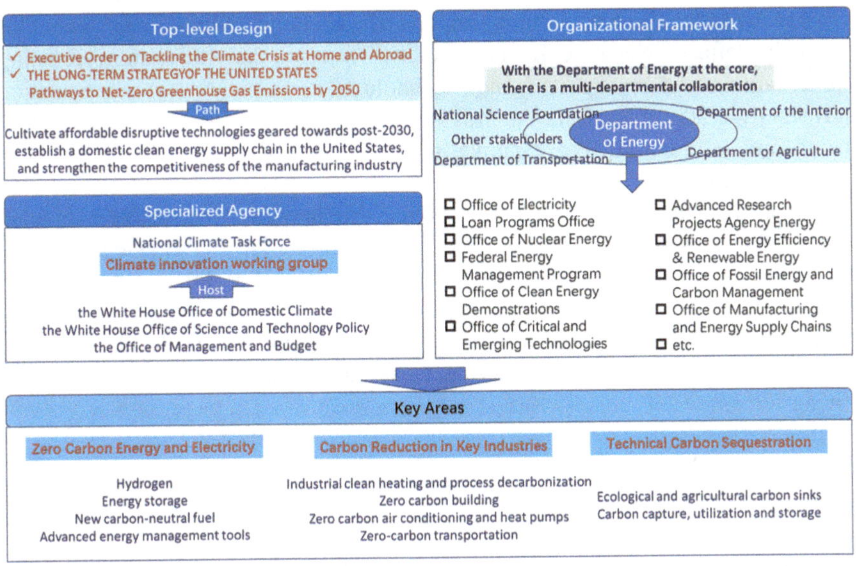


**Fig. 4** Carbon neutrality technology deployment in the United States

## *3.6 International Experiences in Technological Innovation for Green and Low-Carbon Transition*

### 3.6.1 A Renewed Focus on Industrial Policy

**Major economies are increasingly adopting "new industrial policies" to foster the learning-by-doing process for low-carbon technologies, expand the learning pool, and internalize the externalities of green and low-carbon technological innovation.** Since green and low-carbon technologies bring widespread societal benefits, the creators and pioneers of these technologies often do not fully capture the benefits of R&D and green gains. In response, many major economies have welcomed the return of industrial policy over the past decade, recognizing the critical role of policy intervention in effectively allocating market resources.

These new industrial policies share several characteristics: First, they emphasize broad support for R&D and technology diffusion in specific sectors to promote technological innovation and knowledge accumulation. For example, the *Net-Zero Industry Act* in the European Union supports a broad range of low-carbon technologies on its defined "Net-Zero Technology List". Second, they focus on overcoming market failures and constraints on new technology development through regulation and incentives, addressing issues such as the lack of niche markets, fragmented supply chains, and externalities. For example, the U.S. *Inflation Reduction Act* provides subsidies along the entire supply chain, from basic raw materials to manufacturing and domestic consumption. Third, these policies are integrated into broader strategic frameworks focused on creating comparative advantages. The EU's low-carbon technological innovation plan aligns with its broader strategy for global climate leadership and pursuit of low-carbon industry dominance. The basic paradigm for these "new industrial policies" can be summarized as identifying, supporting, and updating strategic sectors in a dynamic process, with technological innovation serving as the driving force.

### 3.6.2 Innovation Entities

**In terms of green and low-carbon innovation policy, developed countries share common goals, focusing on creating national strategic leadership, with enterprises as the main driving force, market-oriented mechanisms, and deep integration of research, industry, and application to achieve high-level outcomes and commercialization**. These policies have three main features: First, they emphasize top-level design and coordination of national scientific and technological forces, with governments playing an indispensable role in leading and organizing technological innovation. This involves coordinating resources across public and private sectors, including research institutions, universities, and enterprises, to create a virtuous cycle between top-down goal-setting and bottom-up demand-driven innovation. Second, they focus on building a new innovation ecosystem that serves carbon neutrality, promotes breakthrough technological advancements, and fosters commercialization. Countries are deploying interdisciplinary and cross-agency large-scale research plans to achieve these goals. Third, they promote public-private partnerships to establish new research and development institutions that provide advanced technologies and solutions for emissions reduction. For example, the UK has proposed the Industrial Decarbonisation Research and Innovation Centre, while the U.S. is focusing on industrial electrification and decarbonization with plans to establish a Department of Defense Manufacturing Innovation Institutes. And Germany has established the Competitiveness of Energy-Intensive Industries as a one-stop platform for government, industry, academia and research, dedicated to uniting all stakeholders to provide consulting, technology and funding for energy-intensive industries.

### 3.6.3 Innovation Environment

**Green and low-carbon transition faces political, economic, technological, and other uncertainties and risks. Developed countries have introduced diverse environmental policy tools to eliminate obstacles to green and low-carbon technological innovation, creating a conducive regulatory and policy environment**. Different countries and regions have developed three primary models in terms of policy tools and their combination: The U.S. primarily adopts supply-side policies, supplemented by environmental and demand-side policies, focusing on financial investment, infrastructure development, and technological innovation, supplemented by planning, subsidies, and tax incentives. Japan adopts a demand-side model, which is mainly based on environment and demand, supplemented by supply, emphasizing policies such as diversified financial measures, green government procurement, and carbon markets. Europe and the UK apply a balanced approach, combining financial, fiscal, carbon market, pricing, and other tools to create a comprehensive policy system that supports green and low-carbon transition.

Developed countries use a balanced approach of regulation, standard-setting, financial support, and fiscal incentives to create a favorable environment for low-carbon technological innovation. By utilizing regulation and standards to guide financial investments and technological development, they promote the scale-up of low-carbon economic activities. Financial support and tax incentives, meanwhile, reduce innovation costs and stimulate market vitality. The balanced application of these environmental policy tools has cultivated fertile ground for the rapid development and application of green and low-carbon technologies. For instance, the UK takes a short-term approach by relying on government intervention and policy incentives, stimulating investment in low-carbon technologies through research and demonstration government funds, green financing, carbon pricing, and the promotion of green products. In the medium and long term, the UK focuses on market mechanisms, gradually building a comprehensive framework that includes carbon pricing, carbon trading, energy efficiency regulations, and international trade agreements. In terms of environmental policy coordination, both the UK and Japan have proposed improving environmental regulatory frameworks to ensure that new decarbonization technologies and policies contribute to achieving environmental goals.

### 3.6.4 Innovation Funding

**Governments in developed countries have adopted diversified funding strategies to support different stages of development for startups**. In addition to direct grants, governments also provide loans and loan guarantees to support business growth. For instance, in the European Union, small businesses can receive up to 70% of the industrial research costs or 45% of the experimental development project costs. This type of funding reduces the financial burden on startups during the technological R&D process. Governments also participate directly in the growth of startups through equity investments. An example of this is the collaboration between the European Investment Bank and the Breakthrough Energy Ventures-Europe fund (BEV-E), which provides long-term, capital-intensive project funding to support the growth of clean energy technology startups.

**In providing funding to startups, governments balance risks and rewards by using reasonable funding terms and strict regulatory mechanisms**. This helps mitigate risks to public funds while ensuring expected economic and social benefits. For instance, the U.S. Department of Energy provides debt financing for large energy projects based on project viability, usually targeting companies that have already secured late-stage venture capital and major customers. This approach reduces financial risk while ensuring effective use of funds. Meanwhile, the European Union has encouraged banks to offer loans to startups by sharing part of the risk through loan guarantees, facilitating more private capital flow to high-risk, high-reward clean energy technology projects.

**Governments often fund startups indirectly through incubators, accelerators, and venture capital funds**. These intermediaries are better equipped to identify, guide, and provide financial support to startups. For example, the European Institute of Innovation and Technology (EIT) through its InnoEnergy Highway program offers not only financial support but also acceleration services tailored to energy technology innovators. This program helps startups by providing consultation and access to market analysis, improving their chances of success even before the business is established. This model of indirect support increases funding efficiency while also offering added value in the form of technical consulting and market insight.

**Developed countries also provide a variety of incentives, such as tax breaks and financial subsidies, to encourage private sector investment in startups**. For example, the UK's Enterprise Investment Scheme (EIS) encourages private investors to fund high-risk startups by offering tax relief, including a 30% income tax reduction for investors and capital gains tax exemptions for investments held for more than three years. These incentives attract substantial private capital to startups, encouraging private investors to take on higher risks. In addition, developed countries have placed a strong emphasis on supporting collaboration between financial institutions and private enterprises through policies and regulations, particularly in areas such as environmental information disclosure and the development of green bond markets. Transparent environmental information disclosure standards and innovative financial products draw more private capital into green technology sectors, fostering innovation and green investments. The above policies aim to create a favorable market environment, enhance the flexibility of the financial system, provide more convenient and broader financial support for green and low-carbon transition, and jointly promote the development of clean energy technologies.

**Governments also support startups through awards and competitions, which not only provide financial backing but also enhance their visibility in the market**. For instance, Germany's Start Up Energy Transition Award (SET) links innovators with stakeholders in venture capital, public and private sectors, increasing the market visibility and appeal of winning companies through media coverage. This broad exposure and synergy encourage greater interest and investment in clean energy technologies, motivating more innovators to participate.

### 3.6.5 International Cooperation

**Developed countries actively promote international cooperation on green and low-carbon technological innovation through existing international platforms, new technical alliances, industry alliances, and bilateral or multilateral cooperation agreements**. For example, the *G7 Industrial Decarbonisation Agenda* calls for strengthened cooperation on market regulation, standard-setting, investment, and procurement between governments and multilateral platforms. One initiative is the creation of a Clean Energy Buyers Alliance, designed to create markets for low-carbon industrial products. The agenda also emphasizes the role of existing multilateral platforms such as the Clean Energy Ministerial, Mission Innovation, and the CCUS Acceleration Consortium. Japan has proposed strengthening cooperation with the United States and Europe on the standardization of key technologies and rules and working with emerging economies to expand markets for green products. The United States has established various climate, energy, or green technology alliances at different levels and across regions. For example, the U.S. is forming a Transatlantic Green Technology Alliance with Europe, and within the Indo-Pacific region, it has set up a climate working group under the Quad (comprising the U.S., Japan, India, and Australia) to collaborate on future key technologies. Through these international cooperation efforts, developed countries aim to coordinate global efforts in market regulation, standard setting, investment, and procurement strategies, promoting technology export and seeking leadership in the global green transition. Simultaneously, they seek to secure key materials and resources from diverse sources, ensuring supply chain security in the face of geopolitical conflicts and resource crises.

# 4 China's Strengths and Challenges in Technological Innovation for Green and Low-Carbon Transition

## 4.1 *Strengths*

### 4.1.1 Massive Market Scale

**China has a vast consumer market, providing broad application scenarios for green and low-carbon technological innovation**. A large consumer market supports the large-scale application of green and low-carbon technologies, enabling these technologies and products to quickly enter the market, receive feedback, and make improvements. This facilitates rapid iteration of technologies, driving down costs and advancing technological progress. According to a recent report Technology Outlook on Wind and Solar Power toward China's Carbon Neutrality Goal by Institute for Carbon Neutrality, Tsinghua University, China's wind and solar installed capacity is expected to reach 1582 to 2130 GW by 2030 and 5496 to 7662 GW by 2060, with wind and solar generating capacity increasing 13-fold by 2060 compared to 2020 levels. By 2060, wind and solar will account for 83% of total installed capacity for all energy sources [24]. In addition, solar PV manufacturing is the technology with the greatest regional concentration. Approximately 80% of global solar PV manufacturing capacity is concentrated in China, with the entire supply chain benefiting from lower production costs. China is expected to increase its share of solar panel production to slightly above 80%, while its share of wafer and polysilicon production may approach 95%. Due to economies of scale and technological innovation, China has been a major driver in the reduction of global solar PV and onshore wind energy costs. As noted by the Executive Director of the International Energy Agency (IEA), Fatih Birol, China's contributions have significantly enhanced access to clean energy technologies and lowered the cost of utilizing green technologies globally [25]. According to data published by the International Renewable Energy Agency (IRENA), between 2010 and 2022, the global weighted average cost of electricity (LCOE) from utility-scale photovoltaic plants dropped by 89%, offshore wind by 59%, and onshore wind by 69% [26]. China's vast market capacity provides enterprises with broad platforms for experimentation and promotion, facilitating a complete industrial chain from R&D to commercialization. This not only accelerates technological iterations but also enhances the competitiveness of enterprises.

### 4.1.2 Economic Growth Potential

**China's economy has shifted from high-speed growth to high-quality growth, yet its economic growth rate remains among the highest globally, demonstrating significant potential. This continued growth will provide ample space for the application of green and low-carbon technologies**. A continuously expanding economy offers sufficient funding and resources for innovation activities, fostering increased investment in R&D and driving breakthroughs in technology. In 2023, China's total GDP was 12.606 trillion yuan, a 5.2% increase from the previous year. Although this represents a slowdown compared to previous periods of rapid growth, the underlying potential of China's economy remains strong, and the long-term trend of positive development has not changed. Additionally, in the first half of 2024, China's industrial added value from enterprises above a designated size increased by 6.0% year-on-year. The green and low-carbon industries showed a particularly strong growth trend. For example, in the first half of the year, production of new energy vehicles (NEVs) increased by 34.3% year-on-year, while the output of complementary products like charging stations and lithium-ion batteries for vehicles grew by 25.4% and 16.5%, respectively [27]. Meanwhile, the combined export value of the "New Three" products—NEVs, lithium-ion batteries, and solar cells—reached 1.06 trillion yuan, breaking the trillion-yuan mark for the first time with an annual growth rate of 29.9%. Notably, one out of every three vehicles China exports is a new energy vehicle [28]. This promising economic growth potential will attract more foreign investors to participate in China's green and low-carbon industries, providing long-term capital support for innovation projects.

### 4.1.3 Strong Organizational and Engineering Capabilities

**China's ability to organize and implement large-scale engineering projects provides a strong foundation for green and low-carbon transition and technological innovation**. First, China has a mature engineering management system and an efficient decision-making mechanism that can quickly mobilize resources and effectively advance large projects. Additionally, the Chinese government's emphasis on infrastructure construction ensures policy and financial support, facilitating the smooth implementation of large-scale green and low-carbon projects. Moreover, China possesses the most complete production supply chain network in the world, with significant economies of scale, allowing companies to reduce costs through bulk procurement and production. This extensive supply chain network fosters upstream and downstream cooperation and accelerates the diffusion and application of new technologies. In regions such as the Yangtze River Delta and the Pearl River Delta, the wind and PV industries have developed comprehensive supply chains. In the wind power sector, China has established an integrated industrial system, including wind turbine sets, special steel for wind power, hubs, towers, main shafts, flanges, gears, bearings, nacelles, special coatings, and other key components [29]. In the PV industry, 80% of China's solar panel production—equivalent to two-thirds of global output—comes from three provinces: Jiangsu (40% of production), Anhui, and Zhejiang (each contributing about 20%) [5]. Furthermore, China's robust technical expertise and a vast pool of engineering talent lay a solid foundation for the development and application of green technologies. With these strong organizational and engineering capabilities, China is well-positioned to accelerate the promotion and application of green and low-carbon technological innovations during the transition to a greener economy.

## *4.2 Challenges*

Despite its achievements, China still faces several challenges in green and low-carbon technological innovation.

**Innovation environment**: Green low-carbon transition involves cross-disciplinary technological innovations. Although China has a large number of researchers, the country lacks top-tier talent with cross-disciplinary expertise in green and low-carbon fields. Furthermore, in the areas of basic research and technology development, China lacks professionals with international competitiveness. The current education and training system has not fully adapted to the rapidly evolving demands of technological innovation, limiting the cultivation and attraction of high-level talent and impeding the enhancement of innovation capabilities.

Many entrepreneurs exploring new technologies and markets are often discouraged by difficulties in obtaining financing, policy uncertainties, and insufficient market demand. Additionally, the high risk of innovation failures and the lack of a robust tolerance mechanism suppress entrepreneurial initiatives.

Although China actively promotes international cooperation in the field of green and low-carbon technologies, the depth and breadth of such collaboration are still limited due to geopolitical tensions and trade restrictions. These limitations hinder technological exchange and transfer and make it harder for China to access cutting-edge global technologies and resources.

**Industrial policy**: Current policies are more focused on supporting mature industries, with relatively less emphasis on emerging green technologies. Policy stability, continuity, and enforcement still need to be strengthened. Moreover, the issue of inefficient subsidies is prevalent, and the fairness and consistency of regional industrial policies need to be improved. The lack of coordination between regional and industrial policies also hampers the progress of industrial transformation and upgrading. Meanwhile, the support and coordination for corporate restructuring and exit are insufficient, and the policy effect evaluation and adjustment mechanism is missing, which affects the accuracy and effectiveness of the policy. Finally, the connection between the old and new policies is not close enough, and the restrictive policy benefits of high-energy-consuming industries need to be transformed into support for emerging green industries to promote the green transition of the industrial structure.

**Financing channels**: The development of green and low-carbon industries requires substantial long-term investment, but China's green financial system is still in its nascent stage. Small and medium-sized enterprises (SMEs) in green technology innovation have limited access to capital and often struggle to obtain long-term, stable funding. Venture capital models are not fully developed, and the current models provide insufficient support for green and low-carbon projects. As a result, many promising innovation projects fail to scale due to a lack of funding. In terms of carbon trading, the auction mechanism, as a way of allocating carbon quotas, can provide price signals and promote market competition, but there is still room for improvement in its design and implementation. The price signal problem is a key challenge in the carbon market. If the carbon price is too low, it may not be enough to motivate companies to invest in emission reduction; while too high a carbon price may increase the operating costs of companies. In addition, with the continuous expansion of the carbon market, the complexity of the market mechanism has increased, which has put forward higher requirements for supervision and information disclosure. The existing supervision system and information disclosure mechanism are not yet sound, which may lead to inefficient use of funds and difficulty in achieving the expected emission reduction effect.

**Consumer demand**: Although public environmental awareness has gradually increased, market demand for green products and services has not been fully unleashed. Price sensitivity and consumption habits have prevented the green consumer market from reaching scale, limiting companies' market expansion and technological promotion. Existing incentive measures, such as tax breaks and subsidies, are mainly directed at production and companies, with few direct incentives for consumers. This reduces consumers' economic motivation to purchase green products.

**Coordination mechanisms**: Green and low-carbon transition involves multiple sectors and industries, but there is insufficient cross-departmental collaboration and policy coordination. Different departments often have inconsistent policy goals and implementation pathways, lacking a unified action framework and standards. This results in inefficient resource allocation and hampers the effective promotion and application of green low-carbon technologies.

# 5 Policy Recommendations for China's Green and Low-Carbon Technological Innovation

## *5.1 Next Steps for the China Council for CCICED in Green and Low-Carbon Technological Innovation*

### 5.1.1 Research on the Green and Low-Carbon Technological Innovation Models and Policies of Developed Countries

**First**, a comprehensive review of green and low-carbon technological innovation actions and deployments in developed countries and regions (including the United States, the European Union, the United Kingdom, Germany, Japan, etc.) should be conducted. This includes analyzing their innovation models and pathways, as well as compiling lists of next-generation green and low-carbon technologies and disruptive innovations prioritized by these economies. **Second**, comparisons should be made between China and developed countries in terms of technological innovation for green and low-carbon transitions, identifying areas where China needs to improve. **Third**, a classification-based evaluation of the policy effects of green and low-carbon technological innovation in developed countries should be undertaken, analyzing the conditions under which different policies are effective. Lessons learned from these experiences can then be used to inform China's policy formulation.

### 5.1.2 Research on China's Near-, Mid-, and Long-Term Roadmaps for Green and Low-Carbon Technological Innovation

**First**, organize multidisciplinary experts and scholars to draft a roadmap for China's green and low-carbon technological innovation, with clear goals for the near term (5–10 years), mid-term (10–20 years), and long term (20–30 years). This roadmap can be divided into several sectors, such as energy transition, industry, transportation, construction, agriculture and land use, circular economy, digital technologies and their integration with green and low-carbon innovations, as well as disruptive and intergenerational technologies. **Second**, develop selection criteria for technologies in each sector, considering factors such as emission reduction potential, economic and social impacts, and international scientific and technological competition. It is also essential to respect the uncertainties of the innovation process and introduce regular evaluations and adjustment mechanisms for the roadmap.

### 5.1.3 Research on the Development of China's Green and Low-Carbon Technological Innovation Policy System in the New Era

Given China's dual carbon goals, there is a need to build a green and low-carbon technological innovation system that aligns with these objectives. Research should focus on the challenges and shortcomings of the current system, examining dimensions such as the innovation environment, fiscal and tax policies, financing mechanisms, and innovation entities. Recommendations should be made to improve the system, particularly in fostering original and disruptive innovations in the green and low-carbon technology field. These efforts aim to enhance China's capacity to provide globally competitive green and low-carbon public goods.

## *5.2 Policy Recommendations for Accelerating China's Green and Low-Carbon Technological Innovation*

### 5.2.1 Cultivating an Environment Conducive to Innovation

First, a coordinated layout for China's medium- and long-term science and technology development plans should be implemented to ensure the long-term stability and consistency of technology-driven carbon neutrality goals. Second, entrepreneurial spirit should be promoted. A social environment that encourages innovation and tolerates failure must be created. It is essential to protect the innovation rights of entrepreneurs through legal means and involve them in planning, policy-making, and setting standards for green and low-carbon technological innovations. Policies such as science and technology innovation vouchers and equipment rental incentives should open up scientific infrastructure to enterprises, fostering shared use of resources. Third, China's basic research should be more open. Cooperation with international third-party organizations in establishing dedicated funds for forward-looking, breakthrough, and disruptive green and low-carbon technologies should be encouraged. Partnerships between domestic and international energy companies, universities, research institutions, and industry organizations should be fostered to form innovation consortia, with the goal of attracting global talent to China for green and low-carbon research and collaboration.

### 5.2.2 Adjusting the Focus of China's New Round of Green Industrial Policies

First, the next round of green industrial policies should be innovation-driven, avoiding over-expansion of production capacity, and ensuring that limited fiscal resources are more directed toward green and low-carbon technological innovation. Second, a public information monitoring platform for the green and low-carbon industry and technologies should be established. This platform should periodically release data and information on capacity utilization, market supply and demand, and technological investment and application. Third, green and low-carbon technologies should be categorized based on their emission reduction benefits, cost, and industrial impact. Regular evaluations should be conducted to adjust and update technology lists, which can then be linked with subsidies, financing, and other policies. Fourth, government procurement of green and low-carbon products should be strengthened, with government agencies and public institutions prioritizing the purchase of green and low-carbon technologies and products.

### 5.2.3 Strengthening Multi-source Financial Governance to Provide Financing Support for Leading Green and Low-Carbon Technological Innovations

First, a financing policy system should be developed based on carbon emissions reductions. Multiple sources of funding, including bank loans, equity investments, and green bonds, should be mobilized to create a multi-tiered, multi-channel green financial system that meets the diverse financing needs of different types and stages of green and low-carbon innovation projects. Second, venture capital should play a more significant role in supporting original and disruptive innovations in green and low-carbon technologies. A financial services system based on venture capital and backed by stock markets should be established. Third, revenues from carbon trading markets, as well as fines and penalties from other carbon reduction regulatory policies, should be used to provide competitive subsidies to support green and low-carbon technological innovations.

## References

1. IEA. Net Zero Roadmap. A Global Pathway to Keep the 1.5 °C Goal in Reach. (2023-09) [2024-09-04]. https://www.iea.org/reports/net-zero-roadmap-a-global-pathway-to-keep-the-15-0c-goal-in-reach
2. IEA. The Role of Carbon Credits in Scaling Up Innovative Clean Energy Technologies. (2024-04) [2024-09-04]. https://www.iea.org/reports/the-role-of-carbon-credits-in-scaling-up-innovative-clean-energy-technologies
3. Zhao, X., Nakonieczny, J., Jabeen, F., et al. (2022). Does green innovation induce green total factor productivity? Novel findings from Chinese city level data. *Technological Forecasting and Social Change, 185*, 122021.
4. Wu, J., Xia, Q., & Li, Z. (2022). Green innovation and enterprise green total factor productivity at a micro level: A perspective of technical distance. *Journal of Cleaner Production, 344*, 131070.
5. IEA. The State of Clean Technology Manufacturing November 2023 Update. (2023-11) [2024-09-04]. https://www.iea.org/reports/the-state-of-clean-technology-manufacturing-november-2023-update
6. IEA. Renewables 2023. (2024-01) [2024-09-04]. https://www.iea.org/reports/renewables-2023
7. IEA. The importance of focusing on jobs and fairness in clean energy transitions. (2021-07-06) [2024-09-04]. https://www.iea.org/commentaries/the-importance-of-focusing-on-jobs-and-fairness-in-clean-energy-transitions
8. EU. The future of European competitiveness—In-depth analysis and recommendations. (2024-09-09) [2024-09-04]. https://commission.europa.eu/topics/strengthening-european-competitiveness/eu-competitiveness-looking-ahead_en
9. Edler, J., Gök, A., Cunningham, P., et al. (2016). Introduction: Making sense of innovation policy. In *Handbook of innovation policy impact* (pp. 1–17). Edward Elgar Publishing.
10. Borrás, S., & Edquist, C. (2013). The choice of innovation policy instruments. *Technological Forecasting and Social Change, 80*(8), 1513–1522.
11. Stucki, T., Woerter, M., Arvanitis, S., et al. (2018). How different policy instruments affect green product innovation: A differentiated perspective. *Energy Policy, 114*, 245–261.
12. Fabrizi, A., Guarini, G., & Meliciani, V. (2018). Green patents, regulatory policies and research network policies. *Research Policy, 47*(6), 1018–1031.
13. Skjærseth J, B. (2021). Towards a European green Deal: The evolution of EU climate and energy policy mixes. *International Environmental Agreements: Politics, Law and Economics, 21*(1), 25–41.
14. IEA. How Governments Support Clean Energy Start-ups. (2022-03) [2024-09-04]. https://www.iea.org/reports/how-governments-support-clean-energy-start-ups
15. Burgess, M. G., Van Boven, L., Wagner, G., et al. (2024). Supply, demand and polarization challenges facing US climate policies. *Nature Climate Change, 14*, 134–142.
16. EU. The EIF and InvestEU. [2024-09-04]. https://www.eif.org/InvestEU/index.htm
17. Ministry of Economy, Trade and Industry. Green Growth Strategy Through Achieving Carbon Neutrality in 2050. (2022-10-17) [2024-09-04]. https://www.meti.go.jp/english/policy/energy_environment/global_warming/ggs2050/index.html
18. OliverWyman. Investment In Clean Energy Startups Is Booming. [2024-09-04]. https:// www.oliverwyman.com/our-expertise/insights/2023/may/investment-in-clean-energy-startups-is-booming.html
19. EU. The European Commission, European Investment Bank and Breakthrough Energy Ventures establish a new €100 million fund to support clean energy investment. (2019-05-29) [2024-09-04]. https://ec.europa.eu/commission/presscorner/detail/en/ip_19_2770

20. Japan patent Office. The Green Transformation Technologies Inventory (GXTI). (2023-10-31) [2024-09-04]. https://stip.oecd.org/stip/net-zero-portal/policy-initiatives/2023%2Fdata%2FpolicyInitiatives%2F99996882
21. Morgan, J. P. Where is sustainability headed in 2024?. (2024-01-29) [2024-09-04]. https://www.jpmorgan.com/insights/esg/sustainability/green-economy-outlook-sustainability-trends-for-2024
22. EU. Ecodesign for Sustainable Products Regulation. [2024-09-04]. https://commission.europa.eu/energy-climate-change-environment/standards-tools-and-labels/products-labelling-rules-and-requirements/sustainable-products/ecodesign-sustainable-products-regulation_en
23. The White House. FACT SHEET: President Biden Signs Executive Order Catalyzing America's Clean Energy Economy Through Federal Sustainability. (2021-12-08) [2024-09-04]. https://www.whitehouse.gov/briefing-room/statements-releases/2021/12/08/fact-sheet-president-biden-signs-executive-order-catalyzing-americas-clean-energy-economy-through-federal-sustainability/
24. Institute for Carbon Neutrality, Tsinghua University. Technology Outlook on Wind and Solar Power toward China's Carbon Neutrality Goal. (2024-01-08) [2024-09-04]. https://www.icon.tsinghua.edu.cn/info/1070/1436.html
25. People's Daily. China's new energy industry helps the global green and low-carbon transition. (2024-04-09) [2024-09-04]. http://en.people.cn/n3/2024/0412/c90000-20156194.html
26. IRENA. Renewable power generation costs in 2022, International Renewable Energy Agency, Abu Dhabi. (2023-08) [2024-09-04]. https://www.irena.org/Publications/2023/Aug/Renewable-Power-Generation-Costs-in-2022
27. National Bureau of Statistics. Tang Weiwei: Industrial production is generally stable and improving, and high-quality development continues to advance. (2024-07-15) [2024-09-04]. https://www.stats.gov.cn/sj/sjjd/202407/t20240715_1955610.html
28. Chinese government website. In the first quarter, the combined exports of electric passenger cars, lithium batteries and solar cells increased by 66.9%—the "new three" lead the foreign trade exports. (2023-05-02) [2024-09-04]. https://www.gov.cn/yaowen/2023-05/02/content_5753916.htm
29. People's Government of Wuxi Binhu District. Wuxi's 13th Five-Year Energy Development Plan. (2018-07-30) [2024-09-04]. http://www.wxbh.gov.cn/doc/2018/07/30/

# Chapter 7
# Beyond GDP

## 1 Why Move Beyond GDP?

The world is facing unprecedented challenges. Climate change has already caused irreversible impacts on natural systems [1], and planetary warming will, according to some, surpass the Paris Agreement target of 1.5 °C in the 2020s without urgent avoidance action [2]. Biodiversity and the ecosystems that support it are deteriorating worldwide [3]. On top of these—and other—environmental challenges are economic and social challenges of similar scope and concern: growing global conflict [4], mounting levels of private and public debt [5], and declining trust and increasing polarization [6] to name a few.

The seriousness of the situation we face today is leading to calls for fundamental change in the way the progress of nations has been defined and measured for the last 75 years, both in China and around the world [7, 8]. Through the lens of the dominant development paradigm, progress has been equated with growth in the output of the market economy as measured by GDP. A growing GDP has been the sine qua non of success. Politicians, investors, business people, and the public have come to see it as the primary touchstone for how well their countries are doing. If GDP is up, a country is said to be moving in the right direction. If growth is weak or, worse, negative, alarm bells will ring, and a change of course will be called for. It is hard to overstate the influence this one measure has had in shaping the modern world.

It has to be said that many nations—not least China—have embraced the growth paradigm with considerable success, at least if one only considers GDP growth itself as the yardstick. Global economic output as measured by GDP has increased more than ten-fold since the 1950s after accounting for inflation [9]. Population, for its part, grew (only) about three times during the same period [10], meaning that many people saw their real incomes increase significantly as a result of growth. The absolute

The content of this chapter is extracted from the main findings of the 2023–2024 CCICED Scoping Study on "Beyond GDP".

 
*Adhering to Green and Open Cooperation to Advance Modernization*,
https://doi.org/10.1007/978-981-95-5547-5_7

number of people living in extreme poverty has dropped from over half of the global population in 1950 to about 10% today [11]. Since opening up in 1978, China has been among the most successful countries in this regard. The Chinese economy has grown nearly 13-fold in a bit more than 4 decades, surpassing what it took the world on average 7 decades to achieve [12]. As a result of this, the number of extremely poor fell from more than 90% of the population in 1978 to almost none today.

In spite of its successes in meeting people's material needs and wants, many find the dominant growth paradigm unfit for purpose in the Anthropocene [13, 14]. Critics of GDP growth as the central goal of government policy point to the challenges outlined above, among many others, as ample evidence of the need for change.[1] Climate change, especially, is seen as a reason to "rethink core assumptions about how we build our economies and our infrastructures, how we travel, how we plan for global pandemics, and even how we eat" [14]. China's "dual carbon" commitment—to achieve peak carbon emissions by 2030 and carbon neutrality by 2060—is seen by leading economists as "a profound shift in the old development narrative regarding both the purpose of development, and what and how to produce" [7].

World leaders, too, increasingly agree that countries need to move beyond GDP as the central measure of progress. The United Nations Secretary-General says—without equivocation—that "global decision-making is fixed on immediate gain, ignoring the long-term consequences of decisions" [16]. If the world wants to achieve sustainable development, it requires "economic analysis based on today's realities, rather than outdated ideas of economic success." This means "we must correct a major blind spot in how we measure progress and prosperity." That is why he is calling for "new metrics that value the life and wellbeing of the many over short-term profit for the few."[2] G7 leaders, for their part, also recognize that GDP alone is "insufficient for measuring success."[3] Text Box 1 outlines in greater detail the reasons why GDP fails in this regard.

The seventeen United Nations Sustainable Development Goals (SDGs) and the numerous related targets and indicators represent a widely agreed upon—if expansive[4]—efforts to move beyond GDP toward an integrated development agenda. In fact, the very last target in the SDG framework—target 17.19—explicitly calls on

[1] It is worth noting that such calls are hardly new. As the famed Yale University Economists William Nordhaus and James Tobin [15]* wrote in a seminal work on the topic 50 years ago, growth was in the 1950s and 60 s "simultaneously the hottest subject of economic theory and research, a slogan eagerly claimed by politicians of all stripes, and a serious objective of the policies of governments." They went on to note how "disillusioned critics [in the early 1970s] indict[ed] both economic science and economic policy for blind obeisance to aggregate material 'progress,' and for neglect of its costly side effects. Growth, it [was] charged, distorts national priorities, worsens the distribution of income, and irreparably damages the environment" (emphasis added). So, if we've made little progress in addressing the GDP's shortcomings in the last 50 years, we cannot claim it is because we were not aware of them.

[2] [7]* United Nations. Secretary-General's Report "Our Common Agenda" [EB/OL]. [2024-09-03]. https://www.un.org/en/content/common-agenda-report/

[3] See, for example, [1]* Global Affairs Canada. The Charlevoix G7 Summit Communique, 2018 [EB/OL]. 2018-06-09-summit-communique-sommet-en.pdf (international.gc.ca).

[4] In total, there are 17 SDG targets and 248 indicators.

governments to "build on existing initiatives to develop measurements of progress on sustainable development that complement gross domestic product."[5] As argued above, this target must not be an afterthought on the road to 2030.

While the SDGs and their indicators have the benefit of political acceptance, they alone cannot take the world beyond GDP. One reason is simply the number of indicators. Recognizing that greater parsimony is desirable in a measurement framework, the United Nations Secretary-General argues that moving beyond GDP means finding 10–20 indicators at most. Some means of paring the SDG indicators down to a more manageable size would thus be needed, and such paring would be challenging.

A second challenge to the SDGs as a measurement framework is that the goals must be reviewed and, likely, replaced by others after 2030. Such transience is undesirable in a measurement framework. Without stability,[6] the statistical systems needed to compile the measures and the political systems needed to act upon them will not have time to evolve.

These two challenges mean that the SDGs offer an important starting point for measurement beyond GDP—with their explicit and inspiring focus on well-being, respect for human life and the planet, and reduced inequalities—but they cannot be its endpoint.

## 2 If Not GDP, Then What?

Though consensus is emerging that countries must move beyond GDP and adopt new indicators of progress, just what those indicators should be remains a matter of debate. Careful thought will be required in choosing them, as GDP remains popular with decision-makers, and resistance to change is likely. As a guide, GDP has many attractive features. It is concise, summarizing a great deal of information in a single number. It is robust, resting on decades of theoretical and practical development. And, for all its shortcomings, it is clearly relevant to the concerns of decision-makers and citizens alike. What GDP measures (income in people's pockets) matters. Moving beyond GDP should not mean its abandonment, then. Rather, what is needed are complementary indicators that provide a foundation for more sustainable policy decisions in a carbon-constrained world, but that also match GDP's conciseness, robustness, and relevance [17, 18].

Some argue that GDP itself simply needs to be modified to address its shortcomings. One of the earliest such proposals was the "measure of economic welfare" of Nordhaus and Tobin [19], which built upon GDP by adjusting it to account for

[5] See [2]* United Nations. Sustainable Development Goal 17: Strengthen the means of implementation and revitalize the Global Partnership for Sustainable Development [EB/OL]. 2015. https://sdgs.un.org/goals/goal17#targets_and_indicators.

[6] Note that stability is not stasis. Measurement frameworks can and must evolve over time. But such evolution should be slow and predictable, neither of which is assured by politically determined frameworks.

household non-market production, the value of leisure time, and welfare-reducing disamenities of urban life. Later efforts like the "index of sustainable economic welfare" and the "genuine progress index" built upon Nordhaus and Tobin's work [20].

While reforming GDP to make it a "better" measure of welfare may be a good idea,[7] reform is not sufficient on its own. Even if GDP were improved in various ways, it would remain inadequate as the core measure of national progress. Notably, the determinants of future well-being will always fall outside GDP's scope, no matter how it is measured. As argued by the Cambridge University economist Sir Partha Dasgupta in his landmark review of *The Economics of Biodiversity*, GDP may be "indispensable in short-run macroeconomic analysis and management [but] it is *wholly unsuitable for … identifying sustainable development*" [15]. Rather, "in order to judge whether the path of economic development [nations] choose to follow is sustainable, [they] need to adopt a system of economic accounts that records an inclusive measure of their wealth" [15]. More recently, Dasgupta and Levin [21] have wondered "how and why GDP has assumed prominence as the measure of long-term economic performance … [when] … we should be studying *the wealth of nations, not the GDP of nations*" (emphasis added).

Dasgupta and Levin call for a "revised economic grammar" [21] in which economics is seen as primarily the study of asset management rather than income growth. Seen this way, development would be a process of managing national wealth to maximize its returns for citizens rather than a process of maximizing national income. Such a change in the grammar of economics is urgently needed, they argue, because the unprecedented improvements in living standards that have been achieved since the start of the 1950s are predicated on the accumulation of produced capital and human capital at the expense of the equally unprecedented depletion and degradation of natural capital. This fact—which may well be a matter of planetary and societal survival today—is entirely invisible when GDP is the primary basis for assessing progress.

A growing number of other experts,[8] including at the United Nations Environment Programme and the World Bank, agree with the argument above. They argue that decision-makers need to focus on managing wealth to ensure the sustainability of development, not simply on wealth as traditionally measured. Rather, in addition to the traditional measures of produced and financial capital, they insist countries also track the evolution of their human, natural, and social capital. This broad portfolio of assets is called either inclusive or comprehensive wealth to reflect the fact that it

[7] If a kind of evangelical zeal for GDP [15]* is, in fact, part of the reason for its abuse as a measure of progress, a case can be made that "tweaking" GDP rather than supplanting it simply gives the evangelists reason to deepen their faith and, therefore, is best avoided. Plus, it is fair to note, none of the GDP-based welfare measures proposed since Nordhaus and Tobin's original work has ever gained a following among decision-makers.

[8] See, among others, Agarwala et al. [3]*;Arrow et al. [4]*;Dasgupta [5, 6, 8, 9]*; Dasgupta and Mäler [10]*; Hamilton and Clemens [11]*; IISD [12, 13]*; Kurniawan and Managi [14]*; Managi and Kumar [16]*; Polasky [22]*; SEI and CEEW [17]*; Office for National Statistics [18]*; UNU–IHDP and UNEP [19, 20]*; UNEP [21]*; World Bank, 2011 and 2018 [23, 24]*; Zenghelis [25]*.

covers all possible types of assets (here, we call it "comprehensive" wealth, though we could as easily have used "inclusive" as the descriptor).[9] The assets that make up the comprehensive wealth portfolio—such as healthy ecosystems, strong communities, educated citizens, efficient buildings, and sound financial holdings—are what countries need to generate well-being for their citizens, both today and in the future. A sound national wealth portfolio is the basis for clean air and water, social trust, a productive workforce, and a vibrant financial sector, among many other elements of well-being. Because the assets of the comprehensive wealth portfolio are long-lived, they can generate well-being far into the future if they are properly managed [18].

Writing about the need for China to adopt a new model of development, Stern et al. [23] and Zhu et al. [7] also note the importance of measuring wealth rather than (or in addition to) GDP. According to the dominant development paradigm, production is a function only of labour and produced capital, while the environment and society are inconsequential. These authors argue that China's twenty-first-century development paradigm—indeed, any country's—cannot be so narrowly focused. There can be no single fixation on income growth as measured by GDP, as this fails to capture important aspects of well-being. Instead, the new development paradigm must be based on a balanced accumulation of all types of capital, recognizing that well-being is shaped by combinations of them all. Though the new paradigm must recognize the importance of all forms of capital, special attention must be paid to natural and social capital. The latter is difficult to substitute for and prone to irreversible thresholds and collapse. Given this, ensuring the quality of the environment and managing social cohesion should be core economic objectives for China and all nations.

Writing about China's need to adopt a modern development paradigm from a slightly different angle, Zhang [8] argues that the way in which consumer utility is conceived in the traditional growth paradigm must be replaced with a view reflecting the broader spectrum of assets in the comprehensive wealth portfolio. As he notes, the objective function in standard economics is the maximization of utility from the consumption of goods and services provided by the market under an assumed set of preferences. The utility gained from the consumption of any good or service produced outside of the market does not enter into this function, nor does the possibility of changing preferences. To address this serious shortcoming, which Zhang argues is partly to blame for economics' failure to address the gravity of environmental and social decline, Zhang proposes a new objective function. In this new function, utility is understood to derive from the consumption of all goods and services, whether

[9] Though both "inclusive" and "comprehensive" are in common use, there is no substantive difference between them. Both terms refer to a measure of wealth that reflects the full spectrum of assets upon which humans rely for inputs into their myriad economic and social activities. There remains some disagreement as to what this spectrum includes; for example, some experts—notably, Professor Dasgupta [25]*—argue that social capital is more of an enabler of other assets while others—notably, Professor Stern and colleagues [26]*, present it as an asset unto itself. However, these debates have nothing to do with which label—"inclusive" or "comprehensive"—is used. Similarly, there remain debates about the best methods and data to use in measuring assets, but those debates transcend the label.

produced within the market or through non-market (environmental or social) activities. Furthermore, consumer preferences are assumed to change with "changing income, identity, knowledge, education, culture, beliefs, institutions, policies, and so on" [8]. He introduces the variable for non-market goods and services in his utility function (see Eq. 9 on p. 61) as a non-rival aggregate product, meaning that it can be consumed by all users (both businesses and individuals) simultaneously. To avoid overuse, this requires a constraint on its use, which Zhang notes could be imposed by limiting either businesses or individuals.[10]

Zhang rightly notes that maximizing wealth—even comprehensive wealth—does not necessarily translate to maximizing well-being (since well-being is a function of what a country does with its wealth, and this will vary from country to country). Thus, measuring comprehensive wealth is a necessary condition for implementing Zhang's conception of a modern development paradigm but not a sufficient condition. We take up this point further below.

## 3 From Income to Wealth[11]

As a means of moving beyond GDP, comprehensive wealth has much to offer. Like GDP, it is concise. Just a few high-level indicators suffice to measure comprehensive wealth. It is also robust, resting on theory and guidance stretching back more than a century (Text Box 2). And, as noted, it is relevant to well-being across all its dimensions—economic, social, and environmental.

Though not everyone will be familiar with comprehensive wealth, most people can grasp its importance intuitively. People understand that their well-being is not determined in the long run by how much they earn today but by their capacity to earn tomorrow and beyond. They also understand that more than money matters to their well-being. What they "earn" more broadly through interacting with family and neighbours, enjoying the benefits of nature, and feeling safe in their communities also counts for much. Even though they might not use the term, they will grasp that it is their assets that determine their long-term earning potential, both monetary and non-monetary. These include both personal assets (a home and property; money in the bank; skills, knowledge, and experience; relations with family, friends, and society at large, etc.) as well as assets shared with other citizens, such as healthy ecosystems and efficient public infrastructure. Together, these personal and shared assets make up each person's comprehensive wealth portfolio—and determine his/her prospects for long-term well-being. People understand that their personal wealth

[10] In regard to the latter, he notes that governments could establish "environmental accounts" with quotas for individuals that, if exceeded, could be topped up with the purchase of additional quotas from the market. Such an idea would likely be effective only in countries where citizens are accustomed to high levels of control over their personal lives, such as China.

[11] This section and the next draw heavily on the arguments raised in Smith et al. [27]*.

portfolios must be maintained if their well-being is to be sustained in the future. This is the essence of "saving a penny for a rainy day."

The same is true for countries—national wealth portfolios determine the prospects for the long-term well-being of countries and their citizens. A country's comprehensive wealth portfolio comprises the individual portfolios of its citizens and businesses plus the public assets owned collectively, such as ecosystems and natural resources, roads and other infrastructure, and institutions. As with individuals, national wealth portfolios must be maintained over time for well-being to be sustained. If wealth is not maintained, a country is eroding its productive base, living off its inheritance rather than building for the future. That is the essence of unsustainability.

## 4 Measuring Comprehensive Wealth

An obvious question is, "What exactly must be maintained to preserve and enhance wealth?" The total number of factories, cars, educated workers, trees, lakes, mineral deposits, corporate shares, savings bonds, and engaged, trusting citizens? In some sense, yes. But this is not a very useful way to think about it. For one, it would be cumbersome to track all these things individually. More importantly, it would be difficult to compare them with one another. How does a factory compare against a hectare of trees in terms of supporting well-being? Are citizens better or worse off if they own more corporate shares but levels of community trust fall?

The problems of measurement and comparability are greatly reduced if all assets are measured using the same yardstick. Practically speaking, this means using monetary values, as no other common yardstick exists. Ideally, the monetary values used should be based on so-called "shadow prices" of assets, though such prices are more of a theoretical than a practical reality.[12] In practice, assets are most often valued using their market price or, where markets don't exist, modelled estimates that reflect what their market price would be. Monetary valuation permits most of the comprehensive wealth portfolio to be measured using just a handful of indicators.

Of course, not all elements of the comprehensive wealth portfolio can, or should, be valued. Social capital, in particular, is likely best measured using non-monetary indicators, at least for now.[13] The same is true of elements of natural capital, such as

[12] Shadow prices (which are sometimes also called "accounting" prices) are the value of the contribution that an additional unit of an asset, other things remaining the same, would make to well-being across generations. They are not readily observed, though in some cases they can be approximated with market prices. They may also be estimated using models, though this is challenging. None of this is a reason not to compile comprehensive wealth measures, however. As Dasgupta and Levin [21]* put it, though "it would be entirely wrong to imagine that [shadow] prices can be estimated, even approximately, for all capital stocks," this does not negate the importance of "[expanded] wealth and the concept of well-being across the generations from their centrality in economic reasoning" (p. 3).

[13] It should be noted that research on the valuation of social capital is being undertaken. See, for example, Hamilton [29]*.

certain ecosystem services, for which no known substitutes exist. Thus, a complete wealth measurement system would comprise monetary indicators of assets amenable to valuation (the majority) plus a limited number of non-monetary indicators related to social and natural capital. As such, it would represent a robust and concise addition to decision-makers' toolboxes. If decision-makers in China—and elsewhere—focused on this small suite of indicators in addition to GDP and a limited number of flow-based indicators of well-being (see Text Box 3), it would lead to more sustainable policy outcomes than today's approach in which achieving GDP growth counts above all else.

## 5 Practical Experience Measuring Comprehensive Wealth

Though no country currently compiles official comprehensive wealth indicators, both the World Bank and UNEP have substantial experience measuring comprehensive wealth. The World Bank has released global reports on the topic under its "Wealth of Nations" series regularly since the early 2000s.[14] UNEP, which uses the term "inclusive wealth," has released four global reports since 2012[15] plus another focused specifically on Pakistan [24]. Though there are many commonalities between the World Bank and UNEP in terms of the scope, methods, and data used to measure comprehensive wealth, differences do exist in their approaches. Chinese leadership could be beneficial in encouraging the agencies to review and align their approaches in the name of converging on a single, global approach. Ultimately, such an approach should be laid out by the United Nations Statistical Commission and adopted as a global statistical standard.

The International Institute for Sustainable Development (IISD) has played a leading role in promoting the measurement of comprehensive wealth for nearly a decade [25]. Its efforts began with two reports on comprehensive wealth for Canada [25, 26]. These reports painted quite a different picture of Canada's development than that offered by GDP alone. While Canada appeared to do well with GDP as the main gauge of success, comprehensive wealth analysis showed that success was resting on a shaky foundation. According to IISD's figures,[16] human capital—Canada's greatest asset[17]—was stagnant in per capita terms from 1980 to 2015. At the same time, natural capital was declining, produced capital was overly concentrated in fossil fuel extraction and residential housing, and financial capital was too reliant on holding gains on foreign assets.

[14] World Bank [23, 24], [30–32]*.

[15] UNU–IHDP and UNEP [19, 20]*; Managi and Kumar [16]*; UNEP [33]*.

[16] IISD drew the figures for its analysis directly from Statistics Canada, Canada's national statistical agency, so the results are considered to be as robust as possible given current concepts, methods, and data.

[17] Human capital is not just Canada's greatest asset but the most important asset in every country.

Building on what it learned from the two Canadian studies, IISD more recently tackled the challenge of measuring comprehensive wealth in countries with less advanced statistical systems. IISD worked from 2020 to 2024 with researchers and experts in Ethiopia, Indonesia, and Trinidad and Tobago to compile comprehensive wealth estimates for those countries. A primary motivation for the project was to determine if comprehensive wealth estimates could be compiled in countries with limited statistical resources using mainly nationally sourced data, following methods and time frames similar to those used in the World Bank's wealth of nations reporting.

China's comprehensive wealth has been the subject of two studies, one at the national/provincial level [27] and the other focused on cities [28]. Both studies took as their starting point the methodology developed by UNEP in its global reports. Fan et al. report an average annual increase in per capita comprehensive wealth[18] rate of 2.3% from 2000 to 2016, which is considerably lower than the growth of per capita GDP over the same period (8.9%). Gender inequality and declines in natural capital were found to hinder the growth of comprehensive wealth. Provincial results varied substantially, with the poorest province (Qinghai) having just 1/18th the comprehensive wealth of the richest (Guangdong). Inner Mongolia had the highest average GDP growth rate among the 30 provinces (15%), but its comprehensive wealth growth rate was only 1.6%.

Looking at the comprehensive wealth of Chinese cities, Cheng et al. [28] report that cities in the eastern part of China are characterized by high levels of comprehensive wealth and increasing sustainability, largely driven by the high numbers and skill levels of their human capital. In comparison, western and northern cities have large amounts of low-skilled human capital, low levels of produced capital, and declining natural capital. Demonstrating the value of comprehensive wealth accounts as modelling tools, Cheng et al. present projections of comprehensive wealth to 2030, based on which they argue northern cities must transition away from their dependence on fossil energy assets as sources of wealth.

The results of Fan et al. and Cheng et al. demonstrate just how different the portrait of the nation can be when painted with the brush of comprehensive wealth rather than GDP. The results suggest that China's current development patterns may not be sustainable, with income growing much faster than the asset base that sustains it, wide regional disparities, and declining natural capital, among other challenges. These results point to the need for comprehensive wealth estimates to be presented with the greatest granularity possible. Distributional analysis by region, by asset, by household type and by industry are all essential to fully understanding comprehensive wealth and its trends. Just as with GDP, aggregate values can mask important differences among groups, allowing instances of gross inequality to go unnoticed by decision-makers. Failure to address wealth inequalities is a growing concern in today's world, where resentment of the "haves" by the "have-nots" can breed polarization.

[18] Following the practice of UNEP, Fan [34]* refer to it as "inclusive wealth."

## 6 Recommendations

To stimulate the move beyond GDP, leadership will be needed. Zhu et al. [7] argue that China's dual carbon commitment is, implicitly, a decision to "accelerate its shift towards a new model of modernization" (p. 2). Assuming this is correct, China is well positioned to play a leadership role in this generational change. To help China fulfill this role, we recommend that CCICED mount a Special Policy Study beginning in 2025 to:

- the NDRC and other central agencies of the Chinese government in discussions of the importance of moving beyond GDP and the value of comprehensive wealth as a tool in the context of 5-year plans and identify opportunities for Chinese leadership in promoting comprehensive wealth as a means of moving beyond GDP regionally and globally.
- a conceptual framework for moving beyond GDP in China based on comprehensive wealth and consider how this framework aligns with existing efforts to measure well-being-related flows in China, including Green GDP, Gross Ecological Product and other prominent indicators.
- pilot comprehensive wealth figures for a region of China (e.g., the city of Ordos) or for a state-owned enterprise, focusing on those elements of the comprehensive wealth portfolio that can be measured using available data.

In the longer term, we believe the Chinese government should engage its decision-makers—and their counterparts in other countries—in discussions about comprehensive wealth and the role it can play in moving beyond GDP. Such discussions should be tailored to both national and regional circumstances. While the basic arguments in favour of using comprehensive wealth to measure progress do not vary from country to country, the context in which the arguments must be made does. Discussions should be held at high levels where policy decisions are ultimately made.

In these discussions, it should be made clear that comprehensive wealth indicators are not intended to replace GDP but to complement it. They should further highlight that measuring wealth is a necessary, but not a sufficient, condition for creating sustainable well-being [7, 8]. The goal would be to enhance decision-makers' toolboxes not just with comprehensive wealth indicators but also with additional flow-based well-being indicators of the sort discussed in Text Box 3 below. Such enhanced toolboxes would ensure that decision-makers keep their eyes firmly on creating well-being not just in the short-term but in the long-term as well and that all dimensions of well-being, and not just economic factors, are kept in focus.

China should also fund the National Bureau of Statistics to undertake the necessary research and data development to measure the country's comprehensive wealth portfolio. Indicators of comprehensive wealth should be published alongside GDP. Advanced national statistics offices—like China's—should be able to produce useful estimates in 2 years with sufficient funding and a concerted effort. These estimates will reveal, among other benefits, where investments in wealth creation will most increase the sustainability of well-being. In China, this may be in better education to build human capital even further and in controlling the degradation of natural capital.

Finally, China should commit to integrating comprehensive wealth indicators into its decision-making processes. This would mean, for example, making wealth measurement and management a core part of the 5-year planning process. It would also mean ensuring that incentives for officials are aligned with wealth—rather than income—maximization. Efforts will be required all along the decision-making chain to ensure that comprehensive wealth indicators are front and centre. Policy analysts will have to craft new analytical tools incorporating wealth indicators, and senior officials will have to resist the instinct to emphasize GDP and think short-term.

The above steps could be taken quickly following the conclusion of a successful Special Policy Study (within 2 years). They would be rooted in sound theory and be well supported by international statistical guidance. Most importantly, they would ensure that today's short-term bias is replaced with a more balanced perspective squarely focused on what matters in the long run—sustainable well-being for all citizens of the world.

## References


1. Pörtner, H.-O., Roberts, D. C., Adams, et al. (2023). *Climate change 2022: Impacts, adaptation, and vulnerability. Contribution of working group II to the sixth assessment report of the intergovernmental panel on climate change.* Cambridge University Press.
2. Hansen, J. E., Sato, M., Simons, L., et al. (2023). Global warming in the pipeline. *Oxford Open Climate Change, 3*(1), kgad008. https://doi.org/10.1093/oxfclm/kgad008
3. Intergovernmental Science-Policy Platform on Biodiversity and Ecosystem Services. (2019). *Global assessment report on biodiversity and ecosystem services of the intergovernmental science-policy platform on biodiversity and ecosystem services.* IPBES Secretariat.
4. ACLED. *Statistics dashboard, country view.* [2024-08-31].
5. IMF. (2023). *Global debt is returning to its rising trend.*
6. Edelman. (2023). 2023 *Edelman trust barometer*: *Global report.* [2024-08-31].
7. Zhu, M., Stern, S., Stiglitz, J. E., et al. (2023). *Embracing the new paradigm of green development: China Carbon Neutrality Policy Framework research report.* Grantham Research Institute on Climate Change and the Environment, London School of Economics and Political Science. Embracing-the-New-Paradigm-of-Green-Development-in-China.pdf (lse.ac.uk).
8. Zhang, Y. (2023). Reshaping the relationship between environment and development: A theoretical framework under the paradigm of eco-civilization and its policy implications. *Social Sciences in China, 44*(1), 44–72. https://doi.org/10.1080/02529203.2023.2192086
9. World Bank.(2022). Maddison Project Database [dataset]. Our World in Data. [2024-08-31]. Gross domestic product (GDP). (ourworldindata.org).
10. UNEP. (2023). *Inclusive wealth report 2023: Measuring sustainability and equity.* United Nations Environment Programme.
11. Moatsos, M. (2021). *Share of population living in extreme poverty, world.* Our World in Data.
12. Feenstra, R. C., Inklaar, R., & Timmer, M. P. Gross domestic product (GDP), 1978 [Penn world table (2021)]. 2015.
13. Waters, C. N., et al. (2016). The Anthropocene is functionally and stratigraphically distinct from the Holocene. *Science, 351*(6269). https://doi.org/10.1126/science.aad2622
14. Oreskes, N. (2024). Calling our times the 'Anthropocene epoch' matters dearly to you. *Scientific American.*

15. Dasgupta, P. (2021). *The economics of biodiversity: The Dasgupta review*. HM Treasury.
16. United Nations. (2021, September 10). *Global Crisis Response 'Too Little, Too Late', Secretary-General Tells General Assembly 'Our Common Agenda' Event, Warning of Instability, Climate Chaos*. UN Secretary-General António Guterres' remarks to the General Assembly.
17. Bizikova, L., Zoundi, Z., & Smith, R. (2022). *Moving beyond GDP to achieve the SDGs. Think20 Indonesia*. Policy Brief.
18. Smith, R., Zoundi, Z., & Bizikova, L. (2022). *Engaging decision-makers in moving beyond GDP and toward wealth*. International Institute for Sustainable Development.
19. Nordhaus, W., & Tobin, J. (1972). Is growth obsolete? In *economic research: Retrospect and Prospect, volume 5, economic growth*. National Bureau of Economic Research. http://www.nber.org/chapters/c7620
20. International Monetary Fund. (2020). *Measuring economic welfare: What and how?* International Monetary Fund.
21. Dasgupta, P., & Levin, S. (2023). Economic factors underlying biodiversity loss. *Philosophical Transactions of the Royal Society B: Biological Sciences, 378*. https://doi.org/10.1098/rstb.2022.0197
22. System of National Accounts 2008 (2009). *European Commission, International Monetary Fund, Organisation for Economic Co-operation and Development, United Nations, World Bank*, United Nations. SNA complete.book (un.org).
23. Stern, N., Xie, C., & Zenghelis, D. (2020). *Strong, sustainable and inclusive growth in a new era for China—Report 2: Valuing and investing in physical, human, natural and social capital in the 14th Plan*. Grantham Research Institute on Climate Change and the Environment, London School of Economics and Political Science.
24. UNEP. (2021). *Inclusive wealth of Pakistan: The case for investing in natural capital and restoration*. United Nations Environment Programme.
25. IISD. (2016, November 30). *Comprehensive wealth in Canada—measuring what matters in the long run*.
26. IISD. (2018, October 26). *Comprehensive wealth in Canada 2018—measuring what matters in the long term*.
27. Fan, J.-L., Li, K., Zhang, X., et al. (2022). Measuring sustainability: Development and application of the inclusive wealth index in China. *Ecological Economics, 195*.
28. Cheng, D., Xue, Q., Hubacek, K. et al. (2022). Inclusive wealth index measuring sustainable development potentials for Chinese cities. Global Environmental Change, *72*, 102417.
29. Hamilton, K., Helliwell, J. F., & Woolcock, M. (2016). Social capital, *trust and well-being in the evaluation of wealth* (NBER Working Paper No. w22556). National Bureau of Economic Research. https://ssrn.com/abstract=2832560
30. World Bank. (2011). The changing wealth of nations: Measuring sustainable development in the new millennium. Washington: The World Bank. https://openknowledge.worldbank.org/handle/10986/2252
31. World Bank. (2018). The changing wealth of nations 2018: Building a sustainable future. Washington: The World Bank. https://openknowledge.worldbank.org/handle/10986/29001
32. World Bank. (2021). The changing wealth of nations 2021: Managing assets for the future. Washington: The World Bank. https://www.worldbank.org/en/publication/changing-wealth-of-nations
33. UNEP. (2023). Inclusive wealth report 2023: Measuring sustainability and equity. Nairobi. https://wedocs.unep.org/handle/20.500.11822/43131
34. Fan, J.-L., Li, K., Zhang, X., Hu, J., Hubacek, K., Da, Y., Liang, X., & Cheng, D. (2022). Measuring sustainability: Development and application of the Inclusive Wealth Index in China. *Ecological Economics, 195*. https://doi.org/10.1016/j.ecolecon.2022.107357

# Chapter 8
# Digitalization and Green Technologies for Sustainable Development

## 1 Introduction

This CCICED Special Policy Study (SPS) aims to provide insight on how to holistically integrate the digital and green transformations at the city level in China. We propose combining well-tested sustainability capabilities with innovative and visionary digital appliances. The study provides policy recommendations to the Chinese government for promoting the coordinated advancement of digitalization and green transformation, thereby contributing more effectively to sustainable development (Fig. 1).

## 2 Digital Capabilities for a Sustainable Urban Future

In the face of the climate crisis, economic turmoil, and social tensions, urban development is encountering increasingly severe challenges. In the meantime, digitalization provides unprecedented opportunities to enhance municipal capacities and expand the boundaries of urban development.

### *2.1 Cities as Complex Technology-Economy-Society Systems in the age of Digitalization*

Cities are highly complex systems comprising numerous subsystems including technology, economy, politics, society, culture, and ecology. Infrastructure and the

The content of this chapter is extracted from the main findings of the 2023–2024 CCICED Special Policy Study (SPS) on "Digitalization and Green Technologies for Sustainable Development".

 
*Adhering to Green and Open Cooperation to Advance Modernization*,
https://doi.org/10.1007/978-981-95-5547-5_8

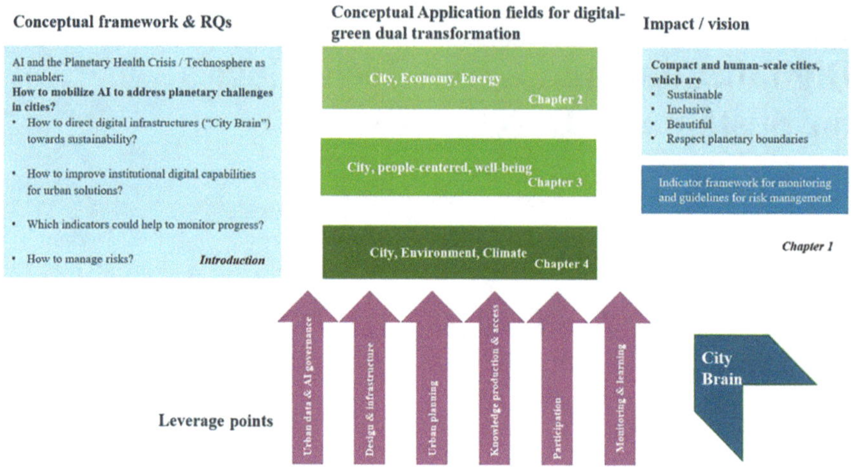


**Fig. 1** Digital-green dual transformation and the sustainable development of cities: A conceptual and analytical framework

ecological carrying capacity present limiting factors for urban development. Meanwhile, the potential for technological innovation and industrial transformation arising from the interaction between technological and economic systems is key to breaking through the borders of urban development.

### 2.1.1 Drivers and Boundaries of Urban Development

The effect of technological innovation on urban development potential, characterized by increasing returns on investment, manifests in three main areas: first, it creates new opportunities for infrastructure investment, thereby reducing costs and enhancing the efficiency of existing infrastructure; second, it improves the efficiency of economic activities, including both production and transaction efficiency; and third, it strengthens the ecological and environmental carrying capacity, as evidenced by the adoption of green energy and the expansion of the sharing economy.

### 2.1.2 Digital Capability and the Expansion of Boundaries in Urban Development

AI is the core engine of the Fourth Industrial Revolution and is set to profoundly impact urban development. As a general-purpose technology, AI signifies a transformation in technological, economic and social systems:

- a substitute for human intellect, AI involves a highly complex technological system.

- is a form of data intelligence that has emerged from cyberspace. By optimizing and controlling physical and social spaces through data and algorithms from cyberspace, AI can revolutionize urban resource allocation.
- can drive structural changes in both the economy and society.
- can enhance ecological and environmental carrying capacities.
- promotes the coordinated and balanced development of urban economies, societies, and ecosystems.
- innovative application of AI can enhance urban resilience and adaptability to environmental changes.

### 2.1.3 The City Brain

The City Brain is a complex technological system composed of interwoven technologies, including the Internet of Things, big data, AI, cloud computing, edge computing, 5G, cyber-physical systems, and digital twins.

Since 2016, several Chinese cities, including Beijing, Shenzhen, Shanghai, Hangzhou, Guangzhou, Chongqing, Chengdu, Qingdao, Tianjin, Nanjing, Zhengzhou, Wuhan, Xi'an, and Jinan, have each launched City Brain initiatives tailored to their own developmental needs. In China, AI companies like Huawei, Alibaba Cloud, China Mobile, and China Unicom are the main innovators in the construction of City Brain.

The development of City Brain initiatives has facilitated the accumulation of digital technologies and capabilities, encompassing both foundational and vertical hardware and software systems. The collaboration between leading tech firms and AI small and medium-sized enterprises (SMEs) creates a network that builds these digital technologies and capabilities.

## *2.2 Pathways towards Urban Sustainability*

### 2.2.1 A Vision for tomorrow's Cities

German Environment Agency's *Tomorrow's Cities* provides a vision of the future of urban living, focusing on creating cities that are sustainable, liveable, and efficient and emphasizing the need for an integrated approach to urban development that balances environmental, social, and economic factors.

### 2.2.2 The Compact City as a Focal Point for Sustainability

An important paradigm of sustainable urban development is the compact city, where distances are short, and the reliance on cars is minimized. The compact city provides multiple benefits. It reduces the need for long commutes, lowers transportation emissions, and promotes physical activity. Furthermore, optimizing building designs for energy efficiency helps significantly reduce energy consumption, leading to lower GHG emissions and operating costs.

#### 2.2.3 New European Bauhaus as an Integrative Sustainable Urban Development Pathway

The New European Bauhaus (NEB) is a process-oriented vision of future urban living with a human-centred theory of change. The NEB aims to integrate cultural and creative aspects into the European Green Deal. The initiative focuses on envisioning and achieving an environmentally sustainable, inclusive, and beautiful urban future.

### *2.3 Nexus of Digitalization and Sustainability: Synergies and Risks*

#### 2.3.1 Potentials of Digitalization for Sustainable City Development

Coordinating digitalization and greening is an inherent requirement of the ecological civilization strategy. First, it indicates that digitalization serves as the technical basis for achieving green development. Second, it denotes that green development offers practical scenarios for the application of digital technologies, thereby promoting the growth of digital technology and the digital economy.

#### 2.3.2 Energy Demand for Digitalization and AI

While digital-driven socio-economic development improves energy efficiency, it also results in a carbon footprint [1]. Statistics in 2021 show that the total electricity consumption of national data centres reached 216.6 billion kWh, with carbon emissions amounting to 135 million tonnes accounting for approximately 1.14% of the country's total $CO_2$ emissions.

#### 2.3.3 Systematic Risks of AI Deployment

AI deployment in smart cities and cloud operations presents significant yet uncertain systemic risks. These technologies can concentrate economic power, widening the gap between the wealthy and the poor, destabilizing social structures, and exacerbating urban tensions [2] (Fig. 2).

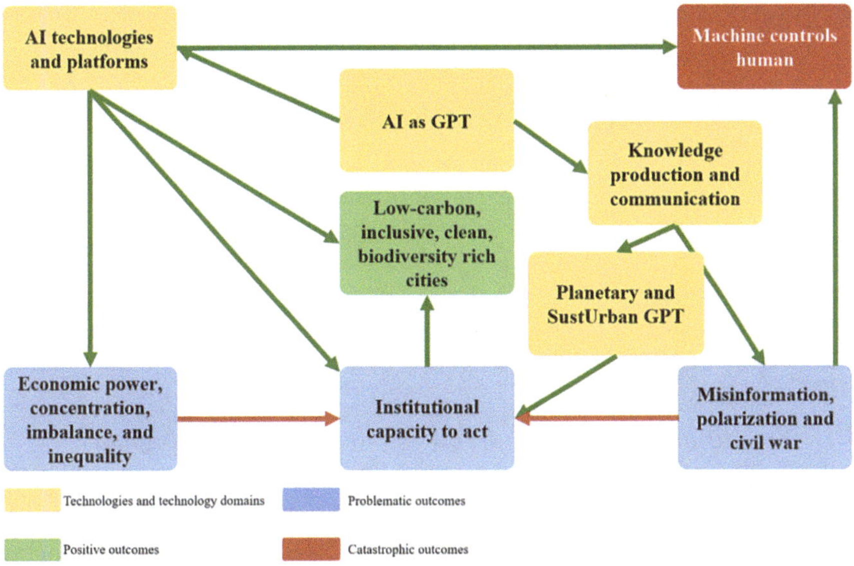


**Fig. 2** The systemic risks of the digitalization of cities

### 2.3.4 AI as both an Enabler and a Solution in the Technology Space

AI super-platform enterprises like Huawei, Baidu, Tencent, and Alibaba are extensively deploying AI in green energy and societal sectors. Huawei's Digital Energy division focuses on transforming energy, intelligence, and mobility in the carbon neutrality process, emphasizing intelligent photovoltaic storage generators and integrating photovoltaics, wind power, and energy storage to construct high-quality, digital clean energy bases.

### 2.3.5 Gender Data Gap and Bias

Gender bias in digital technologies is often the result of the gender data gap, meaning the absence of robust data disaggregated by gender, resulting in an incomplete understanding of reality [3]. Addressing the gender data gap and bias in the context of smart sustainable cities leads to increased urban safety and liveability for all [4].

# 3 Dual Transformation and Sustainable Economic Development

Section 2 aims to shed light on the green and digital transformations of the urban economy, focusing on the economic dimension of sustainable development. After addressing the demand and supply sides of sustainable economic growth at the city level in the first section, Sections 2.2 and 2.3 discuss the implications of the dual transformation for the urban economic structure and employment, as well as for the sharing and circular economies, respectively. The importance of technological innovation is highlighted throughout the section.

## *3.1 Sustainable Economic Growth Through Technological Innovation*

Sustainable economic growth can be significantly advanced through technological innovation, addressing both the supply and demand sides of urban development. On the supply side, investing in renewable energy sources and digitalized energy systems enhances efficiency and reduces carbon footprints, driving cleaner and more resilient energy infrastructures. Simultaneously, on the demand side, cities can leverage digital platforms to optimize resource use, improve public services, and promote sustainable practices among residents.

### 3.1.1 Renewable and Digitalized Energy Systems

By leveraging digital and AI-driven innovations, digitalization can be a key enabler of electrification. In order to enable this transition, governance innovations and new infrastructure will be critical. Modern and digitally integrated cities will be characterized by much higher power and data densities, requiring up-front investments in electricity grid and data connection infrastructure. Leveraging finance will need to be complemented by electricity market reforms to permit incentive schemes and real-time pricing, as well as sharing appliance controls with actors that manage systems reliability and resiliency.

### 3.1.2 Demand-Side Measures in Buildings and Transport

Achieving sustainable cities requires rapid and extensive systemic transformation across various sectors, such as buildings, transport, urban planning, and industry. This transformation involves several key strategies, including "Avoid" strategies, which aim to significantly reduce energy and material demands. "Shift" strategies involve utilizing sharing and circular economy platforms for resource sharing and collaborative consumption. "Improve" strategies focus on enhancing the efficiency of existing technologies, including efficient design, construction, retrofitting, and strategies to minimize energy and material consumption through material substitution and electrification in transportation.

While digitalization plays a moderate role in "avoid" and "improve" measures, such as dematerialization and energy-efficient building envelopes, its most pronounced impact is in advancing high-efficiency smart home technologies, thereby promoting economic growth and reducing environmental impacts (see Table 1).

Digitalization has a high impact on "avoid" measures such as telecommuting and teleworking, which can significantly reduce energy use and emissions, with benefits that extend to the building sector. For shift measures, digitalization substantially improves the effectiveness of public transport, shared mobility infrastructure, and urban freight logistics solutions. When it comes to improving existing systems, electrification of vehicles greatly benefits from advancements in digital and technology, offering high potential for increased energy efficiency and reduced environmental impact (see Table 2). Industry sector strategies, from a material perspective, are discussed in the building sector. Strategies related to supply and structural adjustments will be addressed in the following section.

**Table 1** Impact of digitalization on maximizing the potential of buildings demand-side measures for energy and emissions reduction and economic growth

| Demand-side measures | | Impact of digitalization on maximizing potential |
|---|---|---|
| Avoid | Reducing the floorspace | Medium (through telecommuting) |
| | Behavioural and lifestyle changes | Medium (through new smart appliances and gadgets) |
| | Compact buildings and urban forms | Low |
| | Avoid material waste, repurposing | Low* |
| | Dematerialization, lightweighting design | Medium-high (through 3D printing) |
| | Nature-based solutions | Low (through smart design) |
| Shift | Electrification | High |
| | On-site renewables like solar PV | High |
| | City-integrated distributed generation | High |
| Improve | Energy-efficient building envelopes | Medium-high |
| | Smart home technologies | High |
| | Low-carbon and bio-based materials | Low |

*A higher outcome is envisioned for the future, but there is currently insufficient study to support this

**Table 2** Impact of digitalization on maximizing the potential of transport demand-side measures for energy and emissions reduction, as well as economic growth

| Demand-side measures | | Impact of digitalization on maximizing potential |
|---|---|---|
| Avoid | Telecommuting | High |
| | Urban planning and zoning strategies | Low (through new smart applications and sensors) |
| | Active mobility | Medium (through smart applications and platforms) |
| Shift | Public transport | Medium-high |
| | Shared mobility | Medium-high |
| | Urban freight logistics solutions | Medium-high |
| Improve | Electrification (EVs) | High |

## *3.2 Dual Transformation and Industrial Structure Adjustment*

Sustainable and digital technologies not only enhance performance and efficiency but also reshape the structure of cities. In particular, the dual transition of cities might have a significant impact—positive and negative—on quality of life.

### 3.2.1 Economic Structure Change

The dual transition of cities has been beneficial for the development of new technological segments and thus the creation of new-tech-empowered sectors [5, 6]. Meanwhile, urban decision makers will be forced to develop governance frameworks to adapt to the dual transition [7]. In this new scenario, the application of data and digital technology can enable the improvement of conventional sectors (such as green buildings, waste management, water management, and smart healthcare).

### 3.2.2 Employment Structure

Intensive investments and the expansion of new technologies and infrastructure in cities might result in higher unemployment [8]. Nevertheless, digital appliances will create new job opportunities [9] and facilitate employment expansion [10]. Even more so, the transition to smart sustainable cities will create more green jobs [11], which will produce goods or services that improve environmental protection and natural resource conservation.

### 3.3 The Impetus of Digital Solutions for the Sharing and Circular Economy

The adoption of digital solutions in the Chinese market has delivered on the promises that this so-called "fourth industrial revolution" [12, 13] would provide enormous contributions to the economic development of the country. No country has so far been able to exploit the economic potential of the sharing economy to the extent that China has. At a level of $15 billion in 2014, the Chinese sharing economy is projected to skyrocket to $335 billion by 2025 [14, 15]. The relevance of the sharing economy for the country can also be understood by noticing that China is home to the world's largest company in the sharing economy: as of 2023, Meituan employs about 115,000 full-time people and had total revenues of RMB 276 billion, tripled from the 2019 figures.

Adopting digital solutions has the potential to significantly contribute to the development of both the sharing and circular economies. Digital solutions also have the potential to help deal with the vast production of municipal solid waste (MSW), a characteristic byproduct of rapid urbanization [16].

## 4 Dual Transformation and Sustainable Social Development

Sustainable social development is of paramount importance for achieving a successful dual transformation. Cities are not only agglomerations of physical units but complex systems of entities that are supposed to respond to the needs of their inhabitants.

### 4.1 Ethical Considerations

In the digital age, robust privacy and security frameworks are essential for building trust between governments and citizens. Prioritizing these measures ensures ethical standards, increases public confidence, and supports social inclusion and effective stakeholder engagement.

### 4.2 Digitalization of Government Services

Digitalization of government services refers to the process of transforming traditional, paper-based government operations and services into digital formats that can be accessed and used online. This transformation aims to make government services more efficient, accessible, and user-friendly for citizens, which can have huge impacts on urban infrastructure accessibility.

Accessibility gaps in urban digital infrastructure often manifest as disparities in connectivity, digital literacy, and equitable use of technology. To effectively bridge these gaps, it is imperative to evaluate the role of key stakeholders and the complexities they face in the digitalization process [17].

### 4.2.1 Government Structures and Key Stakeholders

In the digitalization process, from a data perspective, any individual or organization that provides data or is affected by data should become a stakeholder in digital government services. Stakeholders who provide data offer decision-making based on accurate information for the specific implementation of digital government services. While maximizing stakeholder utility, the implementation of government services incentivizes stakeholders to continuously provide relevant data to support subsequent government services. From a societal perspective, the openness and sharing of individual information data will ultimately produce externalities, and whether these externalities are positive or negative is determined by the quality of government service provision.

### 4.2.2 Digitalization and Strategies for Inclusivity

Digital transformation is a multifaceted and intricate process, especially when considering the transformation of urban infrastructure to bridge accessibility gaps. In analysing the role of stakeholders in this digitalization process, it's crucial to understand that stakeholders include not just government entities but also private sector participants, communities, NGOs, and other international organizations. Their collaboration is instrumental in accelerating a stakeholder-driven digital transformation that prioritizes accessibility and inclusivity.

The conclusion that can be drawn is that a stakeholder-driven approach to digital transformation involves concerted efforts aimed at increasing digital literacy, fostering innovation, and implementing policy reforms to support an inclusive digital ecosystem. Strategies should be context-specific and address the unique challenges faced by different demographic groups, ensuring that technological advances do not leave anyone behind but instead work as a bridge that closes the gap between the digitally empowered and the underserved.

### 4.2.3 A Digital Future for Government Services

The benefits of digitalization in government services are multifaceted and expansive, profoundly revolutionizing the way government agencies interact with citizens and manage their operations. Therefore, digital government is not merely a technical upgrade but a fundamental reconceptualization of public sector service design, delivery, and interaction with society. It promises to reshape sectors pivotal to social welfare, including healthcare, education, infrastructure, and broader governance, ensuring that these critical services can adapt, sustain, and thrive in an increasingly digital world [18].

## *4.3 Impact of Digitalization on Social Urban Planning in the Context of Green Transformation*

Social urban planning plays a central role in the dual transformation. Through socially acceptable approaches to urban planning and process design that incorporate the experiences and knowledge of various actors, there is great potential to combine green and digital developments that do justice to the living conditions of all population groups.

### 4.3.1 The Digital Urban Twin as an Instrument for Strengthening the Involvement of Local Communities

Essentially, the extent to which a Digital Urban Twin can serve as an instrument for improving the quality of life for local communities depends on the quality and quantity of the integrated data. The development and use of Digital Urban Twins for social purposes require a holistic approach that balances technological advancements with social equity, privacy, and ethical considerations. By prioritizing stakeholder engagement, data accessibility, and continuous improvement, cities can leverage Digital Urban Twins to enhance urban resilience, sustainability, and quality of life for all residents.

### 4.3.2 The Involvement of Local Communities in Green Smart Cities Design-Processes

Digital technologies have the potential to serve people and their communities and improve living conditions for all (People-Centred Smart Cities, n.d.). For digital technologies to be used in a targeted and people-centred way, everyone needs to be involved. In addition, both digital and climate issues are very closely linked to social issues. These should always be taken into account. The neighbourhood level plays a central role here, not only in terms of the planning itself but also because neighbourhoods are considered to be the nuclei of social cohesion and bring with them key factors for a successful transformation. By involving local communities, local conditions ("Eigenart") can be better understood, and solutions can be human-centred.

One lever to meeting the dual transformation lies in the involvement of communities. A just transformation considers the needs and concerns of the community and responds to needs by enabling involvement in development processes. The goal of transformational processes should be to strengthen the social fabric and improve community well-being. All citizens should be able to benefit equally from progress and equally enjoy the advantages of digital and sustainable development. Without sufficient consideration of social aspects, the dual transformation can lead to certain population groups being cut off from the possibilities of technology and existing inequalities being exacerbated (see Sect. 3.3).

## *4.4 Intersectionality: Women, Migrant Workers, and the Elderly*

In China, the trend across the board is that inequalities are rising and intersecting more strongly. Since the 1980s, China's Gini coefficient rose from roughly 0.25 to over 0.4, making China one of the most unequal countries globally. However, China made huge advances compared to the 1980s and succeeded in raising 770 million Chinese out of absolute poverty.

### 4.4.1 Women

In lockstep with the national trend, gender equality has deteriorated in China over the last 50 years. Urban experiences vary significantly between genders. Women often juggle multiple roles requiring efficient transport systems to meet different needs and usually have less budget for transport [19]. Safety in urban environments is a central concern for gender mainstreaming. A comparative survey in Wuhan and Urumqi found that 70% of women expressed anxiety on public transport [20]. Finally, many women enter the "gig economy", which is an integral part of today's smart cities, offering opportunities such as flexible work and a low bar to re-enter the work force after pregnancy.

### 4.4.2 Migrant Workers

Migrant worker households[1] have two major challenges: first, a strong focus on smart cities may channel resources to the already rich cities. Second, nearly 300 million of China's 734-million-person labour force are migrant workers who may be left behind

[1] Migrant workers in China are generally understood to be workers residing in cities with a rural household registration.

since they work in sectors threatened by the advance of digitalization. Among migrant workers, women are particularly vulnerable with, on average, a higher concentration in low-skilled jobs and lower health levels than men.

### 4.4.3 Elderly Population

Between the years 2020 and 2022, the share of over 60-year-olds in the Chinese workforce rose from 8.8 to 12.8%. The number of people aged over 60 in the total population is expected to rise from just over 20% to approximately 30%. Older rural women are particularly disadvantaged, as their average income covers only 21% of their expenses [21]. Elderly women tend to live longer than men but are more likely to suffer from chronic diseases and generally have poorer health compared to their male counterparts [22]. Women also have higher unmet elderly care needs [23].

The emergence of digital tools, smart cities, and platforms has so far not been a remedy but threatens to leave the elderly further behind. Government interventions to remedy this situation face a tricky challenge brought about by a mixture of expectations and reluctance. On the one hand, the elderly expect the government to intervene and support them, but on the other, many of them are reluctant to turn to people outside the family and, in particular, do not to turn to government programs for help [24].

### 4.4.4 The Value of Inclusive Multi-stakeholder Planning and Digital Tools

This section recommends basing urban design on inclusive multistakeholder planning processes and using digital tools to facilitate the creation of sustainable smart cities.

First, inclusive multi-stakeholder planning. Second, digital technologies can play a pivotal role in integrating gender mainstreaming into urban planning, ensuring cities are equitable and inclusive [4, 25, 26].

To circle back to this section's initial question about whether smart cities, data, and digital technologies will contribute to the globally increasing inequalities or be remedies, it has become abundantly clear that technology alone will not be the solution, but its careful, socially conscious, and—above all, inclusive—use could pave the way for socially equitable results.

# 5 Dual Transformation and Sustainable Environmental Development

Multiple environmental challenges, such as climate change, overconsumption of raw materials, or high land take, illustrate the need for a transformation for sustainable urban development [27]. Digitalization can boost this development. From an environmental perspective, it is necessary to analyze how digitalization can benefit sustainable development, and how technical innovation and digitalization can be linked with sustainability.

## *5.1 Integrative Urban and Environmental Planning and Development for Tomorrow's Cities*

German Environment Agency sets out a vision for Tomorrow's Cities [28] that aims to develop compact and green cities, mixed-use districts, environmentally friendly mobility, and noise reduction. It proposes 10 interconnected packages of measures designed to tackle these issues systematically, addressing core challenges like traffic congestion, air pollution, noise, and the lack of green spaces.

The implementation of these measures needs collaborative efforts between central and local governments, as well as private sector and community involvement. Implementing these measures requires substantial investment, supportive legal frameworks, increased funding from both the central and local levels, and the reallocation of existing funds. As public finance is limited, central measures include eliminating environmentally harmful subsidies and redirecting those funds toward sustainable urban development initiatives. At the municipal level, integrated urban planning is a key mechanism for cities to implement these goals.

## *5.2 Digital Innovations to Re-invent Urban Environmental Protection and Development*

### 5.2.1 Digital Innovations and Urban Environmental Governance

Digital technologies innovate ecological governance pathways and modes to provide systematic environmental services for the public and intelligent environmental decision-making for the government. As a modern way of urban ecological environment governance, digitally oriented smart governance of the ecological environment integrates digital technologies such as big data, 5G, AI etc., into urban environmental protection. The advanced integration between digitalization and greening keeps improving the scientific, precise, and intelligence level of the environmental governance of urban ecology. The key to smart governance of the ecological environment lies in coordinating the relationship between digitalization and greening, promoting their mutual reinforcement and coordinated development. The promotion of digitalization and the use of technologies such as AI to discover the patterns behind data can be an important driving force for ecological environment governance and help achieve its efficient and sustainable effects. The promotion of green development is, on the other hand, an essential part of building a digital ecological civilization.

### 5.2.2 Applying Big Data Analysis and AI in Sustainable Environmental Development

The application of AI and big data technologies brings new opportunities to improve the quality of the urban environment. Applications such as intelligent transport systems, waste treatment, and environmental monitoring can effectively reduce traffic congestion, improve waste treatment efficiency, and optimize environmental pollution control. At the same time, scientific urban planning, refined environmental governance, and smart city management can be achieved through the analysis of big data. The building of eco-environmental smart governance can focus on synergistic efforts of platform construction, data mining, digital integration, policy protection, etc., to form a so-called "one network" or "one map," research data value, develop algorithms, promote in-depth application of digital technology in the field of ecological environment, and elaborate the fundamental and leading role of digital ecosystem technology.

## *5.3 Sustainable Urban Mobility for All*

Traffic areas are key to urban environmental development for three reasons [29]. First, they are comprised of a large share of urban space that could be used for, for example, climate adaptation. Second, the mostly strongly sealed traffic areas increase climate risks themselves. Furthermore, they are mostly used for motorized individual transport as a spatially inefficient and emissions-intensive transport mode that amplifies climate change as the driver for the aforementioned risks. The mobility transition is the process of addressing these three effects, in addition to local environmental problems, traffic safety, or social justice and gender issues associated with traffic. It can be summarized as aiming to avoid unnecessary traffic, shift from motorized individual to public transportation or active mobility, and improve the efficiency of fuel-based individual transportation. The mobility transition as a means of sustainable urban development can benefit from digitalization in a number of ways.

## *5.4 Carbon Emission Reduction, Resource Efficiency, and Zero Pollution in Urban Development*

Digital tools, along with standardized methods and processes, can help to manage the increasing complexity by systematically mapping and evaluating the information required for addressing the challenges with regard to refurbishing the existing building stock and setting standards for energy, materials used, and pollution for refurbishment and new-built structures—and thus contribute to achieving the political goals in the EU of achieving GHG neutrality in the building sector by 2050 as part of the European Green Deal, and in China of peaking $CO_2$ emissions before 2030 and achieving $CO_2$ neutrality before 2060.

## *5.5 Green-Blue Cities for Climate Adaptation, Biodiversity Conservation, and Human Well-Being*

Given ongoing worldwide urbanization and multiple crises, such as climate change and the loss of biodiversity, urban green spaces have been internationally gaining more importance [30–32]. Nature-based solutions incorporating blue (water) and green (vegetation) infrastructure are vital for mitigation, adaptation, human well-being, and urban biodiversity. They provide multiple ecosystem services, such as regulating temperature, water balance, improving air quality, and offering spaces for recreation and social encounters [33]. Therefore, urban green-blue infrastructure is a crucial factor for the quality of life and environment in cities and is as important as technical, digital, and social infrastructure. Digital solutions can support the modelling, planning and monitoring of urban biodiversity, nature-based solutions and green-blue infrastructure in many ways. Remote sensing helps to map the urban green-blue infrastructure as a data-base for green space information systems that support urban planning processes.

### 5.5.1 Quality and Stability of Biological Systems in Urban Areas

As an artificial ecosystem dominated by humans, the vulnerability of urban ecosystems is determined by their artificiality, openness, and incompleteness. Increasing the species quantity and complexity of nutritional structures of different species can enhance the resilience and stability of urban ecosystems. At present, the focus of improving the quality of urban ecosystems is to increase urban biodiversity, reduce hard surfaces, develop sponge cities, increase research on urban metabolic pathways, maintain the stability of urban ecosystems, and enhance their resilience.

### 5.5.2 Urban Park Construction and Management Based on Biodiversity

Park city[2] is to show the real characteristics of the city through digital technology and improve the quality of the city through the integration, deployment, and design of various resource elements so that urban residents can improve their living, business, and travel experiences. The core of a park city is to make full use of digital technology to generate spatial form schemes; establish the interaction path between "pedestrian activities" and "urban spatial form"; coordinate the ecological, cultural, and industrial construction of urban space; promote the integration of industry and city through digital industry and ecological living; and explore the future city where nature and people coexist without boundaries.

## 6 Policy Recommendations

Cities are central to human civilization, generating over 80% of social wealth and 90% of technological achievements. However, the concentration of populations and economic activities in urban areas brings challenges such as social pain points and environmental degradation. As we enter the second quarter of the twenty-first century, technological innovation, particularly in the field of AI, has become increasingly important and opens up new opportunities to better understand urban processes and adapt them to future requirements through targeted impulses. For China, facing the impacts of climate change and geopolitics on the economy, figuring out how to promote the digital-green dual transformation to achieve green economic growth of more than 5% is a challenge. However, it is important to strike the right balance between economic development and the environment and society, without sacrificing the latter two aspects in pursuit of growth. This section lists central operationalized policy recommendations—allocated to six core action areas or leverage points that can assist city officials and policy-makers in integrating the sustainable and digital transformation of cities. This is followed by a set of overarching strategies.

There is no one-size-fits-all solution for the development of smart sustainable cities. The population, development level, economic structure, history, and culture of each city are different, which determines the variations in the paths of smart city construction. In the process of improving the top-level design of smart cities, it is therefore necessary to emphasize adapting to local conditions, adhering to a pragmatic orientation, and implementing the principle of inclusive and sustainable development. At the level of practical policy-making, Chinese cities should strategically identify the most important policy and strategic levers that will enable them to achieve their SDGs. As to this, it should be noted that digitalization is no "silver bullet"; sometimes *smart* simply means well-thought-through. In many cases, technical efficiency

[2] One definition of "park city" can be found on https://www.hanspub.org/journal/paperinformation?paperid = 28,511.

is not sufficient, while vibrant social networks and local communities are crucial for sustainable development at the city level. As the experience during the COVID-19 pandemic demonstrates, these networks and communities can respond quickly to emerging challenges and mobilize various resources. As powerful sources of urban resilience, they should be strengthened in the digital-green dual transformation.

## *6.1 Action Area 1: AI and Data Governance*

Strengthening urban data governance requires addressing both capacity and control. Key interventions include enhancing digital infrastructure, which ensures robust and secure data collection, storage, and processing capabilities. To further strengthen municipal data governance, it is crucial to institutionalize human decision-making and leadership. Ensuring that decision-makers are well-informed and capable of interpreting complex data sets helps maintain transparency and accountability. Additionally, fostering collaboration between various stakeholders, including academia, industry, and civil society, can enhance the governance framework by incorporating diverse perspectives and expertise. This helps, for example, to take different scenarios into account in municipal planning projects or to organize low-threshold participation procedures. Specifically, cities can implement the following policies to develop strong data and AI governance:

- **people first and make their needs the starting point to promote AI application and data governance**
- **control over algorithms, ensuring they serve the public interest and are used ethically and transparently**
- **data governance and optimize data resources**
- **environmental and social risks**

## *6.2 Action Area 2: Energy and Industrial Infrastructure*

Ensure that energy demand and energy supply meet flexibly, supporting the integration of intermittent renewables into the energy system by relying on smart industrial and smart home systems, supplemented by dynamic pricing.

Specifically, cities can implement the following policies to transform their energy sectors and industrial base using digitalization:

- **Target financial structures and flows for the green and sustainable transformation of the urban economy**
- **Develop governance and policy for technology and infrastructure development**

## 6.3 Action Area 3: Smart and Sustainable Urban Planning

Integrate the City Brain technology into urban planning to enhance sustainability by utilizing AI to simulate and predict the impact of interventions on GHG emissions in mobility and buildings, and on climate resilience. Cities should leverage AI for transit-oriented development and climate adaptation strategies, using high-resolution spatial data to guide sustainable urban development and reduce environmental impacts. The use of advanced data-driven methods allows cities to tailor their planning efforts to their unique characteristics, ensuring that urban development aligns with sustainability objectives and enhances the quality of life for residents. The integration of these technologies provides a powerful toolset for addressing the complex challenges of urban climate change mitigation and adaptation. Specifically, cities can seize the means of digitalization for sustainable urban planning by implementing the following policies:

- **Design environmentally sustainable smart cities by following the Tomorrow City vision**

## 6.4 Action Area 4: Social Participation and Inclusion

Offer green and digital literacy programs to equip citizens with an understanding of sustainability and its relation to quality of life as well as with the skills needed to understand algorithmic systems. Implementing participatory design processes and involving community members, experts, and city officials, ensures that algorithms are developed with diverse input, addressing potential biases and aligning with community values as well as contributing to sustainable urban development. Engaging citizens through public consultations, workshops, and forums further demystifies technology, empowering them to actively participate in the governance process. Specifically, cities can ensure digital participation and socially inclusive development by adhering to the following policies:

- **Prevent the exacerbation of inequalities and promote socially equitable planning**
- **Develop governance that is capable and motivated to incentivize change**

## 6.5 Action Area 5: Knowledge Generation and Access

Establish data portals that provide access to sustainability-relevant metrics, enabling evidence-based climate action through predictive AI modelling. By making data accessible and engaging the public, cities can foster transparency and inclusivity in urban sustainability efforts. In addition to data access, it is also about making the methods of data utilization more accessible. Cities of the future have workshops and experimental spaces to enable different actors (citizens, companies, scientists and city employees) to create knowledge with data for their needs. Specifically, cities can generate knowledge and make it accessible by implementing the following policies:

- **Generate and provide access to urban sustainability-relevant data**
- **Support social innovations for technology adoption**

## *6.6 Action Area 6: Monitoring and Learning*

Smart city indices or indicators are important tools for defining a clear long-term vision and direction for the dual transformation, measuring progress, making agile adjustments in the process and evaluating results. Smart City indices or indicators ought to take into account a wide range of criteria to assess a city's digital and green progress. One of the future tasks of cities should be to work out synergy effects between existing monitoring systems of individual disciplines and to identify smart sustainable indicators. Specifically, cities can advance the monitoring and learning process on their way towards a smart sustainable city by implementing the following policies:

- **Develop city-specific indicators that monitor the digital and sustainable transformation**
- **Improve the existing system of evaluation index**

## *6.7 Overarching Strategies*

China's smart city projects have covered all provinces, municipalities, and autonomous regions. All sub-provincial-level cities, above 95% of prefecture-level cities, and more than 50% of county-level cities have planned to develop smart city-capabilities. Recently, the construction of smart cities is accelerating in combination with "new infrastructure buildup" covering, for instance, 5G, big data centres, AI, and industrial internet. China's rapid smart city construction has been driven by a large amount of investment, the scale of which amounted to RMB 875 billion (about USD 120 billion) in 2023. Despite significant investments and achievements, the developments of smart cities and green cities remain on two separate tracks in China. Cities should prioritize sustainability considerations, provide smart and sustainable solutions, and advance key technologies that promote the coordinated advancement of the digital and green transformation.

- **a people-centred "Smart Sustainable City" strategy**
- **digital capabilities for sustainable solutions**
- **key technologies that integrate digital and green innovations**

The City Brain is at the core of smart city operation. Powered by AI, big data and cloud computing, the City Brain can greatly improve the efficiency of urban operations, which lays the foundation for sustainable development. As to this, it will be critical to not only test the City Brain in pilots and isolated application fields, but to collect and analyse evidence of its integration with urban sustainability capabilities. However, it is not enough for a city to have only an efficient brain—it also needs a warm heart. The heart cares about the environment and nature and loves everyone in the city. Having both an efficient brain and a loving heart is the key to the city's successful digital and green dual transformation. In addition, to most effectively serve the people in their pursuit of a better life, both effective markets and capable governments should play a role. Under the command of an efficient brain and a loving heart, the invisible and visible hands shall work together, and, only through collaboration between the two, can we create a sustainable urban future.

## References


1. Kaack, L. H., Donti, P. L., Strubell, E., Kamiya, G., Creutzig, F., & Rolnick, D. (2022). Aligning artificial intelligence with climate change mitigation. *Nature Climate Change, 12*(6), 518–527.
2. Creutzig, F., Acemoglu, D., Bai, X., et al. (2022). Digitalization and the Anthropocene. *Annual Review of Environment and Resources, 47*(1), 479–509.
3. Gauvin, L., Tizzoni, M., Piaggesi, S., et al. (2020). Gender gaps in urban mobility. *Humanities & Social Sciences Communications, 7*, 11. https://doi.org/10.1057/s41599-020-0500-x
4. German, S., Metternicht, G., Laffan, S., & Hawken, S. (2023). Intelligent spatial technologies for gender inclusive urban environments in today's smart cities. *Intelligent Environments*, 285–322.
5. Allam, Z. (2021). Big data, artificial intelligence, and the rise of autonomous smart cities. In *The rise of autonomous smart cities* (Sustainable urban futures). Palgrave Macmillan.
6. Chandler, C. (2023). Contested autonomy in "smart" and "inclusive" innovation: Test-driving transportation technology and policy in Pittsburgh. *Urban Geography, 44*(8), 1608–1627.
7. Ersoy, A., & Alberto, K. C. (2019). Understanding urban infrastructure via big data: The case of Belo Horizonte. *Regional Studies, Regional Science, 6*(1), 374–379. https://doi.org/10.1080/21681376.2019.1623068
8. Lupu, D., Maha, L. G., & Viorica, E. D. (2023). The relevance of smart cities' features in exploring urban labour market resilience: The specificity of post-transition economies. *Regional Studies, 57*(12), 2406–2425.
9. Cortes, G. M., Jaimovich, N., & Siu, H. E. (2017). Disappearing routine jobs: Who, how, and why? *Journal of Monetary Economics, 91*, 69–87.
10. Acemoglu, D., & Restrepo, P. (2018). The race between man and machine: Implications of Technology for Growth, factor shares, and employment. *American Economic Review, 108*(6), 1488–1542.
11. Sulich, A., & Sołoducho-Pelc, L. (2022). The circular economy and the green jobs creation. *Environmental Science and Pollution Research, 29*(10), 14231–14247.
12. Philbeck, T., & Davis, N. (2018). The fourth industrial revolution: Shaping a new era. *Journal of International Affairs, 72*(1), 17–22.
13. Schwab, K. (2016). *The fourth industrial revolution*. World Economic Forum.
14. Li, D., & Schoenherr, T. (2023). The institutionalization of sharing economy platforms in China. *Journal of Operations Management, 69*(5), 764–793. https://doi.org/10.1002/joom.1253
15. Marukawa, T. (2017). Sharing economy in China and Japan. *The Japanese Political Economy, 43*(1–4), 61–78. https://doi.org/10.1080/2329194X.2018.1555666
16. Chen, Y.-C. (2018). Effects of urbanization on municipal solid waste composition. *Waste Management, 79*, 828–836. https://doi.org/10.1016/j.wasman.2018.04.017

17. Deep, G. *Digital transformation in the public sector: opportunities and challenges.* [2023-12-03]. https://ijsra.net/sites/default/files/IJSRA-2023-0995.pdf
18. Bandhiya, I. *Government digitization: transforming public sector.* [2024-02-01]. https://evincedev.com/blog/government-digitization-for-revolutionizing-the-public-sector/
19. Yang, Y. (2017). *Building gender equality into intelligent transport systems in China.* https://blogs.worldbank.org/en/eastasiapacific/building-gender-equality-into-intelligent-transport-systems-in-china. Accessed 22 August 2024.
20. Feng, L., He, Y., & Zhan, P. (2023). Economic independence and living arrangements of older women with agricultural Hukou in China. *China Economic Review, 81.*
21. Chu, L. (2023). Gender disparities in health and Well-being among older adults in China. *Journal of Women & Aging, 35*(3), 299–317.
22. Chen, X., Giles, J., Wang, Y., & Zhao, Y. (2018). Gender patterns of eldercare in China. *Feminist Economics, 24*(2), 54–76.
23. Li, H., & Kostka, G. (2024). Navigating the digital age: The gray digital divide and digital inclusion in China. *Media, Culture & Society, 46*, 1181–1199.
24. Macaya, J. F. M., Dhaou, S. B., & Cunha, M. A. (2021). Gendering the smart cities: addressing gender inequalities in urban spaces. In *Proceedings of the 14th international conference on theory and practice of electronic governance* (pp. 398–405).
25. Hawken, S., Leao, S., Gudes, O., Izadpanahi, P., & Fernandez, E. (2021). Gendering the smart cities: addressing gender inequalities in urban spaces. In *Proceedings of the 14th international conference on theory and practice of electronic governance* (pp. 398–405).
26. Sachs, J. D., Schmidt-Traub, G., Mazzucato, M., Messner, D., Nakicenovic, N., & Rockström, J. (2019). Six transformations to achieve the sustainable development goals. *Nature Sustainability, 2*, 805–814. https://doi.org/10.1038/s41893-019-0352-9
27. German Environment Agency. *Tomorrow's cities. Environmentally friendly mobility, low noise, green spaces, compact housing and mixed-use districts.* [2017-03-24]. https://www.umweltbundesamt.de/publikationen/tomorrows-cities.
28. Glock, J.-P., Hartl, R., Krieg, M., & Becker, J. U. (2021). No climate-resilient society without a resilient transport system. In R. C. Brears (Ed.), *The palgrave handbook of climate resilient societies* (pp. 1147–1175). https://doi.org/10.1007/978-3-030-42462-6
29. Convention on Biological Diversity/CBD. *Decision X/22 plan of action on subnational governments, cities and other local authorities for biodiversity.* [2010-10-29]. Decision X/22 (cbd.int).
30. Convention on Biological Diversity/CBD. *Decision 15/12 Engagement with subnational governments, cities and other local authorities to enhance implementation of the kunming-montreal global biodiversity framework.* [2022-12-29]. Montreal. Engagement with subnational governments, cities and other local authorities to enhance implementation of the Kunming-Montreal Global Biodiversity Framework (cbd.int).
31. European Commission/COM/2013/0249 final. Green Infrastructure (GI)—Enhancing Europe's Natural Capital. [2013-05-06]. https://ec.europa.eu/environment/nature/ecosystems/index_en.htm
32. European Parliament and Council. (2024). *Regulation of the European Parliament and of the Council on Nature Restoration and Amending Regulation (EU)* 2022/869.pdf (europa.eu).
33. Kowarik, I., Bartz, R., Brenck, M., & Hansjürgens, B. *Ecosystem services in the city—protecting health and enhancing quality of life.* Summary for decision-makers. [2017-03-24]. 190507_TEEB_De_Broschuere_KF_Bericht3_Stadt_engl_web.pdf (ufz.de).

# Chapter 9
# Innovative Mechanisms of Sustainable Investment in Environment and Climate

## 1 Introduction

**China's green transformation reflects, to some extent, the role of economies of scale and market competition in continuously reducing the cost of new types of energy.** However, relying mainly on capacity expansion of existing mature technology industries to drive the green transformation will become increasingly challenging. There are three main reasons for this. First, as China's task of reaching its carbon peak, carbon neutrality will become the primary goal of the green transformation. Achieving carbon neutrality will rely on technological innovation. Second, as demand for land resources in the green industry expands, diseconomies of scale in land use will increasingly constrain the advantages of economies of scale. Third, amid the new geopolitical landscape, countries around the world are placing greater emphasis on security over economic efficiency, and the fragmentation of the global market could restrict China's green industry from fully leveraging economies of scale.

**As the focus of the green transformation evolves, green finance has also entered a new stage, focusing on breakthrough innovations in green technology.** Therefore, the key issue lies in how to better enable the capital market to play its pivotal role and provide effective support for the new task of green technology innovation.

---

The content of this chapter is extracted from the main findings of the 2023-2024 CCICED Special Policy Study (SPS) on "Innovative Mechanisms of Sustainable Investment in Environment and Climate".



*Adhering to Green and Open Cooperation to Advance Modernization*,
https://doi.org/10.1007/978-981-95-5547-5_9

## 2 The New Focus of the Green Transition: Disruptive Innovation

China's achievements in its green transformation are mainly attributed to two factors. First, its large population and economy provide an innate advantage in terms of economies of scale for development of the manufacturing industry, especially when promoting mature technologies. Second, investment in the field of green technology has not led to market monopolies, ensuring that while capacity growth is maintained, market competition also remains vibrant. Nevertheless, relying on expansion of capacity based on existing mature technologies to promote the green transformation could face more challenges in the future.

### *2.1 China Focuses on the Supply Side to Promote Green Transformation*

**The approaches to achieving the goals of green transformation vary among countries, particularly among major economies** (see Fig. 1). China, mainly supports the development of green industries such as new energy vehicles from the supply side, with support measures that include government procurement, tax exemptions for purchases, and a whitelist for batteries. In the burgeoning energy storage sector, there are also policy incentives such as renewable energy storage requirements, time-of-use electricity pricing mechanisms, and electricity market mechanisms.

**The different paths taken by China and Europe are both conducive to the global green transformation, but their industrial outcomes are starkly different.** China's green electricity capacity, represented by photovoltaic and wind power, has grown rapidly, leading to a continuous decline in the price of green electricity in China. From 2009 to 2021, the central government subsidized off-grid electricity

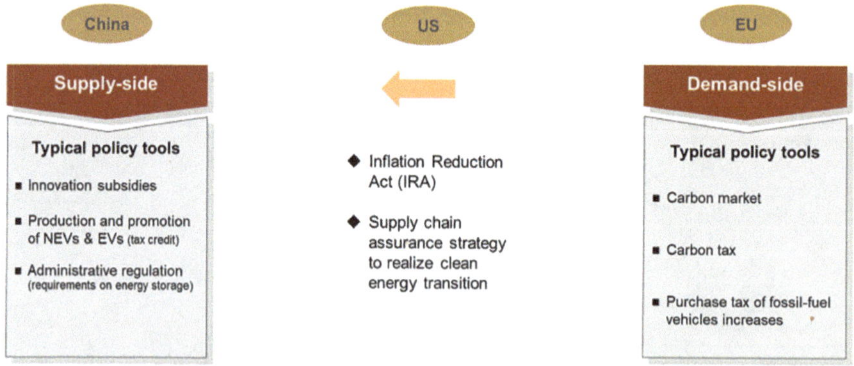


**Fig. 1** Two pathways of public policy. *Source* CICC Global Institute, "The Development of New Quality Productive Forces from the Perspective of Green Transformation: Economies of Scale, Overcapacity, and International Trade," May 2024

prices, and after 2021, local government subsidies replaced central government subsidies. To date, some of China's new energy power generation technologies have achieved grid parity, with unsubsidized generation costs now lower than those of coal-fired power. China's production capacity for new energy vehicles has also grown rapidly; BYD's annual production of electric vehicles has now surpassed Tesla's, ranking first in the world in 2023.

## *2.2 Economies of Scale and Market Competition Effectively Contribute to Reduction of China's Green Premium*

**The success of China's green transition has, to a certain extent, demonstrated the role of economies of scale and market competition in continuously reducing the cost of new energy as part of the green transformation.**

### 2.2.1 Green Economies of Scale

**So far, China's green transition has primarily relied on the continuous expansion of production capacity through incremental innovative technologies.** For mature technologies, expanding production capacity can effectively promote the reduction of the green premium. Economies of scale exist in the expansion of production capacity. This allows larger-scale manufacturing enterprises to benefit more from the positive externalities accumulated in the existing economy. When green economy development is characterized by the expansion of production capacity in related manufacturing industries, economies of scale in the green economy become more pronounced. **China, with its vast population and economy, has a scale advantage in developing manufacturing industries** (Fig. 2).

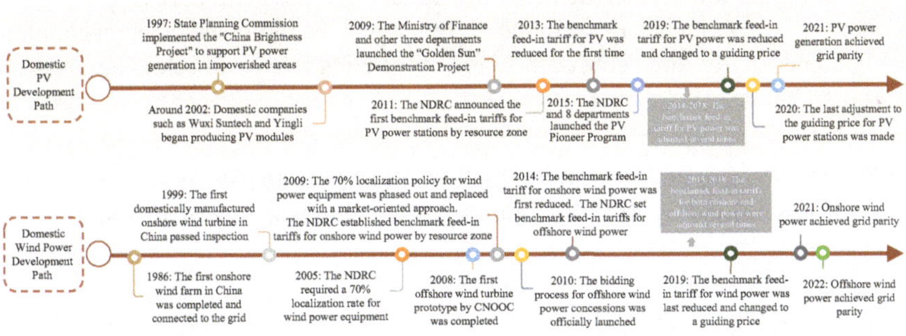


**Fig. 2** Development history of China's PV and wind power industry. *Source* International New Energy Network, International Energy Network, China Youth Daily, Polaris Solar PV Network, China Energy News, WindEurope, CICC Global Institute

### 2.2.2 The Power of Competition

Despite the substantial support provided by the government for the green industry, the market in China remains highly competitive.

**First, the market in China is enormous, and is capable of accommodating and nurturing multiple enterprises and fostering intense competition. Second, local governments in China have strong motivation to spur economic growth and are extensively involved in formulating industrial policies for green transformation and engaging in green investments. Third, China has an abundance of entrepreneurial spirit and accessible entrepreneurial resources.**

These aforementioned reasons ensure that China's capacity expansion is accompanied by ample market competition. The country's economies of scale make the low costs of its green industry possible, while the competitive pressures of the China market make these low costs a reality. The green capacity accumulated by China and its low-cost green manufacturing industry are beneficial to the global green transition and constitute a significant force in driving the global green transition.

## *2.3 New Task in China's Green Transition: Disruptive Innovations in Technology*

### 2.3.1 The New Main Tasks of Green Transformation Require Breakthrough Technological Innovation

China's green transformation has entered a new phase of development, characterized by the coexistence of the initial phase of emerging green technologies and the upgrading and transformation phase of mature green technologies. Compared to the previous phase, which focused on the expansion and large-scale application of mature green technologies, this new phase is centered around breakthrough innovation, thus emphasizing the importance of R&D investment (Fig. 3). The necessity for green technology innovation stems from the current green technologies facing technological bottlenecks, resource constraints, and constraints related to achieving net-zero emissions (Fig. 3).

### 2.3.2 The Economies of Scale in the New Energy Industry Are Constrained by the Diseconomies of Scale Associated with Land Use

**Land, as a factor of production, often exhibits characteristics of diseconomies of scale. It possesses natural exclusivity and competitiveness, and in production, it does not generate the positive externalities typically required for economies of scale.** More importantly, land is heterogeneous, varying in fertility [1] and in locational differences [2]. This heterogeneity leads to diminishing returns to scale for land. This is the characteristic of land's diseconomies of scale. Under the influence

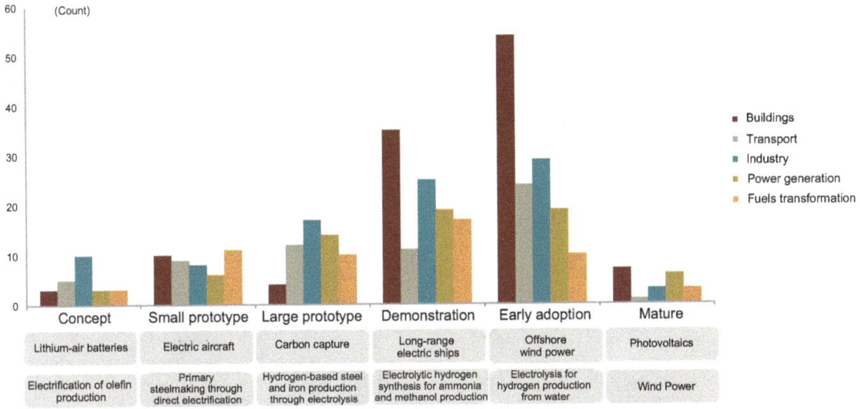


**Fig.3** Number of clean energy technologies analyzed in the ETP clean energy technology guide (2020). *Source* ETP Clean Energy Technology Guide Database, IEA, 2020, CICC Global Institute

of land's diseconomies of scale, economic entities of different spatial scales have asymmetry in their ability to achieve economies of scale. **The green transition leads to a significant increase in the economic demand for land, and the characteristics of land diseconomies of scale are more pronounced in the green transition.**

### 2.3.3 Geopolitical Factors Limit the Scale Advantages of the Green Economy

**In the new geopolitical landscape, countries around the world are placing more emphasis on security rather than economic efficiency, leading to a trend of deglobalization in the world economy. In this context, the green capacity advantage accumulated by China's green transition through fiscal support and policy-based credit may to some extent face a problem of insufficient demand.**

# 3 A New Stage of Green Finance: Enhanced Technology Finance Attributes

## *3.1 Addressing Dual Externalities Requires Green Finance to Possess Dual Attributes*

Green finance encompasses financial activity arrangements aimed at addressing climate change, environmental protection, and pollution prevention.

**Green transformation has dual externalities, manifested as the negative externalities of carbon emissions and the positive externalities of green technological innovation during the transformation process.** On the one hand, carbon-emitting

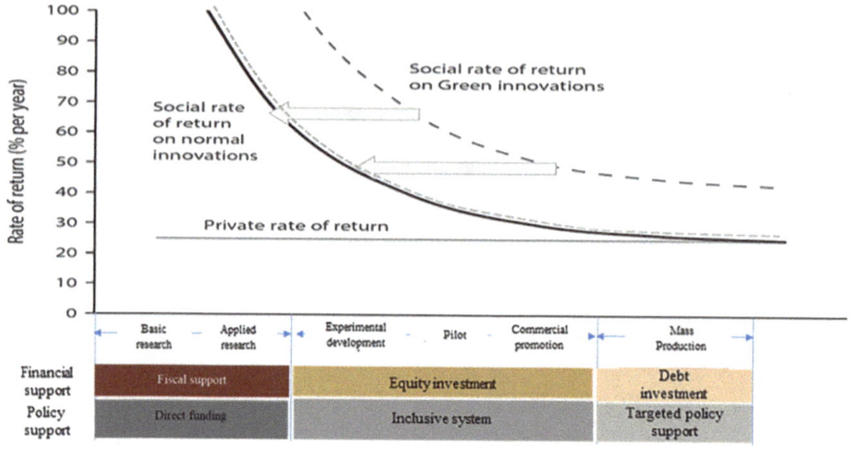


**Fig. 4** To address dual externalities, the financial sector in the field of green technology must provide comprehensive support throughout the entire chain. *Source* CICC Global Institute

economic activities benefit individuals while the resulting climate change and air pollution are costs borne by society as a whole. On the other hand, green technological innovation has positive externalities. Green technology innovation has a positive impact on the environment by promoting carbon emissions reduction and generates new knowledge through R&D of green products and services, leading to knowledge spillover effects that further drive social progress. The positive externalities of green technological innovation represent innovation bringing economic returns to investors and creating public value that is difficult to quantify for society. Therefore, during the process of converging these externalities, from the phase of basic research to the commercial promotion of green technology innovation, the government needs to intervene by providing subsidies, tax incentives, regulations, and other means to establish a market environment, internalize these dual externalities, correct market failures, and attract more investors (Fig. 4).

### 3.2 *Capacity Expansion Financing Is Primarily Characterized by Its Policy-Based Financial Nature*

**When any industry enters the growth stage, it means that capital expenditure is accelerating, making this phase more sensitive to financing costs compared to other stages of the industry life cycle.** Therefore, the main task of green finance is to channel more lower-cost funds into these industries, allowing these industries to expand their production capacity. To achieve this specific goal in industrial development, policies should provide clear direction for capital.

**On the other hand, in order to enhance the transparency of the market and further strengthen the certainty of the green investment environment, China's**

**green financial system also focuses on improving the norms in information disclosure, preventing the risk of "greenwashing", and assisting in the compliant and efficient use of green industry funds.**

## *3.3 To Finance Technology Innovation, the Attribute of Technology Finance Is Enhanced*

### 3.3.1 Green Technology Innovation Is Highly Uncertain

**First, technology routes are diversified, each with its own advantages and disadvantages in terms of subdivision criteria, making it difficult to judge which technology will be the first to make a breakthrough and bring the highest overall benefits in the future. Second, the initial costs of applying technology are high, which may include expenses for new materials, equipment, and infrastructure** (Fig. 5).

**The challenges in commercialization are significant, and the technology may struggle to mature through learning by doing, often remaining stuck at the prototype or demonstration stages for extended periods. This results in a prolonged duration of uncertainty** (Fig. 6). **Third, the lag in institutional and**

**Fig. 5** CCUS technology routes (screenshot). *Source* IEA, China twenty-first Century Agenda Management Center, Global Carbon Capture and Storage Institute, Tsinghua University [3], YANG Pingjian et al. [4], CICC Global Institute

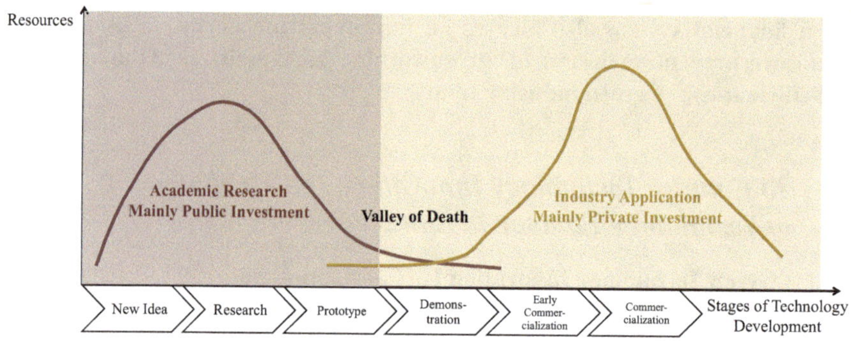


**Fig. 6** The "Valley of Death" for technology development. *Source* CICC Global Institute

**market environments increases the uncertainty of green technology innovation** (Figs. 5 and 6).

### 3.3.2 New Green Finance Policies Should Ensure Stability of Expectations and Inclusiveness of Technologies

**Due to the inherent uncertainties in the new task of green technology innovation for the green transition, it is necessary to establish a policy framework that provides both "long-term stable expectations" and "technological inclusiveness" to support the development of green technology innovation.** Such a policy system should be able to: (1) Create long-term growth expectations, reduce innovation risks, and incentivize enterprises and research institutes to invest in the R&D and application of green technologies so that the track that receives investment has long-term income expectations, thus realizing a win-win situation for both environmental and economic benefits; and (2) it should also respect the autonomy and creativity of science, give sufficient space to the innovative enterprises or teams to develop freely, embrace the plurality of technology routes, and allow society at large to complete the process of selecting the optimal technology. Specifically:

**First, the government should be able to send clear signals to the market that technology is expected to have stable demand in the long term. Second, financial support should focus on longer-term returns rather than short-term profitability, so that there is sufficient space for various technologies to compete freely. Third, from the perspective of the innovation chain, in the process of the diffusion of research and development results from scientific and technology innovation to industrial innovation, it is necessary for all types of institutions in the market to play a concerted role, that is, to build a more complete ecosystem.**

# 4 Capital Markets: A Key Lever in the New Phase of Green Finance

## *4.1 Capital Markets Should Play a Greater Role in Green Finance*

**In China, the rapid growth of green capacity mainly relies on the strong support of debt financing centered on the banking system. This is a phenomenon that is closely related to the structure of China's unique financial system** (Fig. 7).

**As mentioned above, in the early stages of its development, China's green finance system has benefited from clearly defined development paths through policy tools such as the "Guidance Catalogue," which have clarified the boundaries for industry support and implemented targeted funding strategies** (Fig. 8). **This has led to steady growth and gradual expansion in the scale of green bond financing** (Figs. 7 and 8).

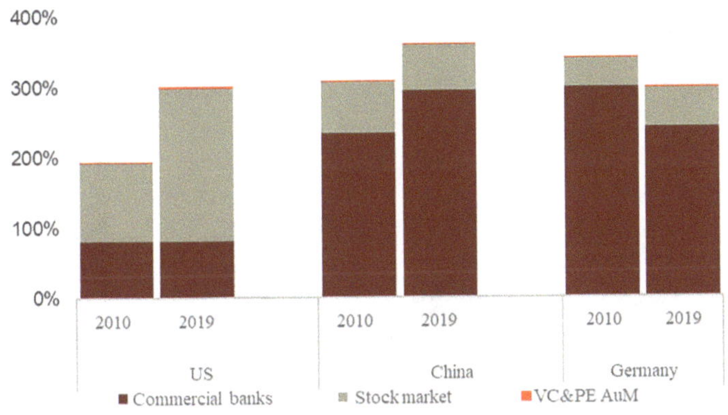


**Fig. 7** Financial structure of China, the US, and Germany: asset size/GDP. *Source* German Private Equity and Venture Capital Association, Japan Private Equity Association, Pitchbook, CICC Global Institute

However, the development of green finance needs to keep pace with the times, continuously innovating and improving financial products and services that match the corresponding stages of development. **Green technology innovation urgently requires equity financing to take on a more core responsibility, thereby promoting the development of green technologies, accelerating their application, and injecting new vitality and momentum into the sustainable development of China's green economy.** The adaptability of equity financing in supporting green technology innovation is primarily reflected in the following three aspects.

**First, the uncertainty of green technology innovation routes and the lack of collateral make equity financing more capable of bearing risks. Second, green technology innovations have a long and uncertain road to commercialization, while equity financing has more patience for the return on investment** (Figs. 9 and 10).

**Third, equity investors are more diverse, which helps in the selection and application of green technology innovations. Therefore, the key to supporting green technology innovation lies in leveraging the central role of capital markets** (Figs. 9 and 10).

## *4.2 The Unique Advantages of Capital Markets: Ecosystem Effect and Selection Effect*

**The capital market has two unique advantages in promoting breakthrough innovation: The ecosystem effect and the selection effect.**

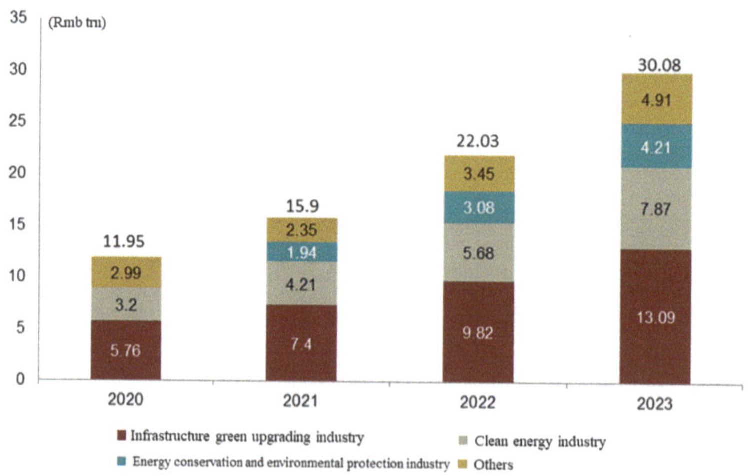


**Fig. 8** Balance of green loans in China (by industry, 2020–2023). *Source* People's Bank of China, CICC Global Institute

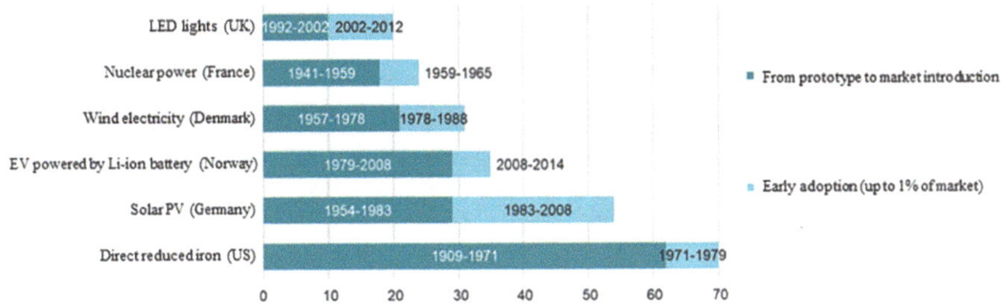


**Fig. 9** R&D and commercialization cycle of typical green technologies. *Source* IEA, CICC Global Institute

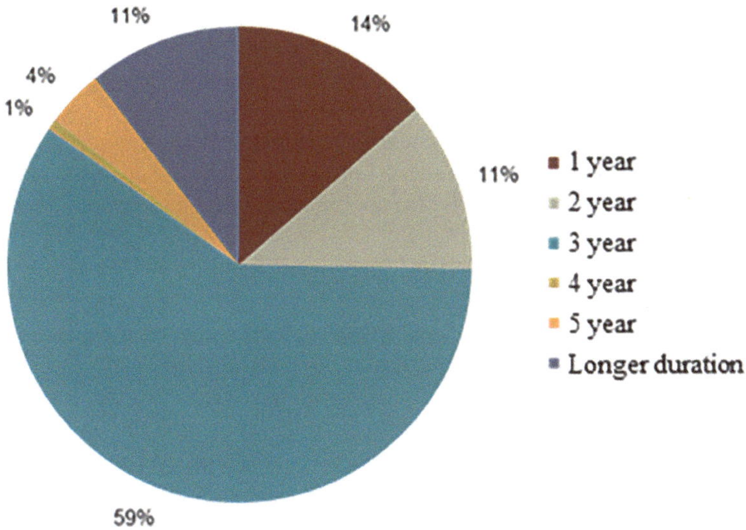


**Fig. 10** Maturity structure distribution of labeled green bond issuance in 2023. *Source* Green Bond Environmental Benefit Information Database, CICC Global Institute

**The ecosystem effect refers to the fact that the decentralized capital market can provide financial support for all fields, which is conducive to the formation of an innovation ecosystem in which all fields promote each other and develop together. The selection effect refers to the capital market's superior ability to "bet" on successful technology routes.**

**In short, the capital market promotes breakthrough innovation by leveraging its ecosystem and selection effects to build an innovation ecosystem that is conducive to mutual development across various fields and technology routes, ultimately achieving sustained breakthrough innovation capability.**

## 5 Challenges Faced by China's Capital Markets in Supporting Green Innovation

In recent years, both the VC market and the stock market in China have encountered several challenges. Specifically, the current challenges facing the domestic capital market in supporting breakthrough innovation are as follows:

### *5.1 Fundraising Side: The Difficulty of Fundraising for Domestic Venture Capital Is Becoming Increasingly Prominent*

The domestic equity market is currently facing severe fundraising difficulties, which have diminished its ability to promote breakthrough innovation in the green sector. Given that fundraising is a vital prerequisite for investment, the continued decline in fundraising in the domestic equity market limits the financial support available for the green sector, hindering its ability to support breakthrough innovation in this area (Fig. 11).

Research shows that the "New Asset Management Regulations" are closely related to fundraising activities in the domestic equity market [5]. Although the "New Asset Management Regulations" have played a positive role in governing financial irregularities and preventing systemic financial risks, they will also likely affect the channels through which bank funds enter the equity market via trusts and other means.

Moreover, the A-share market for IPOs further influences equity investors' willingness to invest in the green sector. Since IPOs are the main exit channel for domestic equity investors, the difficulties in exiting through IPOs in the green sector further

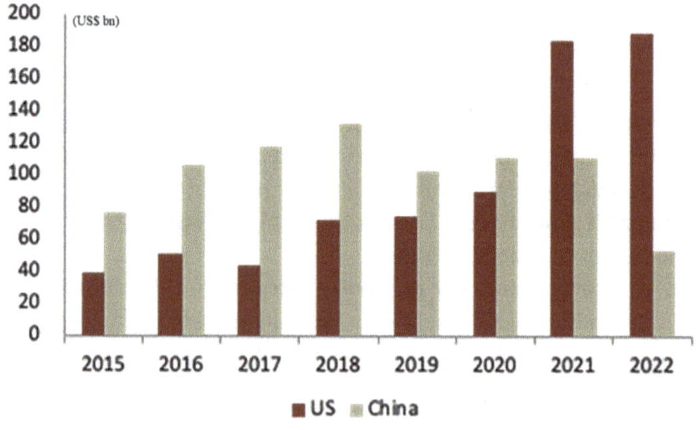


**Fig. 11** Annual fundraising amount comparison between China and US VC markets. *Source* Pitchbook, CICC Global Institute

diminish equity investors' willingness to support early-stage small and medium-sized companies in this field. This situation places greater fundraising pressure on domestic equity funds and investors focused on the green sector (Fig. 11).

## 5.2 *Investment and Management Side: Investment towards Lower Risk and Shorter Time Horizons, with Management Capabilities Needing Improvement*

In the context of the difficulties in fundraising, government-backed funds such as industrial guidance funds have become increasingly important for VC fundraising. Government-backed funds may now account for more than half of the capital in China's VC market, with substantial contributions from local governments, which has significantly alleviated the difficulties in fundraising in the VC market (Fig. 12).

However, industrial guidance funds also face some practical constraints in their investment and operational processes, which makes some of its characteristics unable to perfectly align with the need for promoting breakthrough innovation in the green industry. These constraints are reflected in several key aspects (Fig. 12).

First of all, industrial guidance funds have strong demand for stability, which suppresses their risk appetite as they make investments, making them less conducive to supporting high-risk breakthrough innovations (Fig. 13). Second, industrial guidance funds tend to invest in short-term projects that yield quick returns, which does not align with the long-term capital needs for breakthrough innovation in the green sector. Finally, industrial guidance funds also have problems such as the level of operational specialization needs to be improved, strengthened market segmentation, and poor continuity of capital investment, which limit their function of promoting breakthrough innovation in the green industry (Figs.14 and 15).

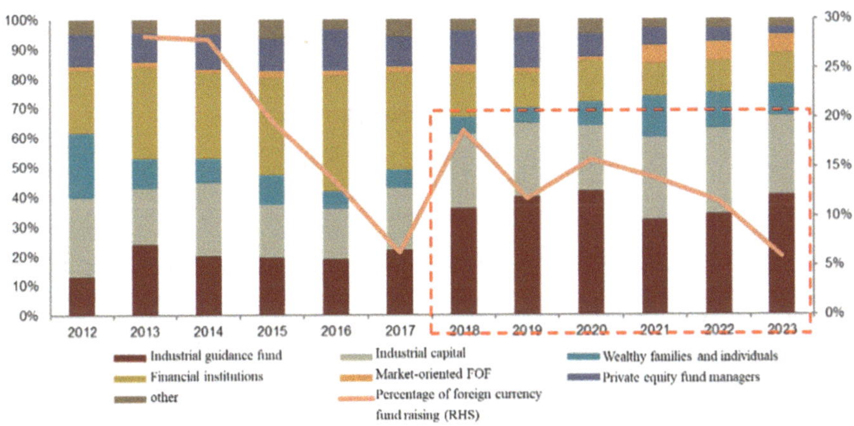


**Fig. 12** The exit of foreign capital since the pandemic has exacerbated the fundraising winter in China's venture capital market. *Source* Zero2IPO, CICC Global Institute

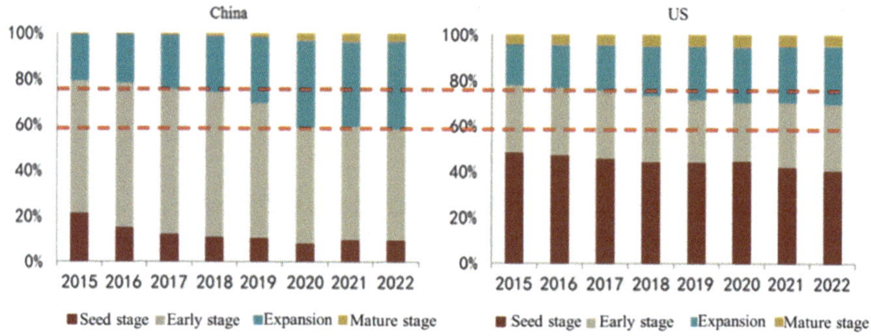


**Fig. 13** Increasing preference for stability in the VC market with the rising proportion of public funds such as industrial guidance funds. *Note* Here, the proportion of investment counts is referred to, with the seed stage corresponding to the angel and seed categories in PitchBook investment types, the early stage corresponding to the early-stage VC category, the expansion stage corresponding to the late-stage VC category, and the mature stage corresponding to the venture growth category. *Source* Pitchbook, CICC Global Institute

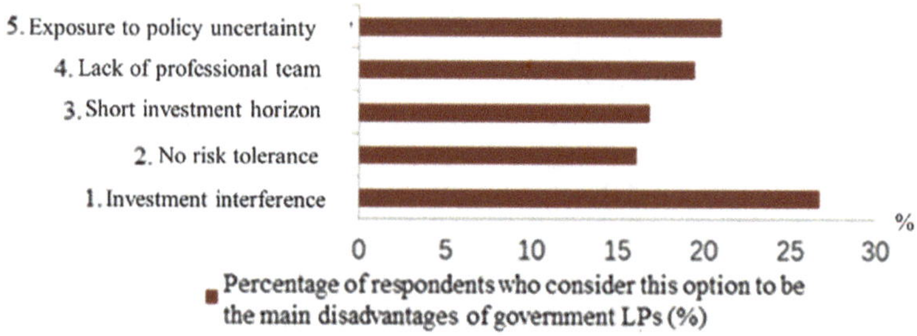


**Fig. 14** Major drawbacks of government LPs as perceived by GPs

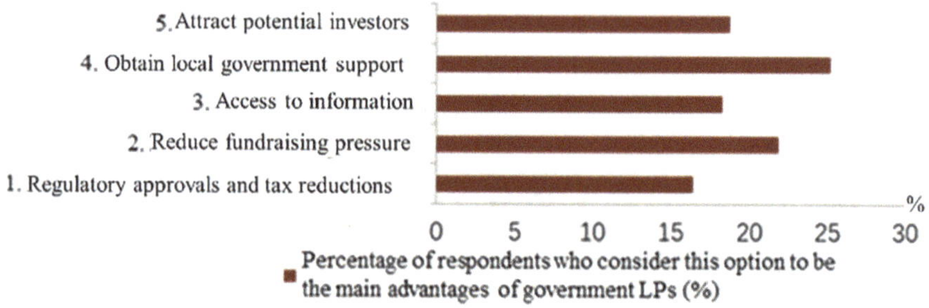


**Fig. 15** Major advantages of government LPs as perceived by GPs. *Source* Emanuele Colonnelli, Bo Li, and Ernest Liu. Investing with the Government: A Field Experiment in China. Journal of Political Economy, 2024, CICC Global Institute.

### 5.3 Exit Side: The Tendency of Investment Homogenization in the Equity Market

As mentioned earlier, promoting breakthrough innovation in the green sector through the capital market relies on effective interaction between the equity market and the stock market. For domestic equity investors, IPOs are one of the most important exit channels, IPOs are also an important lever for the capital market to promote breakthrough innovation in green technology by leveraging ecosystem effects and selection effects (Fig. 16).

However, the continuous aggregation of capital market funds towards the hard technology sector has also led to the green industry equity market facing a shortage of incremental capital. This poses a challenge to the continuous achievement of breakthrough technological innovations in the green industry (Fig. 17).

It is also important to emphasize that there is no direct correlation between a company's current profitability and its future innovation capabilities. Companies engaged in breakthrough innovation activities in the green sector may not yet be profitable and require financial support. Therefore, more efficient exit channels are needed to facilitate continuous breakthrough innovation and technological iteration upgrades in China's green industry (Figs. 16 and 17).

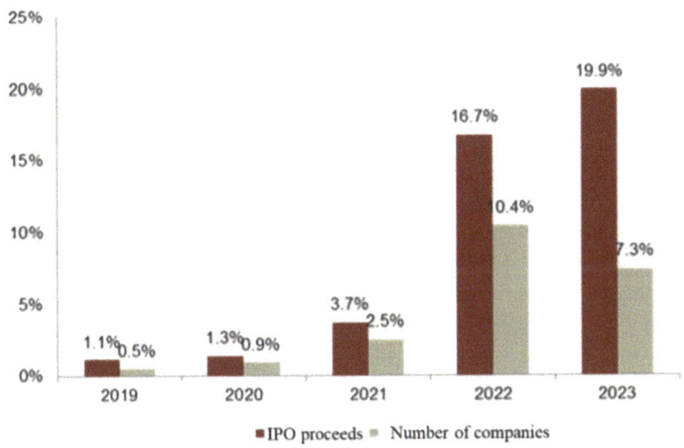


**Fig. 16** The Proportion of semi-conductor and other hard technology companies in A-share IPO fundraising has increased in recent years. *Source* Wind, CICC Global Institute

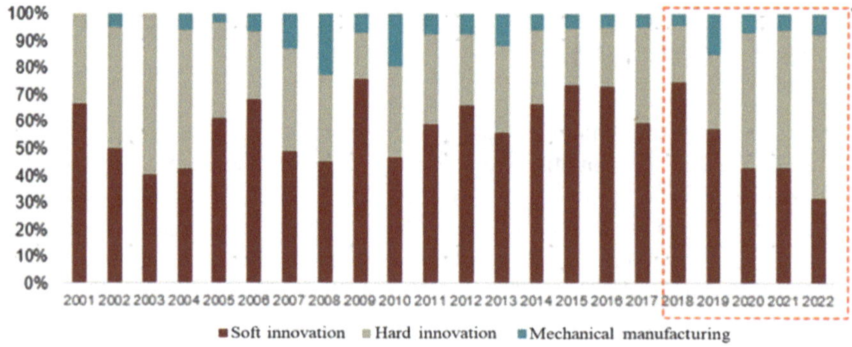


**Fig. 17** Domestic equity market investments increasingly focused on hard-tech sectors. *Note* Hard technology mainly includes semiconductors and electronic equipment, clean technology, and biotechnology/healthcare; soft innovation mainly includes Internet, IT, telecommunications and value-added services, finance, and entertainment media. *Source* Zero2IPO, CICC Global Institute

# 6 Policy Recommendations

## *6.1 Real Economy Policies: Creating Green Demand and Enhancing Green Returns*

Policies supporting green technological innovation in the real economy need to be based on "technology neutrality," which means shifting from a system that selects specific technologies to a technology-neutral policy framework, laying the foundation for the capital market to fully leverage its ecological and screening effects. Specifically, efforts should be made in two aspects: First, by playing an active role in fiscal policy, and second, by improving the carbon market pricing system. Before delving into specific recommendations, it is necessary to make the following clarifications: The construction and improvement of the carbon market is a complex project that needs to be advanced gradually, and its impact may be felt slowly. In contrast, fiscal policy can more directly support green technological innovation by creating market demand for green technologies through government procurement and direct subsidies. At the same time, China's general fiscal support measures for green technological innovation are already relatively comprehensive, including purchase subsidies for green products, financial assistance for trade-ins, and the Ministry of Finance's existing policy of Rmb40bn for green building procurement, all of which can play a role in promoting green innovation. The following will not elaborate further on these points. The policy recommendations in this article will focus on how fiscal policy can directly promote breakthrough innovation.

**Recommendation 1: In the bidding mechanism for green procurement, adopt a "quantitative carbon reduction performance" model instead of a "procurement list" model.**

Currently, China's green procurement system is mainly based on a government procurement list model, and promoting breakthrough innovations with high uncertainty is challenging. A more direct approach might be the "quantitative carbon reduction performance" model. Under this model, first, it is necessary to clearly define the quantitative emissions reduction targets and performance standards that green products need to achieve, providing clear guidance and stable expectations for product suppliers and related investors. Second, it is important to encourage the use of competitive bidding mechanisms as the market ultimately needs green technologies that are both "technically feasible" and "economically feasible." Third, the government could provide additional procurement price preferences for breakthrough technologies to promote continuous innovation by producers and encourage long-term investment.

**Recommendation 2: Establish a green performance auction fund connected to the financial market.**

Financial markets possess an inherent advantage in addressing the information asymmetry issues in the green transition, enabling more effective pricing for corporate green actions. Fully leveraging the role of financial markets can help make the green transition process smoother and more efficient. However, in reality, there is misalignment between financial markets and green transition policies in terms of incentive mechanisms, leading to an inability to effectively mobilize the enthusiasm of financial markets and resulting in low participation of financial markets in the green transition. To address this, it is recommended to use "green auctions" to tackle the externalities and scale constraints in the green transition by combining government incentives with the pricing power of financial markets and the flexibility and innovation of the private sector.

**Recommendation 3: Support the green technology innovation ecosystem, by supporting green technology incubation/acceleration projects, cultivating a team of green technology managers, and improving the green technology innovation service system.**

As mentioned earlier, for breakthrough innovations, technologies may be stuck in the "Valley of Death" at the prototype or demonstration stages, unable to quickly pass market tests to determine whether the technology is viable. Therefore, in addition to incentives for basic R&D and market subsidies for the commercialization phase, fiscal policy should support the market application of technologies at the demonstration stage, promoting market tests to assess whether the technology paths have the potential for commercialization. There should be a shift from supporting green technologies to supporting the green technology innovation ecosystem.

**Recommendation 4: Promote the "two accelerations" in carbon market construction to reduce the green premium for innovative technologies.**

A more binding carbon market can drive a broader range of industries to apply green technologies more vigorously, accelerate the expansion of the carbon market, and expedite the construction of a carbon quota auction mechanism, which is conducive to achieving this goal.

First, accelerating the expansion of the carbon market creates demand for emissions reduction technologies across more industries. Second, accelerating the establishment of a carbon quota auction mechanism will gradually promote the increase of carbon prices and support green technology innovation with the revenue from carbon quota auctions.

## *6.2 Capital Market Policies: Activating Patient Capital and Creating a Diverse Investment Ecosystem*

As previously mentioned, when technological innovation becomes the new main task of China's green transition, equity financing needs to play a greater role; in particular, the capital market should perform its pivotal function effectively. However, compared to the banking system, there is a greater room for improvement in China's capital market, which requires accelerated construction and optimization in order to more effectively support the urgent needs of green technology innovation.

**Recommendation 5: Explore feasible paths for bank funds to support the development of the equity market in capital raising, and guide long-term capital such as funds from wealthy individuals and pension funds into the market.**

First, since most financial resources in China are concentrated in the banking system, there is a need to explore new pathways for guiding bank resources to support breakthrough innovations in the green sector through the capital market in an orderly and moderate manner. Second, starting from the static and dynamic wealth effects, the government shall guide the entry of funds from wealthy individuals and long-term funds such as pensions into the market. Over the past decade, domestic capital has become the main source of fundraising for venture capital in China.

**Recommendation 6: Optimize the assessment methods of industry guidance funds in fund management, establish green VC plans, and value the positive role of outstanding foreign venture capital and CVCs.**

First, optimize the assessment methods of industrial guidance funds to enhance their investment and management capabilities. To address the issues of low risk appetite and short investment horizon of industrial guidance funds, it is necessary to further clarify their role in promoting technological innovation rather than focusing solely on equity finance. Second, create a green venture capital plan to establish a new model of effective cooperation between the government and market-oriented GPs to promote breakthrough innovations in the green sector. Third, increase the intensity of financial opening up to attract high-quality foreign GPs to enter China's equity market. Fourth, optimize the concept of anti-monopoly and pay attention to the positive role of CVC in investment and management.

**Recommendation 7: Create a favorable environment for green sector companies to go public, facilitate diversified mechanisms such as M&As and promote the reform of the exchange company system.**

First, it is necessary to provide more opportunities for software technology companies and unprofitable companies to list. Second, it is important to encourage listed companies to participate in M&A transactions, develop M&A funds, and build diverse exit paths. Third, emphasize the promotion of market-oriented reforms in stock exchanges to enhance their autonomy in increasing inclusivity from the bottom up.

## References

1. Ricardo, D. *On the Principles of Political Economy and Taxation.*
2. Johann Heinrich von Thünen. *The Isolated State.*
3. Yang, P. et al. (2021, July). *Current Status and Prospects of Carbon Capture, Utilization, and Storage (CCUS) Technology.*
4. Chinese Academy of Sciences, Tsinghua University, et al. (2021) *China's annual report on carbon dioxide capture, utilization, and storage (CCUS)* (2021).
5. CICC Research Institute, Research Department of CICC. (2021). Chapter 16: innovative finance: not a solo show of the capital market. *Innovation, the everlasting torch,* Citic Press.

# Chapter 10
# Carbon Pricing

## 1 Part 1: Missing the Paris Targets

**Mounting Impacts, Widening Gaps**: Climate science unambiguously concludes that more ambitious climate mitigation action is needed at scale to meet the UN Paris Climate Agreement goals. The initial UN global stocktake (2023) confirmed a "huge mitigation gap" that makes those goals more likely to be missed. The Intergovernmental Panel on Climate Change (IPCC) confirmed in 2022 that it is "almost inevitable" that the Paris objective of 1.5 °C will be exceeded.

Global greenhouse gas (GHG) emissions increased by 1.1% in 2023 to their highest level ever recorded, measuring 37.4 billion tonnes of $CO_2$ equivalent ($GtCO_2e$), with 65% of that increase coming from coal (International Energy Agency [1].[1] The *State of the Global Climate 2023* [2] reported that 2023 was the warmest year on record, with average global temperatures of 1.45 ± 0.12 °C above pre-industrial levels in that year and the long-term global average ranging from 1.22 to 1.41 °C [3].

In the first half of 2024, China experienced severe flooding in Guangdong, Fujian, Henan, and Hunan provinces. Extreme climate-related events are raising the risk of reaching global tipping points: the *Global Tipping Points Report 2023* cautions that five global tipping points are at risk of being crossed [4]. Once crossed, the risk of

[1] According to the IEA [1], emissions from China in 2023 grew by 565 million tonnes (Mt), of which one third were associated with a decline in hydroelectric power output due to drought and floods.

The content of this chapter is extracted from the main findings of the 2023-2024 CCICED Scoping Study on "Carbon Pricing".



*Adhering to Green and Open Cooperation to Advance Modernization*,
https://doi.org/10.1007/978-981-95-5547-5_10

non-linear, cascading, and irreversible shocks that lie outside of climate scenarios may accelerate quickly, underscoring the unknowns of fat-tailed climate risk.[2]

The UN Environmental Programme (UNEP) [5] reports that China's cumulative historical $CO_2$ emissions between 1850 and 2021 comprised 13% of global GHG emissions. In 2021, reflecting the rapid increase in manufacturing and other capacities, China accounted for 30% of global GHG emissions—they are the largest national source of carbon emissions. Between 2020 and 2023, China's energy consumption growth more than doubled to 4% per year. The largest sources of emissions increases in the past 5 years include coal and heavy industry, including steel and intermediary steel inputs for industrial machinery [6].

**Recommendations**: China's carbon pricing system can play a greater role in the green transition. As the dual control goals shift from carbon peaking to carbon neutrality, carbon pricing should become a primary tool used to accelerate the green transition to low-carbon pathways at both economy-wide and sector-specific levels.

## 2 Part 2: The Growing Role of Carbon Pricing

Measures to close the GHG emissions gap that have been formally adopted at various multilateral levels include the transition away from fossil fuels, the tripling of renewable energy, the doubling of energy efficiency, further increases in climate finance—notably, private sector financing for the next generation of green technologies and innovation—and demand-side consumer actions.

Carbon pricing systems play a pivotal role in supporting these multilateral measures by incentivizing the transition away from fossil fuels and fostering innovation in green technologies.

**The Benefits of Carbon Pricing**: Carbon pricing systems have several key benefits. By assigning a cost to carbon emissions, carbon pricing generates economic incentives that promote investments in renewable energy, energy efficiency, and low-carbon technologies. Additionally, carbon pricing can influence consumer behaviour toward more sustainable choices, thereby advancing comprehensive climate mitigation agendas.

Carbon pricing systems may be applied to larger or smaller emitters across various sectors, including transport, heavy industry, buildings, and households. This broad coverage can promote the substitution away from fossil fuel use and carbon-intensive processes to renewable energy, zero-emission transportation, and other low-carbon pathways.

Moreover, carbon pricing is more cost-effective compared to command-and-control regulations [7].

---

[2] A "fat tail" indicates that the likelihood of very large impacts is greater than what typical statistical assumptions would predict.

**A Growing Number of Carbon Pricing Systems**: Due to these attributes, a growing number of countries have already implemented or are preparing to implement carbon pricing systems. For example, the United Nations Framework Convention on Climate Change (UNFCCC) reports that over 100 countries include carbon pricing in their nationally determined contributions (NDCs). The World Bank's Carbon Pricing Dashboard [8] reports that as of April 2024, 75 compliance carbon pricing instruments have been implemented worldwide, roughly divided between 39 carbon taxes and 36 emissions trading systems (ETSs). Among the largest carbon pricing systems are China's national carbon market, which was launched in 2021 following a few regional pilots, and the European Union (EU) ETS, which was launched in 2005.

The number of implemented carbon pricing systems is increasing across Organisation for Economic Co-operation and Development (OECD) countries and emerging economies. Notable developments in 2023 included the launch of a national ETS in Indonesia covering the electricity sector, the introduction of a carbon tax at the Mexican sub-federal levels in Durango and Guanajuato, a voluntary ETS in Japan, and a sub-federal ETS in Washington State. Additionally, Australia revised its ETS, and Hungary updated its carbon tax. Brazil and India made progress toward introducing enabling legislation for carbon pricing, while Malaysia, Thailand, the Philippines, and Vietnam took steps to establish carbon markets [9]. Moreover, the EU announced plans to include buildings and transport into its ETS.

One barometer of the widening use of carbon pricing is their annual revenues: the World Bank [9] reported in its *State and Trends in Carbon Pricing 2024* that for the first time, the combined revenues from carbon markets exceeded USD 100 billion.

**Climate Externalities**: Carbon markets can help internalize some substantial and increasing externality costs associated with climate change, including flooding, drought, heat waves, wildfires, and other impacts. As noted, science is warning of the crossing of global tipping points.

One proxy to estimate climate-related externalities is through the social cost of carbon dioxide (SC-$CO_2$) methodologies, which estimate the net harm to society resulting from each additional tonne of $CO_2$ emissions. Recent estimates of the SC-$CO_2$ typically range from USD 200 to USD 400/tonne, with some outliers reaching USD 1000/tonne. Rennert et al. [10] estimate a global SC-$CO_2$ of US$185/tonne. In November 2023, the U.S. Environmental Protection Agency (EPA) released a working technical paper with an SC-$CO_2$ price of USD 190/tonne [11].

**Price Levels**:

See Fig. 1.

**Prices Differ Widely**: As Fig. 1 illustrates, there are wide price variations among different carbon pricing systems, ranging from the upper end of USD 167/tonne in Uruguay to well below USD 1/tonne in Baja California, Mexico. Data assessments by the OECD (2024) show that since 2015, carbon prices have diverged, raising leakage risks. Experts of this study have noted that the lack of price convergence complicates international carbon pricing coalitions and suggested developing key parameters for comparing systems and establishing mutual recognition [17]. Several

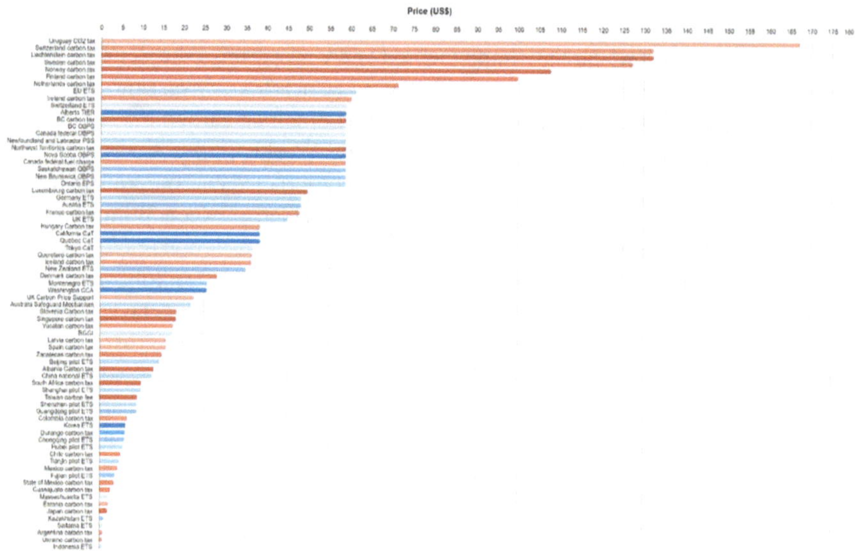


**Fig. 1** ETSs and carbon tax pricing in 2024. *Source* World Bank [9]

experts also noted the importance of this kind of international cooperation work in the context of the EU's Carbon Border Adjustment Mechanism (CBAM). They also noted that the stringency of carbon pricing is more nuanced than the marginal price level alone; it must also consider factors such as the coverage of emissions and compliance flexibilities that lower average costs, such as the allocation of free allowances.

It should be noted that the level of carbon prices is an important—but not the sole—measurement of a carbon market's stringency. Other determinants include the coverage of different sectors or, in the case of ETSs, whether allowances are freely allocated.

**Prices Are Weakened by Fossil Fuel Subsidies**: Fossil fuel subsidies function in the opposite direction of carbon pricing by magnifying environmental externalities, compounding market failures, and weakening the intended price signals and associated investment signals emulating from putting a price on carbon emissions. There are different methods to calculate these subsidies: the International Monetary Fund's (IMF's) assessment [14] includes direct and indirect subsidies coupled with their externality costs, which together were estimated at USD 7 trillion in 2022. The OECD [15–17] estimates covering 82 countries concluded that despite calls for greater climate mitigation ambition, fossil fuel subsidies almost doubled in 2022, reaching USD 1.48 trillion. At the 2009 G20 summit, countries pledged to identify and eliminate inefficient fossil fuel subsidies. Yet little progress has been made to date.

**Optimal Price Levels**: The High-Level Commission on Carbon Prices report (2017) recommended that carbon prices need to be set between USD 50 and 100/t$CO_2$e by 2030 to be consistent with the Paris Climate Agreement's upper-bound 2 °C target. The IMF notes that measures equivalent to a carbon price of USD 75/tonne or more by 2030 are needed [18].

Work by the OECD [15–17] in measuring the average stringency of carbon pricing systems underscores the wide gap between these targets and currently applied rates. The World Bank [9] concluded that in 2024, less than 1% of GHG emissions covered by a direct carbon price are at or above the recommended inflation-adjusted minimum level of USD 63/t$CO_2$e. Another estimate found that 70% of all carbon pricing systems were priced at USD 20/t$CO_2$e or less [19].

China's annual average marginal price for its emissions allowances was around USD 9.6/t$CO_2$e at the close of 2023. Although this marks an increase of 23% from 2022 levels, it is still well below the carbon price levels needed to shift from carbon-intensive to greener pathways set out in the Paris Climate Agreement. Surveys conducted by Environmental Defense Fund China [20] show expectations that China's carbon price will increase to around USD 12.5/t$CO_2$e by 2025 and to USD 18.1/t$CO_2$e by 2030.

**Recommendations**: Carbon prices should rise over time to address increasing climate-related externalities and approach the optimal levels needed to meet the Paris Climate Agreement goals.

Many international jurisdictions, including the EU's ETS, earmark a portion of carbon market revenues into green technology innovation funds. In anticipation of future carbon market revenues from allowance auctions, China should consider establishing green technology innovation funds for specific purposes, such as awarding prizes for clean technology innovation or subsidizing the deployment of new technologies.

## 3 Part 3: Public Support and Equity

**Public Acceptance and Political Economy**: Setting a carbon price level high enough to deliver substantial GHG emissions set by the Paris Climate Agreement while buffering the effects on competitiveness, related leakage risks, and households has been termed the "design trilemma" of compliance carbon markets [21].

Public opinion surveys across many countries broadly show widening support for climate mitigation action. A key finding from multi-country surveys [22] is that support for carbon pricing depends largely on revenue use. Carbon pricing revenues that go into general revenues are unpopular. In contrast, when revenues are earmarked for specific purposes, such as funding green infrastructure and low-carbon technologies, and redistributed to support specific households, notably low-income households or more carbon-dependent households, public support increases. In addition to revenue uses, surveys suggest that public support for policies that are viewed as effective and equitable garner more public support.

**Gender Equity and Wider Fairness Considerations**: Environmental pricing can incentivize consumers and businesses to consider their environmental impact or footprint. However, Gloor et al. [23] have pointed out a troubling trend in which environmental policies and initiatives inadvertently disadvantage women and other marginalized groups. Chalifour [24] examined the gender implications of carbon taxes in British Columbia and Quebec, finding that women are disproportionately affected, mainly due to lower disposable incomes and unpaid caregiving burdens. Women, on average, have less income to pay the additional costs created by the tax. In addition, household and caregiving burdens on women translate into reduced non-working time and flexibility, which limits their ability to change behaviours, such as using less energy-intensive transportation, to reduce tax impacts. Nonetheless, it is worth noting that individual consumption patterns may not accurately reflect the gender impact of consumption taxes, as the individuals may be purchasing goods on behalf of the family [25].

Given these implications, gender-responsive budgeting is a crucial tool for advancing gender equality. OECD countries have increasingly adopted gender budgeting, with participation rising from 44% in 2015 to over 60% in 2023 [15–17]. The OECD report highlights the impact of Canada's system, which enacted the Canadian Gender Budgeting Act in 2018 to ensure all tax and resource allocation decisions consider gender and diversity impacts. Despite progress, UN Women [26] reports that only 26% of 105 countries have comprehensive systems to track gender equality allocations. Strengthening public finance management is recommended to enhance transparency and accountability.

**Recommendation**: Price stringency should be complemented with equity considerations in which eventual carbon price revenues and price pass-throughs are augmented through revenue recycling, rebates, and other schemes that support lower-income and other affected households and groups.

## 4 Part 4: Comprehensive Frameworks, Complementary Policies

**Comprehensive Policies**: Effective carbon pricing systems need to function and interact with a number of complementary climate mitigation measures. Work by D'Arcangelo et al. [27] examined 120 different climate mitigation instruments across 50 countries. Instruments include carbon pricing; regulations such as bans, subsidies, standards; and other measures. Countries were grouped into four clusters: Canada, China, Japan, and New Zealand, for example, are grouped together under category 2, described as relatively diverse measures, with varying stringency levels. By contrast, category 4 countries, described as having diverse and stringent measures, include

France, Germany, Sweden, Finland, and Denmark. Several jurisdictions, such as the EU and Canada, have positioned carbon markets at the centre of comprehensive, whole-of-government[3] measures to achieve carbon neutrality. Three findings are worth noting from this work: (a) well-designed interaction of a limited number of measures can be more effective than dozens of measures with little policy alignment; (b) more stringent measures are associated with lower emissions; and (c) clusters that include carbon pricing are more effective on average compared to those without.

**Methane Emissions**: While carbon pricing can play a primary role in climate mitigation packages, it cannot be the only measure. Therefore, an important design question is the role of complementary measures that work in tandem with—and ideally, unlock—synergies. For example, Canada is developing oil and gas regulations to eliminate routine venting and flaring and enhance leak detection and repair that are hard to price within its carbon pricing system because the extent and emission effects of leaks are difficult to measure, among other factors [28].[4]

**Complementary Measures**: This study's experts also examined complementary mitigation measures to carbon pricing and noted the intentional overlapping of carbon pricing with complementary measures. For example, California's ETS is complemented by other non-market measures led by its low-carbon fuel standard, which in turn has reduced carbon prices in the California market.

Experts underscored the importance of complementary measures to enhance synergies across different instruments. In the case of output-based pricing systems applied in China and Canada, the absence of an emissions cap presents several complex design challenges. For example, one sub-federal system implemented various measures, such as investment tax credits for carbon capture and storage, cheap electricity rates, and specific electricity regulations. These measures have made it easier to generate compliance credits, leading to the potential for an overcapacity of credits that risk glutting the market and undermining the carbon price.

China's current carbon market system is similar to an emitter performance standard in which four benchmarks are applied to the electricity sector. However, no provisions are provided for renewable energy, thereby missing incentives that encourage the shift from fossil fuels to renewable energy sources. Experts noted the role of renewable portfolio standards as an example of an optimal complementary measure.[5]

---

[3] Various governance practices continue to emerge to support policy coherence, including the UK Climate Change Committee carbon budget and France's screening of government-wide budget proposals to determine their compatibility with environmental and net zero goals.

[4] Detecting methane emissions including from fugitive sources, such as from pipeline leaks, continues to improve. UNEP's International Methane Emissions Observatory (IMEO) and its Methane Alert and Monitoring System (MARS) identified over 120 major methane plumes or sources of leaks from the oil and gas sector in 2023. In March 2024, the Environmental Defense Fund launched MethaneSat, to improve the real-time detection of methane emissions from the oil and gas sector. CCICED has made a number of recommendations, including in 2024, for the urgent need to reduce methane emissions.

[5] An OECD paper examines outcomes across various emissions trading systems, including cap-and-trade and tradable performance standards, alongside overlapping policies like renewable subsidies

**Administrative Complexity and Information Requirements**: Compared to carbon taxes, ETSs, especially output-based and performance standards, involve complexity that demands detailed administrative support. They require intensive product knowledge to compare the marginal abatement costs of like products or an apples-to-apples comparison. For example, within the same sector, the average marginal abatement cost could be as high as USD 80/tonne for 20% of the sector's combined emissions but as low as USD 5/tonne for other goods within the same sector. This price differentiation makes benchmarking complex and vulnerable to manipulation, gaming, and capture by firms [29]. Experts suggested that building one or two large net-zero cement or steel plants could flood the market with surplus credits, potentially leading to a price collapse.

**Recommendations**: Ensure comprehensive and coherent policy approaches, particularly between green energy market reforms and the next phase of the national carbon market. Adjust the current carbon pricing system, which is based on emitter performance standards, to include allowances for renewable energy. Additionally, complementary policies such as renewable portfolio standards should be integrated.

## 5 Part 5: China's National Carbon Market

China's national carbon market was launched in mid-2021, following a decade of pilot markets implemented at the provincial or city levels.

**ETS Pilots**: In 2013, China began ETS pilots in Beijing, Chongqing, Guangdong, Hubei, Shanghai, Shenzhen, Fujian, and Tianjin. These pilots evolved from covering roughly 5% of total emissions at their outset to roughly 15% by 2017 [30]. As the national ETS coverage expands to additional sectors, it is expected that the pilot schemes will be integrated into the national system.

**National Market**: In December 2017, China issued the *National Carbon Emissions Trading Market Construction Plan (Power Industry)* (National Development and Reform Commission [31], with the national carbon market officially operating since July 2021. So far covering only the power sector, the national market accounts for

---

and electricity taxes. The study finds that renewable subsidies lower emission prices across all trading systems but have different effects on emissions depending on the performance standard design. Taxing electricity reduces both emission prices and outputs across all systems. With cap-and-trade, overlapping policies like renewables subsidies or electricity consumption taxes can reduce cost-effectiveness. However, under certain tradable performance standards, these measures can decrease distortions and improve overall cost efficiency. The analysis specifically highlights China's system, showing that current policy overlaps can reduce abatement costs by 20–30%, with further 10% savings possible through optimized renewable standards and uniform benchmarks. While cap-and-trade without overlapping policies remains the most cost-effective approach, this underscores the importance of considering the interplay between emissions trading systems and overlapping policies in policy reforms [46].

approximately 5.1 billion tonnes of annual $CO_2$ emissions and represents over 40% of the nation's total emissions.

**Governance**: The Ministry of Ecology and Environment of China (MEE) is the lead national/state authority responsible for issuing opinions, guidance, and rules, as well as overseeing ETS implementation in conjunction with other regulatory authorities. As in other areas of environmental management, provinces are responsible for overseeing the ETS implementation, with provincial departments responsible for identifying entities that are subject to the system, establishing and implementing a monitoring, reporting, and verification (MRV) system that meets MEE state-level specifications, managing provincial registries, and overseeing compliance.

**Policy Coherence**: To date, China's national carbon market has played a relatively limited role in its climate mitigation efforts. The primary drivers of the green transition have been an unprecedented expansion in renewable energy, particularly led by solar and wind, along with regulations and reforms aimed at reducing air pollutants from the transport sector through the adoption of low-cost electric vehicles. Other key measures include energy emission performance standards, feed-in tariffs for renewable energy, green incentives and subsidies, targets for green finance, evolving climate-risk disclosure rules, and the increasing role of public litigation.

In support of China's dual-carbon control goals, approximately 700–800 policies are being implemented at different levels to advance China's green and low-carbon transition. Such comprehensive policies create their own challenges in ensuring policy coherence. There are various examples of policy incoherence—that is, in which differing policies work at cross-purposes. One example of good practice in aligning carbon pricing policies is so-called "feebates," in which taxes applied to carbon-intensive goods such as internal combustion engine automobiles are complemented with rebates or tax incentives to make electric vehicle purchases more affordable [32].

**Next Steps in the National Carbon Market**: The planned next steps include the expansion[6] of sector coverage from the current power sector to other industrial sectors (as set out in Phase 3 of the carbon market plan)[7]; the standardization of GHG MRV requirements for firms; the gradual shift or augmentation of the current system of free allocation of allowances to an auction-based system that will generate public revenues for the first time; and most importantly, the shift from the current intensity-based system to the adoption of an absolute cap.

---

[6] China's *2024 Report on the Work of the Government* listed expanding sectorial coverage of the national ETS as a government task of the year. During a recent interview at the China Carbon Market Conference held in Wuhan in July 2024, Xia Yingxian, Director General of the MEE climate change department, mentioned that the ministry is actively promoting the expansion of China's carbon emissions trading market as soon as possible and the drafting of relevant technical documents is nearly complete.

[7] The NDRC Guidelines identified eight sectors that will eventually be covered under China's carbon market: electricity (including power generation, power and heat cogeneration, and grid distribution), building materials, iron and steel, non-ferrous metal processing, petroleum refining, chemicals, pulp and paper, and aviation. The wider system is expected to regulate 6000 companies [38].

**Revised Energy Law**: In addition to changes underway in the MEE-directed national carbon market, China's energy market, led by the NDRC and the National Energy Agency (NEA), is simultaneously undergoing a series of important reforms. In May 2023, the NDRC introduced a next round of reforms intended to create a unified national power market that would be green, safe and economical by furthering electrification: the 2024 target is to increase renewable energy capacity by 200 GW, which is likely to again be exceeded. The NDRC reforms further call for "reasonable profits" [33, 34] in allowable costs by introducing several market measures. In April 2024, the draft Energy Law was issued for public comment.

These reforms are described as the adoption of a new, more market-oriented system with green power at the centre, increased inter-jurisdictional grid transmission and increased spot trading between jurisdictions, and electricity price reforms that reflect different uses, such as capacity reserves and ancillary services. For 2023–2026, features affecting all industrial and commercial users will include linking feed-in tariffs with feed-in grid loss tariffs and system operating costs; steps toward a flexible national spot market, including a capacity-tariff scheme intended to increase the proportion of renewable energy during non-peak periods compared to tariffs for coal-pricing peak loads; a new peak-shaving market; the introduction of ancillary renewable energy pricing levels; and other measures.

This ongoing shift toward more market-oriented energy prices, intended to favour green electricity, coupled with the next phase of China's national carbon market presents unique opportunities to align these two systems as a primary means to achieve the dual-carbon targets.

**Elevating Carbon Pricing to a Critical Enabler**: *Embracing the New Paradigm of Green Development*—an important 2023 report co-authored by Min ZHU, Nicholas Stern, Joseph Stiglitz, Shijin LIU, Yongsheng ZHANG, and others—concluded that carbon pricing should become the primary policy to realize the dual-carbon climate goals, noting: "A strong and predictable carbon price or its equivalent has a particularly critical role to play in the efficient management of the transition process. Due to the complexity of carbon pricing and multiple market failures, it is important to build a multi-dimensional market mechanism for carbon pricing, including a combination of carbon taxes, carbon emissions trading, carbon derivatives trading, and carbon offset markets."

**Uneven Progress in Climate Targets**: China has set out a number of targets within its dual-carbon control framework. In addition to the overarching goal of carbon peaking by 2023 and carbon neutrality by 2060, other targets to be achieved by 2030 include: reducing $CO_2$ emissions per unit of GDP by 65% based on the 2005 level; reaching 25% of non-fossil fuel primary energy; and increasing installed solar and wind energy to 1200 GW. Some of these targets have been exceeded. For example, China's solar and wind capacity increased by 63% in 2023, adding 297 GW [35]. Some analysts expect carbon peaking to be reached well before scheduled, as early as 2025. However, the mid-term evaluation of progress published by NDRC in December 2023 identified energy and carbon intensity as two out of four indicators that were off track, along with a key air quality target.

**Tradeable Performance Standards**: China's national ETS is based on an output-oriented emissions intensity standard, in which free allowances are allocated based on a facility's verified emission multiplied by the corresponding benchmark factor. Firms can trade emission allowances among covered entities, allowing them to exceed their emission target at low costs. Thus, it more closely resembles a tradeable performance standard (TPS) [36] than a cap-and-trade system, and more specifically, an emitter-based performance standard with differing benchmarks within the power sector.

**Benchmarks and Allocation**: Four distinct benchmarks were employed for the 2021–2022 compliance period: conventional coal plants below 300 MW; conventional coal plants above 300 MW; unconventional coal; and natural gas [37]. In general, coal-fired power plants (both conventional and unconventional) have less stringent benchmarks than natural gas sources.

China's provincial/municipal development and reform commissions noted that setting "reasonable and fair" grandfathering allowance allocation levels was difficult during the pilot periods, given marked increases in the scale of electricity generation [38]. MEE further adjusted allowances to meet fairness criteria.

**Allowances Trading**: Regulations stipulate that only covered entities are permitted to trade allowances through agreement transfer, one-way bidding, or other spot transactions, serving as further price control measures. According to the *2022 China Carbon Pricing Survey* [39], very few covered entities suggested that the benchmarks were "too generous" when asked about their adequacy in encouraging GHG emission reduction. The survey was conducted in October 2022, shortly after the announcement of updated benchmarks for the 2021–2022 compliance cycle, where smaller installations were facing stricter tightening compared to larger ones.

The ETS currently does not allow banking or borrowing. The MEE has indicated that it will gradually introduce a system of banking.

**Intensity-Based System**: The output-based system encourages relative per unit of output efficiency gains but does not encourage wider shifts in fuel choices—for example, from coal to solar, wind, or green hydrogen. While Goulder et al. [40] estimated that the environmental benefits of China's TPS could exceed its cost by a factor of five over the 2020–2035 interval, and these benefits are greater when human health benefits are included, the authors also noted that the cost becomes comparatively greater over time, compared to ETSs based on cap-and-trade features.[8]

In addition, experts of this study have noted that the transition to an auctioning system coupled with more stringent benchmarks and/or the adoption of an emissions cap would help stabilize market expectations and play a key role in carbon price

[8] Among the reasons for this widening cost gap, according to the authors, is that the current free-allowance allocation, intensity-based system implicitly provides a subsidy based on planned production output, since entities are allocated free allowances for each additional unit of production, offering no incentives to reduce production and output. Compared to cap-and-trade systems, as the output-based system increases its stringency, the implicit subsidy characteristic becomes more pronounced, widening the gap between marginal abatement costs and environmental benefits.

discovery, which in turn could help convey important signals to green technology investments.

**Shift From Free Allocation to Auctioning**: While a specific timeline is not provided, the recently released *Interim Regulations on Carbon Emissions Trading Management* [41] indicates a gradual transition[9] in allocation toward a combined method of both free and paid allowances. It underscores the importance of considering market regulation needs[10] in determining the amount of carbon emission allowances and allocation plans.

Anticipating a possible tightening of emission allowances and forward prices, enterprises may be reluctant to sell allowances during this transition period, which may affect trading activities in the national ETS and, in turn, may impede price discovery.

**Price Levels**: S&P Global data showed the annual average trading price for China's emission allowances reached USD 9.48/t$CO_2$e in 2023, up 23.24% on the year. The allowance trade volume was 212 million t$CO_2$e for the year, jumping 316%.

Goulder et al. [40] noted that the stringency of China's carbon market is below an efficiency-maximizing level, in which optimal efficiency is defined as the marginal abatement cost equal to the marginal environmental benefit.

**Stringency and Compliance Flexibility**: China's ETS continues to undergo important developments, including the tightening of disclosure standards, the introduction of carbon offsets, and some tightening of stringency. For example, Guangdong's ETS 2022 updated plan widened the coverage of regulated entities by lowering the threshold from 20,000 t$CO_2$/year or energy consumption of 10,000 tonnes of coal equivalent (tce)/year to 10,000 t$CO_2$/year or energy consumption of 5000 tce/year, while the Shanghai exchange lowered emission factors of power and heat consumption [42].

At the same time, in March 2023, MEE proposed that benchmark values for the 2021 to 2022 compliance period be tightened significantly compared with 2019 to 2020, particularly for coal-fired power plants.

**Coverage and Regulatory Framework on Potential Expansion**: China's national carbon market currently covers approximately 2250 power sector entities, including electricity generation and industrial cooling.

In March 2021, MEE released the *Notice on Strengthening the Management of Enterprise Greenhouse Gas Emissions Reporting*, establishing key actions and timelines on MRV for key emitting entities, not only in the power sector but also petrochemicals, chemicals, building materials, steel, non-ferrous metals, papermaking, and aviation, a necessary precondition for expanding the sectoral scope.

---

[9] During the China Carbon Market Conference in July 2024, Zhao Yingmin, Vice Minister of Ecology and Environment, highlighted the focus on exploring and implementing a mix of free and paid carbon quota allocation methods, with a gradual increase in the proportion of paid allocations.

[10] In the events of abnormal price fluctuations, the Ministry of Ecology and Environment may take measures such as conducting open market operations and adjusting the use of Chinese Certified Emissions Reductions (CCERs).

In February 2024, the State Council of China released the *Interim Regulations on Carbon Emissions Trading Management* (effective on May 1, 2024), which provides a legal framework for the operation of China's national ETS. It signals a potential near-term expansion to more key emitting industries soon, by mentioning carbon accounting in the aviation sector.

**Reporting and Registries**: Carbon information disclosure can help enterprises manage their carbon assets and identify risks, including stranded-asset risks. Effective reporting is a cornerstone of effective government regulation and policy-making.

Multiple stipulations regarding information disclosure are set out in the *Interim Regulations on Carbon Emissions Trading Management*. The policy requires regular disclosure of annual GHG emission reports of key emitters. Firm-level reporting is an important pillar in ensuring the transparency of the national ETS, including providing a guarantee for fair and transparent carbon prices.

System registries, including the national carbon emission allowance registry and clearing platform, are administered by China Carbon Emissions Registration and Clearing Co., Ltd., while the Shanghai Environment and Energy Exchange maintains the allowance trading platform.

The trading and clearing platform for the relaunched carbon offset system, the CCER system, is operated by the Beijing Green Exchange, while the CCER registry is maintained through the National Center for Climate Change Strategy and International Cooperation (NCSC).

In addition, those enterprises that are or will be required to undertake full-scale climate information disclosure can align their MRV systems with the evolving climate-related financial risk disclosure standards and mandatory rules.

**Carbon Offsets**: Following a 6-year pause, the CCER market was relaunched in January 2024. The CCER's initial coverage is forestation, mangrove cultivation, solar thermal power, and grid-connected offshore wind power projects. As of January 2024, entities under the ETS are permitted to offset up to 5% of emissions under the CCER. Various guidelines have been issued related to MRV and other carbon offset characteristics. One question will be the average price of carbon offsets vis-à-vis the price of allowances. Rules regarding the classification of a carbon offset as a financial product are pending. The first certified carbon offset credits tied to the ETS are expected in 2024.

Currently, only spot allowances are allowed to be traded. However, MEE has identified carbon derivatives as a likely product in the future.

**Market Stability Measures**: A common feature of most ETSs, including China's, is the use of various direct measures, such as floor and ceiling prices, and indirect measures, such as allowance withdrawals and retirements, to smoothen price volatility.

In 2021, MEE announced the option of cushioning large swings in trading prices with stability mechanisms such as permit buybacks, auctioning, or adjusting the rules related to carbon offsets under its CCER. Under the Beijing pilot ETS, the Beijing Green Exchange limited price increases and decreases when they exceeded or fell below 20% of a reference price. The Shanghai pilot ETS introduced rules whereby day-trading could be suspended if prices exceeded set parameters (10% and 30%), as well as an annual auction reserve [43].

**Distributional Effects**: Carbon pricing can affect households, regions, firms, and others differently, and extensive literature exists on measuring impacts and options to offset potentially negative effects.

To date, China's carbon market does not pass forward costs to households, while energy and electricity prices are set separately by NDRC. A recent IMF study examined how imposing a USD 50 per tonne carbon price[11] could lead to price hikes in coal, electricity, natural gas, and road fuel, observing that the burden on average households' consumption would predominantly stem indirectly from increased prices for general goods and services, rather than directly from energy prices. The study noted that recycling carbon pricing revenues could alleviate roughly 85% of the burden on households. The study compared various policy options for protecting low-income households, emphasizing the effectiveness of targeted income support while discussing the challenges and trade-offs associated with other approaches, such as broad-based income support or compensating for energy price increases.

Additionally, strategies for mitigating the impact on firms' competitiveness were outlined, including revenue recycling to provide industrial rebates, tradeable emission rate standards, border carbon adjustments, or, ideally, international coordination among major emitters.

## 6 Part 6: Canada's Carbon Pricing System

Carbon pricing is a key component of meeting Canada's targets of reducing GHG emissions to 40–45% below 2005 levels by 2030 and reaching net-zero by 2050. The policy aims to put a price on GHG emissions across Canada as an efficient way to incentivize emissions reductions and spur innovation across the economy. Some 78–80% of Canada's emissions are covered by a carbon price. Environment and Climate Change Canada's energy and economic modelling estimates that up to one third of projected emissions reductions to 2030 would come from carbon pricing.

Canada's approach to carbon pricing is complex, primarily because of the flexibility afforded to subnational jurisdictions to implement their own carbon pricing systems but also because of the interaction of the instrument with the hundreds of non-pricing climate change policies and measures at play across the country (e.g., vehicle efficiency and emissions regulations, a clean fuel standard, clean electricity regulations, etc.).

One component of Canada's system, the federally regulated Output-Based Pricing System (OBPS), may provide relevant insights for China. Given its design, the OBPS

[11] Source: IMF staff using Climate Policy Assessment Tool

has an output-based design that differs from ETSs that set an absolute cap on emissions across specific sources. The OBPS could serve as a useful example for China to look at when considering carbon pricing expansion to sectors susceptible to carbon leakage.

The OBPS sets specific output-based standards for a given activity (i.e., a set level of GHG emissions per unit of output). Compliance obligations reside only on emissions that exceed these performance standards. By limiting compliance costs to emissions above a facility's output-based standard, the OBPS helps maintain international competitiveness and minimize carbon leakage. This is particularly important given that Canada and the United States—a jurisdiction with no carbon pricing at the federal level—share one of the largest trading relationships in the world.

However, some trade-offs must be considered. These trade-offs include the lack of an absolute cap and therefore emissions reduction certainty from the program, one of the principal advantages identified in the literature for a quantity-based—rather than a price-based—approach. This type of approach also has implications for exemptions or rebates from carbon border charges, which are generally based on the explicit carbon price paid per tonne of embedded emissions within exported goods compared to the domestic effective carbon price. Carbon pricing policies designed to mitigate carbon leakage through free allowance allocation or benchmarking result in lower carbon costs per tonne of embedded emissions than the headline marginal carbon price. Meanwhile, the effective carbon price for emissions-intensive, trade-exposed goods within the EU will converge on its marginal carbon price as free allowance distribution is phased out between 2026 and 2034. This will increase the delta (i.e., border charge) with carbon pricing policies that result in lower average carbon prices.

**Price Predictability**: At the federal level, Canada has looked to address "price predictability" issues through a minimum national carbon price schedule, applicable to all carbon pricing systems in Canada. As early as 2020, the government proposed carbon price increases of CAD 15/tonne $CO_2e$, rising from CAD 65/tonne $CO_2e$ in 2023, and by CAD 15/year to CAD 170 in 2030. This price trajectory was confirmed in August 2021.

Canadian systems must set their minimum carbon pollution price according to this schedule. For example, Canada's fuel charge rises in accordance with the price schedule each April, recently hitting CAD 80 per tonne of $CO_2e$. The maximum compliance price under OBPSs must also be at least equal to the minimum national carbon price, which sets a consistent price signal across the country for large industrial emitters. Cap-and-trade systems must have declining annual GHG emissions caps that correspond, at a minimum, to the projected emissions levels that would result from the application of a price-based system using this schedule.

Legislating and publishing the price trajectory in advance gives businesses and households the opportunity to estimate future costs and make informed decisions on investments such as low-carbon technologies, zero-emissions vehicles, heat pumps, or business/home energy efficiency improvements. It also serves to limit economic impacts where the economy can adjust more readily to gradual and predictable price changes.

In addition to publishing the carbon price schedule through 2030, the government is committed to publishing information about how revenues collected in the federal carbon pricing systems are recycled.

In provinces and territories where the federal benchmark applies, and the federal fuel charge is collected at the federal level, revenues are returned to households through the Canada Carbon Rebate. The amount of the Canada Carbon Rebate that households will receive is also published ahead of time on the Canada Revenue Agency webpage. The quarterly rebate values for 2024–2025 were announced in February 2024.

While the carbon price schedule helps ensure transparency and predictability, total carbon costs are much more difficult to predict for large industrial emitters in Canada, where output-based pricing applies. The price of complying with the policy is determined by market dynamics, given the policy parameters, with the minimum national carbon price only providing certainty regarding the highest-value compliance option. Market dynamics—determining the price of compliance units—are specific to each industrial carbon pricing system. Though all systems are modelled to have a net demand for compliance units, market factors may differ in individual markets, thereby affecting total compliance costs (e.g., banking, market liquidity, policy interactions).

Moreover, in industrial pricing systems, the marginal carbon price is not representative of the average cost of emitting, whereas Canadian output-based pricing systems and cap-and-trade systems are designed to lower average costs to prevent carbon leakage for emissions-intensive and trade-exposed industrial sectors. Given the flexibilities noted above, variation in average cost, both across and within jurisdictions, may impact cost projections. Predictability around average costs is important for long-term capital decisions as well as inter-provincial carbon leakage.

Some stakeholders have noted that Canada's pricing systems are unlikely to spur the types of large-scale clean investments needed to decarbonize industry if longer-term price predictability is not strengthened [44]. The government is aware of price predictability concerns for large industrial emitters in Canada's carbon pricing landscape (recognized in *Canada's Emissions Reduction Plan for 2030*), including the federal OBPS. The 2023–2030 update to Canada's federal benchmark commits to an interim review of the policy by 2026, which may include consideration related to carbon price predictability.

Canada is concurrently considering mechanisms such as carbon contracts for difference (CCfDs) for large low-carbon infrastructure projects. CCfDs can be designed to support low-carbon projects by guaranteeing a certain carbon price, essentially subsidizing any delta between an agreed future "strike price" and the market price of compliance units. The 2023 Fall Economic Statement announced that the Canada Growth Fund will allocate CAD 7 billion to investment offerings for industrial decarbonization, including contracts for difference as well as other forms of price assurance/support.

**Flexibilities**: ***Subnational flexibilities***: In 2016, Canada's federal, provincial, and territorial First Ministers agreed to a pan-Canadian approach to reducing GHG emissions, including carbon pricing. At the time, several Canadian provinces already had carbon pricing in place, and the Canada-wide approach committed to recognizing these existing systems. For example, the province of British Columbia implemented a broad-based carbon tax in 2008, and the province of Quebec implemented an economy-wide cap-and-trade system in 2013, which linked the following year to the state of California through the Western Climate Initiative.

To expand the application of carbon pricing across the country, the Government of Canada introduced a federal "benchmark" to set minimum national stringency standards, with this stringency increasing over time. Provinces and territories are required to implement an explicit price-based system—such as a carbon levy on fossil fuels or a hybrid system with an OBPS for industry—or a cap-and-trade system.

The federal government, at the same time, designed a federal backstop, a carbon pricing system that applies in provinces/territories that request it or that do not implement a system that meets the federal benchmark. The federal pricing system has two parts: a regulatory charge on fossil fuels, known as the fuel charge, and a performance-based system for industries, known as the OBPS. This approach saw economy-wide carbon pricing applied across Canada by 2019, providing flexibility for a variety of different approaches.

The federal standard aims to ensure that all systems in Canada are comparable and effective while giving subnational jurisdictions the flexibility to design their own systems tailored to their unique regional circumstances. The benchmark requires that carbon pricing be applied to a broad set of emission sources across the economy and that carbon price increases should occur according to a minimum national carbon price schedule. (see *Price Predictability* above).

***Sectoral flexibilities***: The output-based standards under the federal OBPS are mostly set according to the production-weighted average emissions intensity of all facilities producing similar products across Canada, adjusted for the level of carbon leakage risk. Producers are liable for their emissions above the standards and receive tradeable credits when their emissions are lower than the standard. These credits can be banked or sold to other producers for compliance purposes. Provinces and territories looking to align with the federal approach have the flexibility to set these standards according to their own preferences and unique circumstances as long as the marginal carbon price signal is maintained. This ensures a flexible pricing approach, with the various standards reflecting the current status of emissions intensity for the product/activity.

One drawback to this approach is that it promotes improvements in plant efficiency but does not promote shifting to new manners of production. For example, while the federal output-based standard for coal-fired power plants promotes more efficient practices, it does not incentivize switching over to a lower-emitting fuel such as natural gas. Instead, Canada is developing a Clean Electricity Regulation to help transition toward a net-zero electricity grid [45]. Another drawback to the approach is that regional and sector-specific flexibilities add to the significant complexity of the policy. The standards are determined in a similar way as regulatory development, requiring significant sectoral knowledge.

***Revenue flexibilities***: Where the federal system applies, revenues remain in the jurisdiction of origin. Jurisdictions with their own systems or that request the federal system can use carbon pricing proceeds according to their needs, including to address impacts on vulnerable populations and sectors, and/or to support climate change and clean growth goals. More specifically, 90% of revenue collected through Canada's federal fuel charge is returned to households through the Canada Carbon Rebate as direct payments. Most households receive more money back in rebate payments than they pay directly through the fuel charge. Lower-income households benefit the most, as these households traditionally spend less on energy while receiving the same rebate.

The federal OBPS chooses an approach that foregoes revenue collection on emissions under the specific standards described above, instead choosing to mitigate the risk of carbon leakage and adverse competitiveness impacts for emissions-intensive and trade-exposed industries. Compliance obligations arise on the portion of emissions above specific benchmarks—and where compliance units are not surrendered—and proceeds are transferred to subnational governments, including through funds that support industrial decarbonization and clean electricity. Other policy design options for the use of revenues may be more appropriate in other national contexts, such as the establishment of social programmes or the lowering of distortionary taxes, such as income taxes.

## 7 Conclusion

After 4 years, China's national carbon market will enter its next critical phase, notably with the introduction of emission caps for covered sectors, as well as the gradual phase-in of an allowance auctioning system. According to the Work Plan for Accelerating the Establishment of a Dual Control System for Carbon Emissions published by the State Council of China in August 2024, China will set targets for total emission volumes in the 15th Five-Year Plan period, beginning in 2026. Initially, these total volume targets will be secondary to intensity targets. However, after China reaches its peak emissions—currently expected before 2030—total emissions will become the primary focus. The carbon market will function within a wider context of the transition from carbon peaking to the beginning of carbon reduction measures toward achieving carbon neutrality by 2060. The study envisages a greater role of carbon pricing, a market-based approach to incentivize the shift to carbon-neutral pathways at the firm level and among private investors. As this shift in carbon markets continues, the study highlights several observations and findings, particularly the importance of identifying distributional effects among households, sectors, and regions and designing the early use of carbon pricing revenues from allowance auctioning to support equity outcomes.

## References


1. International Energy Agency (IEA). (2024). $CO_2$ emissions in 2023. Retrieved from https://iea.blob.core.windows.net/assets/33e2badc-b839-4c18-84ce-f6387b3c008f/CO2Emissionsin2023.pdf.
2. World Meteorological Organization. (2024). State of the global climate 2023. World Meteorological Organization. Retrieved from https://library.wmo.int/records/item/68835-state-of-the-global-climate-2023
3. Hausfather, Z. (2023, July 14). Analysis: What record global heat means for breaching the 1.5C warming limit. Carbon Brief Retrieved from https://www.carbonbrief.org/analysis-what-record-global-heat-means-for-breaching-the-1-5c-warming-limit/.
4. University of Exeter. (2023). *The Global Tipping Points Report 2023*. Retrieved from https://global-tipping-points.org/about/
5. UNEP. (2023). Broken record: Temperatures hit new high, yet world fails to cut emissions (again) (Emissions Gap Report). Retrieved from: https://wedocs.unep.org/bitstream/handle/20.500.11822/43922/EGR2023.pdf?sequence=3&isAllowed=y
6. Myllyvirta, L. (2024, May 20). China's manufacturing pushed emissions sky high. What's next? *Dialogue Earth*. Retrieved from https://dialogue.earth/en/climate/chinas-manufacturing-pushed-emissions-sky-high-whats-next/
7. Flues, F., & Van Dender, K. (2020). Carbon pricing design: Effectiveness, efficiency and feasibility (OECD taxation working papers, No. 48). OECD Library. Retrieved from https://www.oecd-ilibrary.org/taxation/carbon-pricing-design-effectiveness-efficiency-and-feasibility_91ad6a1e-en
8. World Bank. (2024a). *Carbon Pricing Dashboard*. Retrieved from https://carbonpricingdashboard.worldbank.org/
9. World Bank. (2024b). *State and Trends of Carbon Pricing 2024*. Retrieved from https://openknowledge.worldbank.org/server/api/core/bitstreams/253e6cdd-9631-4db2-8cc5-1d013956de15/content
10. Rennert, K., Errickson, F., Prest, B. C., et al. (2022). Comprehensive evidence implies a higher social cost of $CO_2$. *Nature, 610*, 687–692. https://www.nature.com/articles/s41586-022-05224-9
11. Environmental Protection Agency. (2023, November). Final report of the social cost of greenhouse gas emissions (Docket ID No. EPA-HQ-OAR-2021-0317). Retrieved from https://www.epa.gov/system/files/documents/2023-12/epa_scghg_2023_report_final.pdf
12. OECD. (2024). *Pricing Greenhouse gas Emissions, 2024*. Gearing up to Bring down Emissions. OECD Paris.
13. Delbeke, J. (Ed.). (2024). *Delivering a climate neutral Europe* (1st ed.). Routledge. https://doi.org/10.4324/9781003493730
14. Parry, I., Black, S., & Vernon, N. (2023). Climate change: Fossil fuel subsidies surged to record $7 trillion. *IMF Blog*. Retrieved from. https://www.imf.org/en/Blogs/Articles/2023/08/24/fossil-fuel-subsidies-surged-to-record-7-trillion
15. OECD. (2023a). Effective carbon rates 2023: Pricing greenhouse gas emissions through taxes and emissions trading. OECD Series on Carbon Pricing and Energy Taxation. OECD Publishing. https://doi.org/10.1787/b84d5b36-en.
16. OECD. (2023b). OECD best practices for gender budgeting. *OECD Journal on Budgeting, 23*(1). https://doi.org/10.1787/9574ed6f-en
17. OECD. (2023c). *OECD inventory of support measures for fossil fuels 2023*. OECD Publishing. https://doi.org/10.1787/87dc4a55-en
18. Black, S., Parry, I., & Zhunussova, K. (2022, July 21). More countries are pricing carbon, but emissions are still too cheap. IMF Blog. Retrieved from https://www.imf.org/en/Blogs/Articles/2022/07/21/blog-more-countries-are-pricing-carbon-but-emissions-are-still-too-cheap
19. Fleurence, L., Fetet, M., & Postic, S. (2023, November 22). *Global carbon accounts in 2023*. Institute for Climate Economics (I4CE). Retrieved from https://www.i4ce.org/en/publication/global-carbon-accounts-2023-climate/

20. Environmental Defense Fund. (2023). 2023 Highlights: ETS Newsletter Issue 21. Environmental Defense Fund. Retrieved from https://www.edf.org/sites/default/files/2024-01/Issue_21_ETS_Newsletter_2023%20Highlights.pdf
21. Böhringer, C., Fischer, C., & Rivers, N. (2023). Intensity-based rebating of emissions pricing revenues. *Journal of the Association of Environmental and Resource Economists, 10*(4). https://doi.org/10.1086/723645
22. Dechezleprêtre, A., et al. (2022). Fighting climate change: International attitudes toward climate policies. OECD Economics Department Working Papers, No. 1714. OECD publishing, Paris. https://doi.org/10.1787/3406f29a-en
23. Gloor, J. L., Bajet Mestre, E., Post, C., & Ruigrok, W. (2022, July 26). We can't fight climate change without fighting for gender equity. Retrieved from https://hbr.org/2022/07/we-cant-fight-climate-change-without-fighting-for-gender-equity
24. Chalifour, N. (2010). A feminist perspective on carbon taxes. *Canadian Journal of Women and the Law, 21*(2), 171. Retrieved from https://papers.ssrn.com/sol3/papers.cfm?abstract_id=1684097
25. OECD. (2022). *Tax policy and gender equality: A Stocktake of country approaches.* OECD Publishing. https://doi.org/10.1787/b8177aea-en
26. UN Women. (2023). Strengthening public finance management systems for gender equality and women's empowerment: Promising practices and remaining gaps. Retrieved from https://www.unwomen.org/sites/default/files/2023-06/Strengthening-public-finance-management-systems-for-gender-equality-and-womens-empowerment-en.pdf
27. D'Arcangelo, F., Kruse, T., & Pisu, M. (2023). Identifying and tracking climate mitigation strategies: A cluster-based assessment. OECD economic department working papers no. 1786. OECD. Retrieved from https://one.oecd.org/document/ECO/WKP(2023)39/en/pdf
28. Environment and Climate Change Canada. (2023). *Minister Guilbeault announces Canada's draft methane regulations to support cleaner energy and climate action* (news release). https://www.canada.ca/en/environment-climate-change/news/2023/12/minister-guilbeault-announces-canadas-draft-methane-regulations-to-support-cleaner-energy-and-climate-action.html
29. Canadian Climate Institute. (2024). The state of carbon pricing in Canada. Climate Institute. Retrieved from https://climateinstitute.ca/reports/the-state-of-carbon-pricing-in-canada/
30. Oppermann, K., Zhang, J., Child, A., Nierop, S., Ramstein, M. S., Long Khanh Lam, Wong, L., et al. (2017). State and trends of carbon pricing 2017. The World Bank. Retrieved from https://documents.shihang.org/zh/publication/documents-reports/documentdetail/468881509601753549/state-and-trends-of-carbon-pricing-2017
31. National Development and Reform Commission of China (NDRC). (2017). National Carbon Emissions Trading Market Construction Plan (Power Industry). Retrieved from https://www.ndrc.gov.cn/xxgk/zcfb/ghxwj/201712/t20171220_960930.html
32. Parry, I. (2021). The critical role of Feebates in climate mitigation strategies. In F. Caselli, A. Ludwig, & R. van der Ploeg (Eds.), *No brainers and low-hanging fruit in National Climate Policy* (pp. 217–244). Center for Economic Policy Research.
33. National Development and Reform Commission (NDRC). (2023a, May 15). Notice on Provincial Grid Transmission and Distribution Tariffs and Related Matters for the Third Regulatory Period. Retrieved from https://www.ndrc.gov.cn/xwdt/tzgg/202305/t20230515_1355748.html
34. National Development and Reform Commission (NDRC). (2023b, December 27). Mid-term Evaluation Report on the Implementation of the "14th Five-Year Plan for National Economic and Social Development and the Long-Range Objectives Through the Year 2035" of the People's Republic of China. Retrieved from https://www.ndrc.gov.cn/fzggw/wld/zsj/zyhd/202312/t20231227_1362958.html
35. International Renewable Energy Agency (IRENA). (2024). *Renewable capacity statistics 2024.* Retrieved from https://www.irena.org/Publications/2024/Mar/Renewable-capacity-statistics-2024
36. Fischer, C. (2001). Rebating environmental policy revenues: Output-based allocations and tradeable performance standards (Resources for the Future Discussion Paper No. 01-22). Retrieved from https://core.ac.uk/download/pdf/9308407.pdf

37. Ministry of Ecology and Environment of China (MEE). (2023). Implementation Plan for the Setting and Allocation of National Carbon Emissions Trading Allowance Total for 2021 and 2022 (Power Industry). Retrieved from https://www.mee.gov.cn/xxgk2018/xxgk/xxgk03/202303/W02023031 5687660073734.pdf
38. Pizer, W. A., & Zhang, X. (2018). China's new national carbon market. *AEA Papers and Proceedings, 108*, 463–467. https://doi.org/10.1257/pandp.20181029
39. Slater, H., et al. (2023). 2022 China carbon pricing survey. Retrieved from https://www.cet.net.cn/storage/files/2022%20CCPS%20Report-EN.pdf
40. Goulder, L., et al. (2024). China's Nationwide $CO_2$ emissions trading system: A general equilibrium analysis (Resources for the Future, 24-02). Retrieved from https://media.rff.org/documents/WP_24-02.pdf
41. State Council of China. (2024). Interim regulations on carbon emissions trading management. https://www.mee.gov.cn/zcwj/gwywj/202402/t20240205_1065850.shtml
42. World Bank. (2023). *State and Trends of Carbon Pricing 2023.* World Bank, Washington, DC.
43. International Carbon Action Partnership (ICAP). (2023). *Emissions trading worldwide: Status Report 2023.* Retrieved from https://icapcarbonaction.com/system/files/document/ICAP%20Emissions%20Trading%20Worldwide%202023%20Status%20Report_0.pdf
44. Clark, J., Bernstein, M., Beugin, D., Shaffer, B., & Wadland, J. (2022). Closing the carbon-pricing certainty gap. Retrieved from https://cleanprosperity.ca/wp-content/uploads/2022/10/Closing_the_Carbon-Pricing_Certainty_Gap.pdf
45. Government of Canada. (2024). Clean electricity regulations. https://www.canada.ca/en/services/environment/weather/climatechange/climate-plan/clean-electricity-regulation.html
46. Fischer, C., Qu, C., & Goulder, L. H. (2024). *Rate-based emissions trading with overlapping policies: Insights from theory and an application to China.* Policy Research Working Paper; PLANET. Washington, D.C.: World Bank Group. Retrieved from http://documents.worldbank.org/curated/en/099733008212419312/IDU12092fd8d16e4014fe21b9b115230cfba2a5f

# Chapter 11
# Circular Economy

## 1 Objectives and Approach

The high demand for raw materials in a largely linear economy is one of the main causes of the climate crisis, pollution, and biodiversity loss. According to UNEP [1], over 55% of greenhouse gas emissions (GHG) and 40% of the health effects of particulate matter are caused by the extraction and processing of material resources. The transition to a circular economy is crucial for the achievement of global and national climate and environmental goals, as it reduces the use of primary resources and the associated emissions as well as the pressure on ecosystems and biodiversity.

The concept of circular economy has been at the center of attention in China for several years, which is also reflected in the work of the China Council for International Cooperation on Environment and Development (CCICED) to date.[1] Furthermore, China's 14th Five-Year Development Plan sets out key tasks and indicators to measure progress in the implementation of the circular economy in China.

The goal of this study is to support CCICED in the identification of relevant areas for further research and policy implementation in the field of circular economy in China. The study has the following specific objectives:

1. Identify and prioritize environmentally relevant sectors for a circular economy in China,
2. Define effective and appropriate circular economy measures in the identified sectors,

[1] In this regard, the CCICED discussion paper The New Era of Green Development—China's Green Transition to 2050 (2017), the Special Policy Study (SPS) Sustainable Consumption and Green Development (2013) and the SPS Major Green Technology Innovation and Implementation Mechanisms (2020) are few examples that emphasise the importance of a circular economy for reducing the environmental impact of production and consumption in China.

The content of this chapter is extracted from the main findings of the 2023-2024 CCICED Scoping Study on "Circular Economy".



*Adhering to Green and Open Cooperation to Advance Modernization*,
https://doi.org/10.1007/978-981-95-5547-5_11

3. Analyze the feasibility of modelling the ecological and socio-economic potentials of implementing a comprehensive circular economy in China (building on the experiences of the study "Modell Germany Circular Economy")
4. Define research areas for potential Special Policy Studies (SPS) on circular economy, and
5. Derive recommendations for the implementation of a comprehensive circular economy in China.

The study is based on a comprehensive literature review on the current status of circular economy implementation in China as well as interaction with and continuous feedback from international and Chinese circular economy experts (selected by CCICED), e.g., through regular meetings and an online consultation. Furthermore, interviews with circular economy experts in government, the private sector, academic institutions, and civil society in China, as well as a one-week field visit to Beijing from 13 May 2024 to 17 May 2024, also contributed substantially to the study.

The overarching goal of the study is to contribute to the development of policy recommendations to be considered in the 15th Five Year Development Plan and in the reform of the Circular Economy Promotion Law in China. The study also seeks to strengthen the bilateral cooperation between China and Germany in the field of circular economy as part of the Sino-German Environment Partnership [2].

## 2 Current State of Circular Economy Implementation in China

China has been the first country to introduce Circular Economy (CE) as a basic principle of its policy [3]. In 2005, a policy framework including principles, goals and key tasks was proposed ("Opinions on accelerating the development of circular economy"), followed by the 11th Five-Year Development Plan (FYDP), which mainly focused on resource productivity (especially energy). The Circular Economy Promotion Law (CEPL) was subsequently published in 2008, focusing on 3R strategies (Reduce, Reuse, Recycle), and later revised (2017) to place a greater emphasis on the circularity of industrial systems. Originally, the focus of CE in China was primarily on resource efficiency and end-of-life waste management. With regard to circular economy, the 14th Five-Year Development Plan (FYDP) (2021–2025) prioritizes resource productivity, recycling and waste utilization by setting national targets and clearly highlighting the importance of CE as an approach to tackling climate change and achieving carbon neutrality by 2060.

The interplay between resources and carbon neutrality has played an increasingly important role in the design of circular economy policies ([4]): The key targets of China's green low-carbon transition—known as the 'dual carbon goals' and also issued in the two key documents: "*Opinions of the Central Committee of the Communist Party of China and the State Council on Completely, Accurately, and Comprehensively Implementing the New Development Concept to Well Implement the Work of Carbon Peaking and Carbon Neutrality* (the Opinion) and Action Plan for Peaking Carbon Emissions before 2030—are to peak $CO_2$ emissions before 2030 and achieve carbon neutrality before 2060 [5]. Furthermore, the NDRC published the '*Guiding Opinions on Coordinating Energy Conservation, Carbon Reduction, and Recycling to Accelerate the Updating and Renovation of Product Equipment in Key Fields*' in 2023 (Chinese version cited in [5]). Therein, NDRC emphasizes the importance of 'large sorting, processing, and trading centers of wastes' including incentives for consumers such as trade-in and cash pledge, as well as green procurement and green labeling. Wind turbines, photovoltaic solar panels, and power batteries are mentioned, the recycling and reuse of which will be further standardized' [5]. Furthermore, CE in China includes a range of environmental indicators (e.g. energy and water consumption and pollutant emissions) and addresses key issues of growth and development [3, 6]. A recent study shows the major contribution that a CE can give to carbon reduction in China [7].

The National Development and Reform Commission (NDRC) is the leading authority in the development of CE regulations, other ministries are in charge of their respective programs [8]. CE projects are implemented at micro (enterprise/product), meso (industrial symbiosis, mainly through eco-industrial parks) and macro (provinces/country as a whole) levels [3, 9].

Another important driver is commodity security [8]: Due to its enormous growth since 1990 and its increasing global relevance in the production of goods, China has a high demand for commodity imports, which leads to a strong dependence on commodity prices and supply chain disruptions. This shows that a future CE development pathway could significantly reduce China's dependence on imports in addition to the ecological effects.

## 2.1 Achievements

China's resource productivity (i.e., the ratio of gross domestic product (GDP) to resource consumption) doubled between 1990 and 2015. According to Vlieger de Oliveira and Mahut [9], resource productivity increased by 26% between 2015 to 2020. As resource productivity represents the ratio of the production result (output) to resource input, this progress can be interpreted as a significant achievement towards a circular economy. At the same time, it only partially reflects the actual success in environmental and resource protection: the increase in resource productivity is mainly due to economic growth, as absolute resource consumption, including the environmental impacts thereof, increased over the same period [8].

In China, the circularity rate for non-metallic minerals increased from 2.7% to 5.8% between 1995 and 2015 from an input socioeconomic cycling rate perspective, reflecting the increasing share of secondary materials. Furthermore, circularity of non-metallic minerals in terms of 'output socio-economic cycling rate' (OSCr) rose from 7.2% to 17%, reflecting progress in industrial solid waste management [8, 10].[2] In comparison, 11.5% of the material resources used in the EU in 2022 came from recycled waste materials: Between 2010 and 2022, the rate increased from 10.7% to 11.5%. [11].

A report by EMF [12] estimates that a CE development pathway could reduce China's greenhouse gas emissions by 23% by 2040 and reduce the consumption of non-renewable resources by between 8% and 71% by 2040, depending on the focus area. For Germany, a study by Oeko-Institut [13] found that the measures in the ambitious scenario could reduce $CO_2$ emissions by 26% and raw material consumption by 27% compared to a business-as-usual in 2045.

## 2.2 *Existing Challenges for a CE in China*

Besides achievements and a tradition in the field of CE policies, there are also challenges that hinder its implementation. First of all, as CE is a multi- and cross-sectoral topic, there are coordination challenges among various ministries and agencies. This challenge is addressed by the establishment of the interministerial joint meeting mechanism by the Chinese government for coordinating the development of the circular economy. Annual meetings within this mechanism ensure a continuous dialogue and cooperation on policy coordination.

Further examples are challenges of implementing CE measures in resource-intensive industries that are fundamental to the economy, or arranging financial support [9]. It is noteworthy mentioning that the stated challenges are inherent to a multi- and cross-sectoral topic of a CE, and not necessarily restricted to China. Similar challenges can be identified in many other countries, including EU.

Coordination difficulties occur also at company, sector, and regional levels [8]. Uneven development is also an obstacle that relates to the asymmetry between sectors; while high-tech and export-oriented manufacturing sectors are more advanced in greening their operations, other sectors (e.g. agriculture, construction) are lagging behind [9]. There are also regional imbalances; for instance, the CE development level of 31 Chinese provinces was assessed by Fan und Fang [6] in 2017. The authors found that there are still large differences between regions and only 23% of the provinces are relatively efficient. Such implementation gaps exist due to "economic imperatives at the provincial level, bonus systems for policymakers if they meet their respective planning goals, a tax system favoring growth and budget issues over environmental concerns, and prevalence of environmentally-intensive 'zombie industries' that are kept alive due to social concerns" [3].

[2] In 2020, a comprehensive utilization rate of 86%, 56% and 50% was achieved for crop straw, bulk solid and construction waste respectively, as well as the utilization of 54.9 million tons of wastepaper and 260 million tons of scrap steel [9]. In 2015, resource recycling saved nearly 200 million tons of standard coal and reduced emissions of 9 billion tons of wastewater and 1.15 billion tons of solid wastes [6]. It is worth mentioning that eco-innovation has been an essential aspect.

Another obstacle is the failure of CE projects after initial success, implementation gaps and path dependencies [8]. Due to limited research funds, comprehensive evaluations of CE in China are still rare, as is quantitative systemic research on CE performance, especially concerning the differing circumstances of Chinese provinces [6]. Further challenges include the high level of investment required in recycling facilities, and the lack of public awareness of CE [14].

Bleischwitz et al. [8] also highlight implementation gaps between pioneers and transforming majorities, emphasizing that SMEs are lagging behind the large companies in moving towards a CE, as well as difficulties in creating a boosting market for secondary materials.

## 2.3 Raw Material Supply for Future Technologies

By expanding waste collection and recycling infrastructures and increasing the use of secondary raw materials, a circular economy contributes significantly to the security of raw material supply, independence from raw material imports and the reduction of the environmental impact of primary raw material extraction and processing.

On the one hand, more than a third of all materials (around 31.3 billion tons) were extracted in China in 2020, on the other hand, the country was the largest net importer of materials (over 2000 million tons) in the same year [1]. Recycling can be a reliable and sustainable supply of secondary raw materials, and product design can consider the use of secondary raw materials as well as design for recycling.

Due to its role in raw material extraction and processing, China is a hotspot country for greenhouse gas emissions, harmful particulate matter emissions from industrial sources and other impacts [1]. The further development of critical technologies implies additional raw material demand and environmental impacts. For example, China's consumption of rare-earth minerals grew at an average annual rate of 7.5% between 2004 and 2014, increasing its share of global consumption from 43 to 70%; and the production of rare-earth end-use products grew by about 70% between 2005 and 2015 [15].

## 2.4 Gender Aspects of a Circular Economy

Globally, women play a big role in the waste management sector and in circular economy (CE). With regards to gender aspects of CE, gender-specific behaviour in consumption, e.g., in terms of expanding products' lifespan through repair, and in waste segregation and disposal can be distinguished from gender-specific jobs in production and manufacturing.

With regards to waste management, a global online survey [16] found that women reported "a sense of purpose working in the waste sector". However, women are predominantly represented in the low-value areas of waste treatment and are under-represented in product design, management and the development of advanced technologies. In addition, female entrepreneurs in the waste sector reported difficulties in obtaining loans. In this study, interviewees reported that recycling and transport are dominated by male workers in China. However, since the aspect of waste segregation has become more prominent, female workers are also beginning to play a role: It has been reported that women are working in the position of teachers or multipliers who provide training and raise public awareness about waste segregation. This observation is supported by UN Women [17]. An ongoing EU project could serve as best practice to accelerate this trend: '***Girls Go Circular***' supports "schoolgirls, and more broadly, any student, to develop their digital and leadership skills while learning about the circular economy and finding solutions for a sustainable future."

In terms of gender-specific behaviour when buying and handling items, according to UN Women [17] and others, women make more sustainable consumption decisions and are more sensitive to ecological, environmental and health issues than men. Research has shown that they are more likely to make environmentally friendly purchases, minimise waste and recycle.

From a broader point of view, gender aspects in CE could be considered together with other aspects such as the formalization of informal sectors under the umbrella of "just transition".

# 3 Sectoral Perspectives

## *3.1 Selection of Environmentally Relevant Sectors in China*

A global, Multi-Regional Input-Output (MRIO) model was used for an initial screening of the relevant industrial sectors. For this study, sector-specific, consumption-based environmental impacts were estimated for three different impact categories: raw material consumption (RMC), global temperature potential (GTP) and land use (LU). This was done using EXIOBASE v3.8.1 [18, 19] in conjunction with the python package 'pymrio' [20]. Version 2022 of EXIOBASE was used for this study. The resulting uncertainties, which add to the inherent uncertainties of MRIO tables in general, were considered acceptable for an initial, economy-wide screening. However, the figures given in this section are not to be understood as precise calculations, but rather serve as rough estimates. The estimates for RMC, GTP and LU were validated against literature values and expert judgement.

The RMC in China for 2022 amounts to roughly 28.5 Gt, with 77% originating from abiotic resources such as minerals, metals and fossil fuels and 23% stemming from biotic resource extraction such as food and fodder crops or wood products. The most relevant individual sectors were the construction industry contributing with 36.7%, the health and social work sector (14.5%) and cattle farming (5.8%).

Together, the 20 most important sectors account for around 89% of Chinese raw material consumption.

The GTP of Chinese consumption was estimated at 12.8 Gt $CO_2$-eq for 2022. The most relevant sectors were again construction, health and social services and cattle farming. The top 20 industries contribute to 88% of China's GHG emissions, with the construction sector alone accounting for around 43% of emissions.

Cattle farming accounts for the largest share of the land use footprint at 18.7%, followed by the construction sector (13.3%) and health and social work (7.6%). Together, the top 20 sectors account for around 82% of Chinese land use.

A longlist of sectors was compiled considering the results of the MRIO analysis as well as the expected future technologies and material demands (see chapter "Raw material supply for future technologies"). It included: Construction, (Manufacture of) vehicles and batteries, (Manufacture of) wearing apparel/textiles, Plastics/packaging, Electric and electronic equipment and machinery, Health sector, Hotels and gastronomy, Metallurgy, Agriculture and farming, Renewable energy installations, Digitalization and Artificial Intelligence, Retail & trade, Robotics and Additive manufacturing and 3D-printing.

## *3.2 Proposed Sectoral Focus for Special Policy Studies (SPS)*

For the purpose of a more targeted approach to be pursued under the potential SPS on circular economy, the following criteria were applied for further sectoral prioritization:

- Environmental relevance,
- Policy relevance: Synergies with the development in the EU product policy (desk research),
- Technology relevance: Future trends and technology development,
- Feasibility of implementation,
- Specific interest shown by the experts during the interviews (expert judgement),
- Complexity of the sector.

Following the evaluation of the sectors based on the above-mentioned criteria, the **following sectors were recommended to be focused on in a possible SPS: (1) Plastics (incl. packaging), (2) Batteries, (3) Renewable energy technologies, (4) Textiles and (5) Construction.**

In addition, a cross-cutting topic was identified for modelling the impacts of a circular economy in the industry sectors to demonstrate the contribution of the circular economy to achieving the climate neutrality target of China. This topic is discussed further in Chap. 4.2.

## *3.3 Specific Development and Challenges for Selected Sectors in China*

This chapter provides a rationale for the prioritization of four sectors for a possible sectoral SPS.

### 3.3.1 Plastics/Packaging

China is the largest manufacturer, user, and exporter of plastics in the world. In 2007, China introduced its first ban on plastic, making it one of the earliest countries globally to address plastic pollution. The average annual growth rate of China's packaging production between 2018 and 2020 was 8.7%. Of the 60 million tons of plastic waste generated in 2020, approximately 35% were landfilled, 37% incinerated, and only 26.7% recycled [21]. Building a circular value chain for plastics is important for China's plastics objectives as well as for carbon emissions and broader environmental goals [21]. In the 14th Five-Year Plan, China has set the following targets for the plastics sector: (1) Reduce single-use plastic production and consumption at source, (2) Improve plastic waste collection and recycling, (3) Establish the whole chain management system of plastic pollution, (4) Reduce plastic waste leakage [21]. Since 2016, the efforts to address plastic pollution have increased significantly, covering a wide range of policy options, such as economic incentives, building recycling facilities, enhancing resource utilization infrastructure and implementing awareness campaigns. In recent years, there has been a further diversification in the policy instruments used, the types of plastics addressed and the aspects of the value chain considered, as well as a clear intention to build a circular plastics value chain [22]. In the coming years, a stronger regulatory focus on initiatives to eliminate plastic pollution is expected (with the aim of reducing the plastic leakage into the environment, especially into the oceans) [22]. There is also a trend towards the use of biodegradable plastics, which is not without problems [23]. In addition, there is a strategy to promote bamboo as a substitute to plastic [24]. The boom of e-commerce and increase of (food) deliveries has led to a significant increase in packaging waste. These sectors have therefore been the focus of several studies and political measures Other policy priorities were dealing with agricultural plastic film, marine litter and microplastics [25].

Overall, a thorough assessment to comprehensively evaluate the effectiveness of policies for addressing plastic pollution in China is required. While some experts suggest that China's comprehensive plastic pollution control policies operates within a closed-loop logic system, including source reduction, ecodesign principles, process substitution, and end-of-life recycling, and has been very effective in reducing plastic pollution, other sources suggest that the plastic packaging industry is not fully integrated into current policy initiatives (e.g. initiatives for zero-waste cities or waste sorting schemes) and the focus has mostly been on downstream measures (such as recycling) [21, 22]. Although regulations are in force in some regions, there is a lack of implementation and penalty (e.g. for SUPs used in food delivery; [26]). High attention given to downstream measures is, however, justified, considering the challenges related to the waste management in China's specific context.

In addition to packaging, chemicals in plastic products also give rise to health concerns, e.g., due to plasticizers (phthalates, bisphenols, e.g. BPA, BPS, BPF), flame retardants, poly- and perfluorinated alkyl substances (PFAS), etc. In 2019, the European Environment Agency issued a warning about plastic children's toys imported from China that contained dangerous chemicals, for example.

### 3.3.2 Batteries

It is expected that the number of private vehicles in China will continue to increase [27]. The stock of electric vehicles is also increasing significantly [28]. There may be restrictions on the sale of fuel cars, as some cities in China have already introduced [27, 29]. Such restrictions could reduce the current growth of the car stock in China, but enhance the share of electric vehicles and thus batteries [27, 30]. The risks of metal supply (apart from lithium) can be reduced by cobalt-free battery technology in combination with efficient recycling systems [31]. The supplementary capacity of second-life EV batteries may be useful for China's prospective novel energy storage applications [31]. However, Wesselkämper et al. [30] argue that there is a conflict between circularity and a longer product life, so that the benefits of a second use (reuse) need to be evaluated depending on the prioritization of objectives.

The switch from combustion engines to battery-powered electric vehicles poses new challenges for the treatment and recycling infrastructure and capacities. In macroeconomic terms, this transition will lead to a lower demand for primary raw materials, as less fossil fuels will be needed. In the vehicle/battery sector, however, critical and strategic metals in particular will initially be in high demand. Cars are getting bigger and heavier, both in terms of combustion engines and electric vehicles.[3] Currently there is little incentive to take ecological assessments into account in the design of vehicles. The recycling rate of end-of-life vehicles is low. In 2018, the recovery rate for end-of-life passenger vehicles was less than 18% of the quantity scrapped [32].

### 3.3.3 Renewable Energy Installations

China has set itself ambitious targets for the expansion of renewable energies, and capacity is expanding rapidly [33]. An increased volume of EOL wind turbines and photovoltaic modules are therefore expected in the coming years [34, 35]. The waste

[3] The International Council on Clean Transportation (2023): Trends of New Passenger Cars in China: Air Pollutant and CO2-Emissions and Technologies, 2012–2021, Yuntian Zhang, Hui He and Zhinan Chen, https://theicct.org/wp-content/uploads/2023/03/China-PV-trends_final-v2.pdf; accessed: 13.09.2024

streams may be unevenly distributed across the country, e.g. for PV waste. Wang et al. [36] find that there will be a concentration in the northern or north-western regions. In terms of timing, it is found that over 80% of the total cumulative PV waste will be generated between 2040–2050. In 2023, the Chinese authorities published guidelines to promote the recycling of decommissioned wind power and photovoltaic equipment [37]. The publication of standards for the recycling of wind turbines was announced for January 2024 [38].

There are challenges for both PV and wind turbines. Currently, there is a lack of technologies and capacities for cost-competitive, high-quality recycling of wind turbine rotor blades [35]. This also applies to large-scale infrastructures for the reuse and recycling of solar photovoltaic modules, which are hampered by high costs for the recycling of the modules and therefore low economic benefits [39]. Zhou et al. [40] state that innovating, or improving the existing PV recycling process, and encouraging low-energy and low-emission technologies could help to reduce recovery costs and increase recovery income, thereby stimulating market interest. Furthermore, there are other bottlenecks in PV recycling: the current concentration of PV waste recycling in small enterprises and self-employed people (without environmental protection measures and with corresponding environmental risks and material losses), the wide dispersion of PV modules in China, and the lack of clear policy guidance for PV panel decommissioning (which results in difficult access to recycling, complex cross-regional flows, inconsistent recycling technologies, unsmooth circulation routes and other problems) [40].

Furthermore, heat pumps are considered an important technology for the decarbonization of heating in China [41–43]. Circular design and circular business models are therefore necessary for this technology to help to reduce resource consumption.

### 3.3.4 Textiles

In 2014, China accounted for 55% of global textile production and generated 70% of the world's synthetic fibres. China is not only "the world's largest fibre producer, but it is also the world's largest textiles exporter, and the largest textiles machinery manufacturer" [12]. China's rapid economic growth and urbanisation coupled with a growing middle class, "tripled the domestic demand for textiles and will grow threefold yet again over the next decade" [12]. It is predicted that China's clothing sales (2016: 6.5 kg) may rise to 11–16 kg per person by 2030, which would put China on a par with North America (EMF 2018). Overall, the linear 'take-make-dispose' approach dominates the production and consumption of textiles. There is a trend towards the relocation of textiles manufacturing plants from the east of the country to western and central regions to counteract rising wages and production costs [12]. Textiles are also an e-commerce trend, "with an online shopping adoption rate of almost 60%" [12].

Of China's textile waste (20–26 million tons annually), only a small proportion is recycled [44]. Government announced in 2022 that the recycling rate for used textiles should be 25% by 2025. By 2030, a rate of 30% and three million tons of recycled fibres are to be achieved.[4] The Chinese state has promoted the recycling of used textiles in recent years, for example through tax concessions and financial support. However, there is still a lack of a comprehensive recycling system and efficient utilisation of textile waste [45, 46].

In 2018, the textiles industry was the third largest water consumer in the country, consuming approximately three trillion litres each year [12]. In addition to water consumption, there is also the water pollution caused by the dyeing of textiles and other treatment processes. In the last decade, however, research has focussed intensively on the treatment of textile wastewater [47].

The prevailing market conditions and the cost structure have resulted in short-life textile products being imported and offered at low prices mainly by bulk buyers from countries with lower labour costs and low labour and environmental standards [48]. There is an excessive focus on the export sector, and a lack of effective control over the import and export of textile raw materials and products [48]. In addition, enterprises lack their own financial resources as well as access to bank loans, which are necessary to upgrade manufacturing [48]. As in all other parts of the world, circular business models are largely unable to compete with this global supply market. There are low prices for new goods, short-lived designs and a limited willingness or ability to repair textiles and use second-hand goods (low societal appreciation of textiles). A lack of take-back systems and low economic efficiency in sorting hinder the optimization of the collection and recycling of used textiles. Thus, there is a lack of incentives for technical innovations and investments at the end of the life cycle of textiles and ultimately for a circular design.

### 3.3.5 Construction

The urbanization trend in China is at an unprecedented pace, resulting in substantial demand for new housing and infrastructure. In 2016, the built-up area under construction and completed built-up area accounted for half of the world's new construction; and by 2030, China's urban population will reach one billion people ([12]). At the same time, the average lifespan of a building is only 25–30 years, which is due to the use of low-quality materials to reduce construction costs ([12]). The building sector is characterized by a high consumption of primary materials (e.g., China produces and consumes around 55% of cement globally), resulting in significant greenhouse gas emissions. GHG emissions from concrete, steel, bricks and aluminium account for around 90% of the GHG emissions from buildings [49]. China now accounts for more than 18% of total $CO_2$ emissions in the global building sector (having surpassed the EU in 2011) [12].

Due to the fast-advancing urbanization and short lifespans of buildings, about 2.36 billion tons of construction and demolition waste (CDW) were produced annually in

[4] State Council of the People's Republic of China (2022): China to up its textile recycling capability, https://english.www.gov.cn/statecouncil/ministries/202204/20/content_WS625f649fc6d02e5335329a8f.html; Accessed: 13.09.2024

the period 2003–2013 [50]. This corresponds to 30 to 40% of the total amount of waste generated in China [51]. Only about 5% is recycled or reused [51]. Furthermore, most of this share is reused for road gravel instead of being utilized in the construction industry. Therefore, the construction sector faces multiple challenges, in particular the fact that the preservation and continued use of existing buildings and infrastructure is not prioritized over resource- and space-intensive new construction. Most existing regulations target treatment rather than the reduction of CDW, and few regulations require CDW reduction during the architectural design stage [51]. Furthermore, there is an under-developed market for reused or recycled CDW products [51]. As there is little awareness of reuse, deconstructability and recycling during the construction of buildings, it is often not possible to provide unmixed material streams from the secondary raw material store.

# 4 Recommendations for Potential Special Policy Studies (SPS)

A comprehensive long-list of potential topics for SPS had been developed based on desk research, exchange with the international and Chinese experts in the CE Scoping Study working group and direct interviews with several stakeholders before and during the field visit of Oeko-Institute's experts in Beijing. Applying the criteria mentioned in '3.2 Proposed sectoral focus for Special Policy Studies (SPS)', a short-list had been selected and was submitted to the experts for their assessment.

Figure 1 shows experts' ranking of the short-list of potential topics for SPS. The two topics with the highest ranking are "CE in industry decarbonization" and "Extended Producer Responsibility (EPR)" in terms of its relevance followed by one or more topics per selected sector.

## *4.1 Sectoral SPS*

Based on the experts' estimation of relevance and the selection of environmentally relevant sectors, the six most important sectoral SPS topics were proposed (Table 1).

## *4.2 Cross-Sectoral SPS*

Given China's ambitious policy goal of achieving carbon neutrality by 2060 and the high level of interest shown by the experts, it is recommended to model the impact of the circular economy to achieve industry decarbonization and climate neutrality in China. All experts consulted believe that such a study would be highly relevant

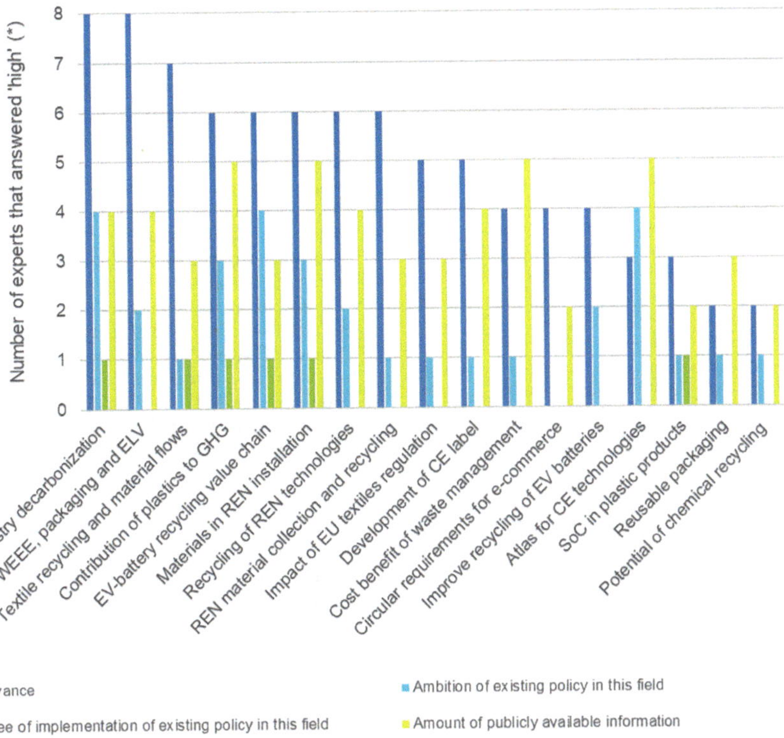


**Fig. 1** Expert ranking of possible SPS topics. *Note* (*) Experts had four options to answer ('high', 'medium', 'low', 'I don't know'). *Source* Expert consultation in the course of this study

(Fig. 1). According to the experts, a comprehensive and professional study on this topic does not yet exist in China. They are of the view that CE is an established policy area in China, whereas the link to decarbonization and climate neutrality has only recently emerged. Hence, policy implementation has so far focused on a few topics such as industrial energy efficiency. Examining both topics together in a quantitative analysis, showing the full range of circular economy measures and the impact of their implementation, could provide a 'boost' that seems to be significant.

To assess the feasibility of a comprehensive CE modelling, several macroeconomic modelling studies and their respective data basis for China were analyzed. In general, computable Global Equilibrium Models (CGE), Integrated Assessment Models (IAM), Environmentally-Extended, Multi-Regional Input-Output models (EE-MRIO) and Dynamic Material Flow Analysis (MFA) are all suitable model classes to evaluate circular economy measures on a national scale. A universal bottleneck for all model classes remains the availability of data, which is a particular challenge for an economy as diverse and rapidly changing as the Chinese economy. Experts point out that publicly available information in English is poor, while the availability of information in Chinese is somewhat better. For bottom-up modelling,

**Table 1** Six most relevant sectoral research questions for SPS on CE

| Research question | Considerations based on experts' feedback |
|---|---|
| Analyze the contribution of the plastic industry (incl. packaging) to GHG emissions/carbon budget in China and assess the potential for reduction | • Urgent need to decarbonize the plastic industry as well, move away from fossil-based feedstocks and replace them with bio-based and especially secondary resources<br>• Plastic packaging and packaging in online deliveries should be included<br>• The macro data on plastic consumption in China is available, obtaining this data will be difficult for the SPS itself<br>• Beyond the contribution to GHG emissions, efforts should be made to analyze in detail the impact of the plastics industry on other types of pollution (e.g. particulate matter or chemical pollution) and the impact on biodiversity, e.g. through modelling studies such as The PEW charitable trusts; systemiq [52] that could be built upon and applied specifically to China |
| Map the value chain of EV-battery recycling in China and analyze the material and economic flows as well as costs and benefits of EV battery recycling in the value chain | • The EV-battery recycling sector in China is developing rapidly [30]. The current battery recycling capacity amounts to more than 500,000 metric tons. The 14th FYDP sets the goal of establishing a complete battery recycling system by 2025<br>• China is improving its regulatory framework to better trace and manage the battery value chain. The Ministry of Industry and Information Technology (MIIT) has introduced guidelines to improve traceability and transparency in the production and recycling of batteries [53, 54]<br>• Such a study could focus on a better understanding and identification of optimization potential, sustainable technology adoption, standardisation, environmental management and system integration (reuse in the energy transition) |

(continued)

**Table 1** (continued)

| Research question | Considerations based on experts' feedback |
|---|---|
| Policy instruments to improve the collection and recycling of raw materials from renewable energy technologies (REN) | • This is a topic where decarbonisation and CE go hand in hand. Experts say that policy should focus on CE in manufacturing in general rather than in the manufacturing of REN, and that measures that contribute to decarbonising REN production could be identified and their potential explored<br>• The focus of such a study should be determined based on literature review and existing studies focussing on China [55–57]. There is also increasing publicly available information on material demand [58]. The choice of topic should take into account that strategies that minimize the need for future energy infrastructure and thus improve strategic autonomy and reduce pressure on critical mineral supplies should be prioritized., UNEP [1], for example, shows that with smart city design, and optimized access to service it would be possible to reduce global energy demand by 27% by 2060. This study could also include the collection and recycling of raw material from renewable energy installations<br>• China requires all renewable energy projects, including wind, solar, and hydro installations, to undergo rigorous environmental impact assessments (EIA) before approval<br>• Out of the three different suggestions for potential SPS related to REN, exploring policy instruments such as design-for-recycling, take back obligations, deposit return schemes etc. seems to be of highest importance, aspects such as, reuse, repurposing and remanufacturing should also be analyzed closely<br>• Potential to look on reuse (e.g. rare earth elements) or carbon fiber remanufacturing / reuse<br>• Experience from Germany shows that typical instruments (deposit returns, technical restrictions) don't work due to dynamic technology development. Hence, requirements for future contracts and management shall be developed considering the dynamic market structure of REN |

(continued)

**Table 1** (continued)

| Research question | Considerations based on experts' feedback |
|---|---|
| Map the technologies for textile recycling and analyze the market and material flows of waste textiles, economic operators, environmental and economic potential of introducing fibre-to-fibre recycling in China | • Given the large textile production capacities in China, it would be worthwhile to increase the output quantities and qualities of textile recycling as there is a huge market for the uptake of secondary fibre-to-fibre recyclate<br>• In order to not only look at end-of-pipe solutions such as recycling. The topic could be explored in combination with a focus on ecodesign, in particular, the impact of the EU textiles regulation on the fashion and textile industry in CN, and provide a knowledge transfer on the EU Ecodesign for sustainable products regulation |
| Feasibility study on implementing extended producer responsibility (EPR) in waste electrical and electronic equipment (WEEE), packaging waste and end-of-life-vehicles (ELV) in China. | • There have been studies for EPR in China in general [53, 54] and in some sectors (e.g., batteries, [59])<br>• Future EPR research for China should focus on overcoming barriers such as standardizing recycling systems, and upscaling access to recycling infrastructure<br>• Understanding the Chinese-specific context will be critical in designing implementable and effective EPRs |
| Assess the feasibility of circular public procurement in the building and construction sector in China and draw conclusions for circular public procurement in general | • It is generally acknowledged that public procurement has great market power, which can be used to reduce environmental impact and test new approaches. However, there are still challenges to making procurement not only sustainable but also circular, which can be studied in the building sector<br>• The environmental impact of the building sector provides a solid rationale for the selection of potential SPS, e.g., Chap. 3.1, e.g., construction and demolition waste (CDW) accounts for 30% to 40% of the total amount of waste in China [51]. Furthermore, it is selected due to its large relevance for public procurement<br>• As potential SPS should build on existing studies, e.g., Huang et al. [60], and select the scope based on the existing, relevant waste categories. One of the potential study's objectives could be the identification of measures to reduce CDW of relevant waste categories from public real estate to provide best practices<br>• The aim is to learn from these best practices examples which standard procedures and monitoring framework could be derived for circular public procurement |

This sectoral research question for a potential SPS was developed at the end of the study based on the feedback from the final expert workshop. Therefore, it is not explicitly included in the long and short lists of potential SPS topics in the appendix process-based LCA is a theoretical option. It is well suited for the assessment of individual products or materials, provided the data is available, but does not translate well to systems as large as the Chinese economy. This is due to the fact that the system boundary is inevitably cut off, which means that a significant part of the economy is not covered. However, sub-sectors and key processes that influence environmental impacts, such as cement or steel production, can be quantified with a process-based LCA. The LCA results could be used for comparison with the macro-scale modelling results. All methods have their individual drawbacks and benefits, therefore a hybrid approach of several model classes should be used to analyse systems as complex as the Chinese economy, as it was done, for example in work done by Circle Economy [61] and UNEP [1].

The feasibility analysis of the CE modelling reveals that comprehensive macroeconomic modelling for China is theoretically possible. However, quantitative models can only complement qualitative research if data is available. A coarse-grained analysis of individual simplified scenarios can be carried out using macroeconomic Input-Output models, but stand-alone results have limited explanatory power and may therefore require complementary modelling approaches. A simple approach to estimating circularity potential is stock modelling, as this only requires time series data on sales, waste streams and average lifespans.

Given the availability of resources and data, it is recommended to focus on the most relevant sectors for a targeted CE modelling in the context of China.

## 5 Policy Recommendations

Based on desk research, the exchange with international and Chinese experts in the CE Scoping Study working group and direct interviews with various stakeholders before and during the field visit of Oeko-Institut experts in Beijing, eleven overarching policy recommendations were identified for consideration in the 15th Five Year Development Plan and the reform Circular Economy Promotion Law in China. Based on feedback from Chinese and international experts (online consultation), the following three main policy recommendations are presented:

Recommendation 1—Continue setting numerical targets and indicators including a resource productivity target to enable responsible consumption of raw materials in China.

China's increase in resource productivity (see 2 Current state of circular economy implementation in China) shows its considerable progress in relatively decoupling the resource consumption from economic growth.[5] For the 15. FYDP, it is recommended to continue setting numerical targets for resource productivity, adding waste reduction targets in relevant sectors, mainstreaming the implementation of waste hierarchy principles (R-Strategies), defining a clear hierarchy of different recycling options based on their environmental impact and recycling, and targets for recycled content at sectoral and material-level. As resources and their consumption patterns are so diverse, progress in resource policy cannot be monitored using a single indicator.

Generally, targets can contribute to actively transparently communicating the necessary, structural social change and setting a clear direction. Of course, social needs must not be forgotten in the process. Similar to the discussion on compensating for the social consequences of climate protection measures, resource policy should also be accompanied by social measures.

Recommendation 2—Implement ambitious framework-setting instruments for creating a level-playing field for Circular Business Models.

In view of China's increasingly important role as manufacturer of final products, more and more strategic decisions are being made in China on the circularity of global material flows. The mix of instruments to enforce circular products and technologies in the mass market should include the following[6]:

(a) Minimum ecodesign product standards (using the example of the EU Ecodesign for Sustainable Products Regulation). As the design of a product has a major influence on the later impact of that very product on the environment, eco-design is a crucial prerequisite for more circular value chains.
(b) Ensure that all legal requirements at all levels of government are consistently and effectively geared towards circular public procurement. Experts estimate that due to the sheer volume of public projects in China, mandatory circularity standards would have a major impact.
(c) Mandatory Extended Producer Responsibility (EPR) schemes for selected sectors, such as batteries, textiles, plastics and electronics. Extended producer responsibility is key to the transformation towards a circular economy—not only because it generates funding for waste management infrastructure, but also because it leads to economic incentives for companies to improve the circularity of their systems. In 2016, China introduced its EPR system, which obliges producers to assume responsibility for waste. It covers a wide range of products,

[5] Relative decoupling: "In relative decoupling the growth rate of the environmentally relevant parameter (such as resources used or environmental impact) is lower than the growth rate of the relevant economic indicator (for example GDP)", UNEP [1].

[6] As for the policy tools, such as the Extended Producer Responsibility (EPR) system and eco-design, related documents, such as (State Council Document [2016] No. 99) and (Ministry of Industry and Information Technology Document [2013] No. 58), serve as a good starting point for further analysis.

including electronics, batteries, vehicles, and packaging materials. Implementation is progressing, but there are still other challenges, such as the complex institutional design, the introduction of EPR, as recycling technology and public awareness are not the same as in other parts of the world. It is understandable that in a large country with regional differences and increasing consumption, the set-up is complex. However, building upon existing ambitious policies, best practice in other countries, scientific studies (see potential topics for SPS), and in combination with awareness trainings and incentives, EPR will greatly promote circularity in China.

(d) Financing mechanisms to mobilize private capital for circular investments (e.g., loans, transformation bonds, public-private partnerships, start-up financing, etc.). Experts say that currently, the diffusion of CE strategies and technologies to SMEs is seen as a weak point in the overall implementation of CE in China. So fiscal instruments would likely have a great impact, but it might prove difficult to implement them and ensure that they reach the SMEs that need them most. There also seems to be great potential for policy measures for SMEs.

For some of these instruments, there are already policies or pilot projects in China, for others not. They need to be (further) developed and will ultimately require consistent implementation.

Recommendation 3—Link Circular Economy to China's industrial decarbonization and climate neutrality targets.

Experts agree that linking CE to climate targets and implementation strategies would be very effective to combat climate change. Tools such as the UNDP/UNEP/UNFCCC's toolbox for integrating CE into NDCs could be used. The link between CE and the goal of climate neutrality has been recognized in several high-level policy documents in China. However, according to experts, there is a gap in how CE is translated into targets for specific sectors to guide implementation, and a roadmap for the future development of CE in China is lacking.

## 6 Conclusion

There is broad agreement on the relevance of the circular economy in China as well as on the successes and challenges related to its implementation. Political recognition of circular economy in terms of its contribution towards high-level policy goals, such as carbon neutrality, has already been explicitly stated in the important government opinions and action plans. Against this background, this study has built upon the existing knowledge on circular economy in China and has highlighted several areas of interventions and research, keeping in mind the environmental relevance of the proposals as well as future technology development and interests of a wide range of actors. Thus, all recommendations for potential Special Policy Studies (SPS) are justified from one angle or another. From the meta perspective, the link between CE and the dual goal of carbon neutrality in China by 2060 was supported by the majority of interviewees and consulted experts to be a positive one. Consequently, this is reflected in the cross-sectoral SPS recommendation on a study on modelling CE's contribution towards decarbonization and carbon neutrality in China. However, depending on the budget and time schedule for the SPS, a CE modelling for selected relevant sectors can also be considered, as it is recommended in the sectoral SPS on plastics.

Modelling CE at the economy or sectoral level is just one of many other identified topics in this study. Improving the information, knowledge and policy instruments for environmentally relevant sectors, such as textiles, EV-batteries and renewable energy installations, offer substantial research opportunities and thus have been recommended at the level of sectoral SPS. This also applies to the construction sector and enormous potentials of improvement that can be channelled in this sector using the instrument of (circular) public procurement, which is already well established in China.

In general, the range of topics identified in this study can be combined in different levels of detail to advance the discourse of CE in China. For instance, the instrument of (circular) public procurement can also be further developed for other sectors, such as textiles and plastics/packaging. The selection of potential SPS topics can also be guided by the availability of data. This may apply to the recommendations on plastics or textiles.

In addition, the study makes three high-level policy recommendations for consideration in the 15th Five Year Development Plan and the reform of China's Circular Economy Promotion Law. Based on the experiences of the authors, experts and interviewees as well as scientific studies, the combination of proposed targets and instruments represents a promising policy mix for mainstreaming CE in China.

It is important to emphasize that all sectoral and policy recommendations made in this study support the focus of bilateral cooperation agreed upon between China and Germany in the field of circular economy [2]. The framework of the Sino-German Environment Partnership provides a solid basis for the follow-up of the identified set of potential research questions and topics of collaboration. The collaboration can benefit from the mutual exchange on success and challenges in implementing CE in both the countries.

## References

1. UNEP—United Nations environment Programme. (Ed.). (2024). Global resources outlook 2024: Bend the trend, pathways to a livable planet as resource use spikes. In collaboration with international resource panel (IRP). Online available at http://unep.org/resources/Global-Resource-Outlook-2024. Last accessed on 27 Aug 2024.
2. Bundesministerium für Umwelt, Naturschutz, Nukleare SIcherheit und Verbraucherschutz and State Commission for Development and Reform (Ed.). (2024). Sino-German action plan on

circular economy and resource efficiency, 16 Apr 2024. Online available at https://www.bmuv.de/en/download/sino-german-action-plan-on-circular-economy-and-resource-efficiency. Last accessed on 28 Aug 2024.

3. OECD; EC—European Commission (2019): Ekins, P., Domenech, T., Drummond, P., Bleischwitz, R., Hughes, N., Lotti, L. The Circula Economy: What, Why, How and Where. Background paper for an OECD/EC Workshop on 5 July 2019 Within the workshop series "managing environmental and energy transitions for regions and cities". Paris. Online available at https://www.oecd.org/cfe/regionaldevelopment/Ekins-2019-Circular-Economy-What-Why-How-Where.pdf. Last accessed on 22 Mar 2024.
4. Wang, Y., Guo, C., Chen, X., Jia, L., Guo, X., Chen, R., Zhang, M., Chen, Z., & Wang, H. (2021). Carbon peak and carbon neutrality in China: Goals, implementation path and prospects. *China Geology, 4*(4), 720–746. https://doi.org/10.31035/cg2021083
5. UNCTAD—United Nations conference on trade and development. (Ed.). (2023). China's policy strategies for green low-carbon development, perspective from south-south cooperation. UNCTAD/GDS/2023/6. Online available at https://unctad.org/system/files/official-document/gds2023d6_en.pdf. Last accessed on 28 Aug 2024.
6. Fan, Y., Fang, C. (2020). Circular economy development in China-current situation, evaluation and policy implications. *Environmental Impact Assessment Review, 84.* Online available at https://www.sciencedirect.com/science/article/abs/pii/S0195925520301943. Last accessed on 22 Mar 2024.
7. CACE—China Association of Circular Economy (Ed.). (2023b): Zhanqiang, G., Junxia, L., Lijia, J., Yang, W., Xiangci, D., Yingying, Fan, Jingjing, W. Research report on circular economy supporting peak carbon dioxide emissions. In collaboration with Jiarong, Z., Liyang, Z., Zhanqiang, G., Kai, Z., Jisheng, L. et al.
8. Bleischwitz, R., Yang, M., Huang, B., Xu, X., Zhou, J., McDowall, W., Andrews-Speed, P., Liu, Z., Yong, G. (2022): The circular economy in China: Achievements, challenges and potential implications for decarbonization. In: *Resources, conservation and recycling* (Vol. 183). Online available at https://www.sciencedirect.com/science/article/pii/S0921344922001951?via%3Dihub. Last accessed on 8 Mar 2024.
9. Vlieger de Oliveira, S., Mahut, C. (Ed.). (2023a): Chen, R. China's circular economy policies: Review and reflection, a circular economy vision (circular press, 4). Circular innovation lab. Copenhagen, Denmark. Online available at https://circulareconomy.europa.eu/platform/sites/default/files/2023-08/China%E2%80%99s%20Circular%20Economy%20Policies_%20Review%20and%20Reflection.pdf. Last accessed on 22 Mar 2024.
10. Wang, R., Deng, Y., Li, S., Yu, K., Liu, Y., Shang, M., Wang, J., Shu, J., Sun, Z., Chen, M., & Liang, Q. (2021). Waste electrical and electronic equipment reutilization in China. *Sustainability, 13*(20), 11433. https://doi.org/10.3390/su132011433
11. Eurostat (Ed.). (2023). EU's circular material use rate slightly up in 2022. Online available at https://ec.europa.eu/eurostat/de/web/products-eurostat-news/w/ddn-20231114-2. Last accessed on 27 Aug 2024.
12. EMF—Ellen MacArthur Foundation (Ed.). (2018). The circular economy opportunity for urban and industrial innovation in China. Ellen MacArthur Foundation, ARUP, McKinsey & company, United Nations Conference on Trade and Development. Online available at https://www.ellenmacarthurfoundation.org/urban-and-industrial-innovation-in-china. Last accessed on 8 Mar 2024.
13. Oeko-Institut (Ed.). (2023). Prakash, S., Löw, C., Antony, F., Dehoust, G., Stuber-Rousselle, K., Liu, R., Gascón Castillero, L., López Hernandez, V., Hurst, K., Köhler, A., Schön-Blume, N., Loibl, A., Sievers, L., et al. Modell Deutschland Circular Economy, Modellierung und Folgeabschätzung einer Circular Economy in 9 Sektoren in Deutschland. Im Auftrag des WWF Deutschland. Oeko-Institut in Zusammenarbeit mit Fraunhofer ISI und FU-Berlin. Online available at https://www.oeko.de/fileadmin/oekodoc/MDCE_Modellierung.pdf. Last accessed on 8 Mar 2024.
14. Vlieger de Oliveira, S., Mahut, C. (Ed.). (2023b). Chen, R. China's circular economy transition: Challenges and solutions ahead, a circular economy vision

(circular press, 6). Circular innovation lab. Copenhagen, Denmark. Online available at https://circulareconomy.europa.eu/platform/sites/default/files/2023-08/China%27s%20Circular%20Economy%20Transition_%20Challenges%20and%20Solutions%20Ahead.pdf. Last accessed on 22 Mar 2024.
15. CSIS—Center for Strategic and International Studies (2021): Nakano, J. The Chinese Dominance of the Global Critical Minerals Supply Chains. Online available at https://www.jstor.org/stable/resrep30033.4.
16. ISWA Women of Waste Task Force. (Ed.). (2023): Godfrey, L., Tsakona, M., Nitzsche, G., Khaled, D., Garcés-Sánchez, G. Findings of the WOW! Global survey II, mapping the status of women in the global waste management sector. Rotterdam. Online available at https://www.iswa.org/wp-content/uploads/2023/10/23-10-31-Findings-of-the-WoW-Global-Survey-II_Endorsed.pdf?trk=article-ssr-frontend-pulse_little-text-block. Last accessed on 27 Aug 2024.
17. UN Women. (2019). Women can be the engines and souls of the circular economy UN Women (Ed.). Online available at https://eca.unwomen.org/en/news/stories/2019/03/take-five-women-can-be-the-engines-and-souls-of-the-circular-economy. Last updated on 2 Mar 2019, last accessed on 28 Aug 2024.
18. Stadler, K., Wood, R., Bulavskaya, T., Södersten, C.-J., Simas, M., Schmidt, S., Usubiaga, A., Acosta-Fernández, J., Kuenen, J., Bruckner, M., Giljum, S., Lutter, S., Merciai, S., et al. (2018). EXIOBASE 3: Developing a time series of detailed environmentally extended multi-regional input-output tables. *Journal of Industrial Ecology, 22*(3), 502–515. https://doi.org/10.1111/jiec.12715
19. Stadler, K., Wood, R., Bulavskaya, T., Södersten, C.-J., Simas, M., Schmidt, S., Usubiaga, A., Acosta-Fernández, J., Kuenen, J., Bruckner, M., Giljum, S., Lutter, S., Merciai, S., et al. (2021). EXIOBASE 3.
20. Stadler, K. (2024). Pymrio Documentation, Release 0.6.Dev.
21. EMF—Ellen MacArthur Foundation (Ed.). (2022). Towards a circular economy for plastics in China: Opportunities and recommendations. Ellen MacArthur Foundation; Tsinghua University. Online available at https://emf.thirdlight.com/file/24/LfW5VnNLfqHQH0PLfagDLGCCcjJ/Towards%20a%20circular%20economy%20for%20plastics%20in%20China%3A%20Opportunities%20and%20recommendations.pdf. Last accessed on 26 Apr 2024.
22. Fürst, K., & Feng, Y. (2022). China's regulatory respond to plastic pollution: Trends and trajectories. *Frontiers in Marine Science, 9*. https://doi.org/10.3389/fmars.2022.982546
23. Greenpeace East Asia. (Ed.). (2020). Biodegradable plastics: Breaking down the facts, production, composition and environmental impact. Online available at https://www.greenpeace.org/static/planet4-eastasia-stateless/84075f56-biodegradable-plastics-report.pdf. Last accessed on 3 May 2024.
24. The State Council of the People's Republic of China. (2023b): Drive for bamboo to replace plastic deepens The State Council of the People's Republic of China (Ed.). Online available at http://english.www.gov.cn/news/202311/08/content_WS654b27a8c6d0868f4e8e112a.html. Last updated on 8 Nov 2023, last accessed on 2 May 2024.
25. Liu, C., & Liu, C. (2023). Exploring plastic-management policy in China: Status, Challenges and Policy Insights. *Sustainability, 15*(11), 9087. https://doi.org/10.3390/su15119087
26. Lu, D. (2024): Igniting a reuse revolution in China's war against plastic waste|break free from plastic. Online available at https://www.breakfreefromplastic.org/2024/03/01/igniting-a-reuse-revolution-in-chinas-war-against-plastic-waste/. Last updated on 1 mar 2024. Last accessed on 2 May 2024.
27. Gan, Y., Lu, Z., Cai, H., Wang, M., He, X., & Przesmitzki, S. (2020). Future private car stock in China: Current growth pattern and effects of car sales restriction. *Mitigation and Adaptation Strategies for Global Change, 25*(3), 289–306. https://doi.org/10.1007/s11027-019-09868-3
28. Statista. (2024). Total electric vehicle stock volume in China from 2009 to 2021 Statista (Ed.). Online available at https://www.statista.com/statistics/1050044/china-electric-car-stock/. Last accessed on 3 May 2024.

29. Liu, M., Chen, X., Zhang, M., Lv, X., Wang, H., Chen, Z., Huang, X., Zhang, X., & Zhang, S. (2020). End-of-life passenger vehicles recycling decision system in China based on dynamic material flow analysis and life cycle assessment. *Waste Management, 117*, 81–92. https://doi.org/10.1016/j.wasman.2020.08.002
30. Wesselkämper, J.; Dahrendorf, L.; Mauler, L.; Lux, S.; Delft, S. von (2024): A battery value chain independent of primary raw materials: Towards circularity in China, Europe and the US. *Resources, Conservation and Recycling* 201, p. 107218. DOI:https://doi.org/10.1016/j.resconrec.2023.107218.
31. Hu, Z., Yu, B., Daigo, I., Tan, J., Sun, F., & Zhang, S. (2024). Circular economy strategies for mitigating metals shortages in electric vehicle batteries under China's carbon-neutral target. *Journal of Environmental Management, 352*, 120079. https://doi.org/10.1016/j.jenvman.2024.120079
32. Liu, F., Zhao, F., Liu, Z., & Hao, H. (2020). The impact of purchase restriction policy on car ownership in China's four major cities. *Journal of Advanced Transportation, 5*, 1–14. https://doi.org/10.1155/2020/7454307
33. Climate Cooperation China (2023) 2022 energy statistics show rapid development of renewable energy in China, Deutsche Gesellschaft für international Zusammenarbeit Climate Cooperation China (Ed.). Online available at https://climatecooperation.cn/climate/2022-energy-data-released/. Last accessed on 15 Apr 2024.
34. Ali, A., Malik, S. A., Shafiullah, M., Malik, M. Z., & Zahir, M. H. (2023). Policies and regulations for solar photovoltaic end-of-life waste management: Insights from China and the USA. *Chemosphere, 340*, 139840. https://doi.org/10.1016/j.chemosphere.2023.139840
35. Yang, J., Meng, F., Zhang, L., McKechnie, J., Chang, Y., Ma, B., Hao, Y., Li, X., Pender, K., Yang, L., Leeke, G. A., & Cullen, J. M. (2023). Solutions for recycling emerging wind turbine blade waste in China are not yet effective. *Communications Earth & Environment, 4*(1). https://doi.org/10.1038/s43247-023-01104-w
36. Wang, C., Feng, K., Liu, X., Wang, P., Chen, W.-Q., & Li, J. (2022). Looming challenge of photovoltaic waste under China's solar ambition: A spatial–temporal assessment. *Applied Energy, 307*, 118186. https://doi.org/10.1016/j.apenergy.2021.118186
37. The State Council of the People's Republic of China. (2023a). China issues guidelines on recycling wind-power, photovoltaic equipment The State Council of the People's Republic of China (Ed.). Online available at http://english.www.gov.cn/news/202308/17/content_WS64dddc1cc6d0868f4e8dea32.html. Last updated on 17 Aug 2023, last accessed on 19 Apr 2024.
38. Ng, E. (2024): Sustainability: China proposes first standards for recycling wind turbines to manage imminent surge of retired equipment South China morning post (Ed.). Online available at https://www.scmp.com/business/article/3249032/sustainability-China-proposes-first-standards-recycling-wind-turbines-manage-imminent-surge-retired. Last updated on 19 Jan 2024. Last accessed on 19 Apr 2024.
39. IEA—International Energy Agency. (Ed.). (2022). Komoto, K., Held, M., Agraffeil, C., Alonso-Garcia, C., Danelli, A., Lee, J.-S., Fang, L., Bilbao, J., Deng, R., Heath, G., Ravikumar, D., Sinha, P. Status of PV Module Recycling in Selected IEA PVPS Task12 Countries. Report IEA-PVPS T12–24: 2022. Online available at https://iea-pvps.org/wp-content/uploads/2022/09/Report-IEA-PVPS-T12-24_2022_Status-of-PV-Module-Recycling.pdf. Last accessed on 16 Apr 2024.
40. Zhou, Y., Wen, J., Zheng, Y., Yang, W., Zhang, Y., & Cheng, W. (2024). Status quo on recycling of waste crystalline silicon for photovoltaic modules and its implications for China's photovoltaic industry. *Frontiers in Energy*. https://doi.org/10.1007/s11708-024-0923-y
41. GIZ—Deutsche Gesellschaft für international Zusammenarbeit. (2024). German heat pump experience helps China to tap the potent, White Paper on Heat Pumps for Carbon Neutrality (2021) released, Sino-German Energy Partnership Deutsche Gesellschaft für international Zusammenarbeit (Ed.). Online available at https://www.energypartnership.cn/home/german-heat-pump-experience-helps-china-to-tap-the-potential-of-emission-reduction/. Last updated on 19 Apr 2024, last accessed on 19 Apr 2024.

42. IEA—International Energy Agency. (Ed.). (2024b). The future of heat pumps in China. International energy agency; Tsinghua University. Online available at https://iea.blob.core.windows.net/assets/217e820f-3344-4144-9d2c-cc444e080421/FutureofHeatpumpsinChina.pdf. Last accessed on 15 Apr 2024.
43. Su, C., & Urban, F. (2021). Carbon neutral China by 2060: The role of clean heating systems. *Energies, 14*(22), 7461. https://doi.org/10.3390/en14227461
44. CACE—China Association of Circular Economy (Ed.). (2023a). Report on the Development of the Comprehensive Utilization of Textile Waste in China (2020–2022).
45. Akter, A. (2024). China to tackle textile waste through recycling textile today (ed.). Online available at https://www.textiletoday.com.bd/China-to-tackle-textile-waste-through-recycling. Last updated on 13 July 2024, last accessed on 28 Aug 2024.
46. Fashion United. (2022). China will 25 Prozent seines Textilabfalls recyceln Fashion United (Ed.). Online available at https://fashionunited.de/nachrichten/business/china-will-25-prozent-seines-textilabfalls-recyceln/2022041446124. Last updated on 14 Apr 2022, last accessed on 28 Aug 2024.
47. Huang, R., Yan, P., & Yang, X. (2021). Knowledge map visualization of technology hotspots and development trends in China's textile manufacturing industry. *IET Collaborative Intelligent Manufacturing, 3*(3), 243–251. https://doi.org/10.1049/cim2.12024
48. Dudin, M., Lyasnikov, N., Kahramanovna, D., Kuznecov, A. (2015). Chinese textile industry: Sustainable development challenges and competitiveness issues in economic environment dynamics. *Fibres & Textiles in Eastern Europe, 23*(4), 14–18. Online available at https://www.researchgate.net/publication/283646243_Chinese_Textile_Industry_Sustainable_Development_Challenges_and_Competitiveness_issues_in_Economic_Environment_Dynamics. Last accessed on 3 Jun 2024.
49. UNEP—United Nations Environment Programme. (Ed.). (2023). Building materials and the climate: Constructing a new future. Nairobi. Online available at https://wedocs.unep.org/20.500.11822/43293. Last accessed on 27 Aug 2024.
50. Zheng, L., Wu, H., Zhang, H., Duan, H., Wang, J., Jiang, W., Dong, B., Liu, G., Zuo, J., & Song, Q. (2017). Characterizing the generation and flows of construction and demolition waste in China. *Construction and Building Materials, 136*, 405–413. https://doi.org/10.1016/j.conbuildmat.2017.01.055
51. Huang, B., Wang, X., Kua, H., Geng, Y., Bleischwitz, R., & Ren, J. (2018). Construction and demolition waste management in China through the 3R principle. *Resources, Conservation and Recycling, 129*, 36–44. https://doi.org/10.1016/j.resconrec.2017.09.029
52. The PEW Charitable Trusts; Systemiq (Ed.). (2020). Breaking the plastic wave, A comprehensive Assessment of Pathways Towards Stopping Ocean Plastic Pollution. Online available at https://www.systemiq.earth/breakingtheplasticwave
53. Yang, L., Fujikawa, K. (Ed.). (2023). Empirical research on environmental policies in China: China towards Decarbonization and recycle economy. Springer, Singapore. Online available at https://link.springer.com/chapter/10.1007/978-981-99-5957-0_2. Last accessed on 28 Aug 2024.
54. Zhao, Y., Peng, B., Elahi, E., & Wan, A. (2021). Does the extended producer responsibility system promote the green technological innovation of enterprises? An empirical study based on the difference-in-differences model. *Journal of Cleaner Production, 319*, 128631. https://doi.org/10.1016/j.jclepro.2021.128631
55. Li, J., Shao, J., Yao, X., & Li, J. (2023). Life cycle analysis of the economic costs and environmental benefits of photovoltaic module waste recycling in China. *Resources, Conservation and Recycling, 196*, 107027. https://doi.org/10.1016/j.resconrec.2023.107027
56. Sanchez Molina, P. (2023). China plans recycling system for wind turbines, solar panels PV magazine (Ed.). Online available at https://www.pv-magazine.com/2023/08/18/China-plans-recycling-system-for-wind-turbines-solar-panels/. Last updated on 18 Aug 2023, last accessed on 28 Aug 2024.
57. Shao, J., Li, J., & Yao, X. (2023). Net benefits change of waste photovoltaic recycling in China: Projection of waste based on multiple factors. *Journal of Cleaner Production, 417*, 137984. https://doi.org/10.1016/j.jclepro.2023.137984

58. IEA—International Energy Agency. (Ed.). (2024a). Global critical minerals outlook 2024. Online available at https://www.iea.org/reports/global-critical-minerals-outlook-2024. Last accessed on 28 Aug 2024.
59. Jiang, C., & Zhang, Y. (2023). Does extended producer responsibility system promote green technological innovation in China's power battery enterprises? *Sustainability, 15*(16), 12318. https://doi.org/10.3390/su151612318
60. Huang, B., Gao, X., Xu, X., Song, J., Geng, Y., Sarkis, J., Fishman, T., Kua, H., & Nakatani, J. (2020). A life cycle thinking framework to mitigate the environmental impact of building materials. *One Earth, 3*(5), 564–573. https://doi.org/10.1016/j.oneear.2020.10.010
61. Circle Economy (Ed.). (2023): Fraser, M., Haigh, L., Conde Soria, A. The Circularity Gap Report 2023, A circular economy to live within the safe limits of the planet. Circle Economy, Deloitte, Amsterdam. Online available at https://www.circularity-gap.world/2023#download. Last accessed on 17 Apr 2024.

# Chapter 12
# Sustainable Trade and Supply Chains

## 1 Introduction

A global transition toward sustainable trade and supply chains is essential in the age of climate change and energy decarbonization. Low-carbon technologies (LCTs) and the related minerals are at the forefront of this transition, playing a critical role in reducing greenhouse gas emissions and promoting sustainable development. The international trade of LCTs, including a wide range of products and technologies from renewable energy, energy-efficiency, energy storage and electric vehicles, is pivotal in accelerating the global adoption of climate friendly development, especially in the Global South. Considering the important role of China in the international LCT market, this report aims to provide a comprehensive analysis of LCT trade, focusing on countries' bilateral trade with China, its environmental impacts, and the economic factors shaping these trends.

The current research progress in this area is as follows: In terms of low-carbon technology trade, relevant studies involve aspects such as technology diffusion, policy support, and international cooperation. Environmental regulations and policy support are crucial for low-carbon technology trade, and research in China has received considerable attention. Regarding the impact of trade on the environment, some literature has investigated the direct impact of trade on the environment, such as how international environmental agreements promote the diffusion of clean technologies and how trade affects environmental quality. There are also studies that discuss the indirect impact of economic growth on the environment, such as the "Environmental Kuznets Curve" hypothesis. In the aspect of gravity models, this model is an important tool in international trade research. Tinbergen and others were the first to apply it, and it has since been continuously expanded and applied to various fields. Some studies have also discussed its theory and estimation.

---

The content of this chapter is extracted from the main findings of the 2023-2024 CCICED Special Policy Study (SPS) on "Sustainable Trade and Supply Chains".



*Adhering to Green and Open Cooperation to Advance Modernization*,
https://doi.org/10.1007/978-981-95-5547-5_12

In terms of data sources, to identify trade in Low Carbon Technology (LCT) Goods we rely on a classification published by IMF in the report Data for a Greener World: A Guide for Practitioners and Policymakers [1]. The list comprises 124 six-digit product codes classified according to the Harmonized System (HS) nomenclature. The United Nations Statistical Division's Commodity Trade Statistics (Comtrade) database records bilateral trade transactions for customs and tariff purposes and is the primary source of data for this study. The period of analysis of this study is from 1992 to 2022, as trade data for China is only available starting in 1992.

## 2 An Overview of Global Low Carbon Technology Trade

The global trade in LTCs is increasingly pivotal for the world's transition toward carbon neutrality.

Like the global economy, the structure of world trade of LCTs has changed dramatically over the past three decades. Specifically, high-income countries, which have traditionally dominated the export market, saw their share decrease from 93.8% in 1992 to 81.4% in 2007, and further to 62% in 2022. In contrast, developing countries, particularly China, have expanded their presence in the export market significantly. China's export share has seen a steady rise from 1.1% in 1992 to 12.2% in 2007.

While high-income countries continue to lead in LCT imports, there is a gradual diversification in global LCT import dynamics. In the global South, China's share in the LCT import market initially increased, rising from 5.1% in 1992 to 14.3% in 2007. However, this trend reversed in the following years, and China's import share dropped to 8.6% by 2022. This decline is largely attributable to the development of its domestic LCT industry, reducing reliance on imports as local production capacities and technologies improved.

As shown in Fig. 1, China emerges as the primary beneficiary of structural changes in the global LCT market. Historically, China was a net importer of LCT products up until the late 2000s. However, post-2008, China transitioned to a net exporter (Fig. 2).

In contrast, high-income countries, which were collectively net exporters of LCTs until the late 2000s, experienced a reversal in global trade. As China's clean technology industry expanded, these countries transitioned to net importers. For the group of developing countries, the trade deficit has worsened over the years. Despite this growing deficit, the broader implications suggest a complex interplay of factors. These economies are likely to experience increased trade deficits of LCTs due to insufficient domestic production capacity and the urgent need to meet environmental targets. However, this also indicates a dependency that could stifle local industry development unless balanced with supportive policies to build domestic LCT sectors.

The total value of the export market grew from $71 billion in 1992 to $1119 billion in 2022. The clean power generation technologies, mainly consisting of solar and wind, outstripped all other sectors in the early 2000s and continued to expand to $281 million in 2022 (Fig. 3).

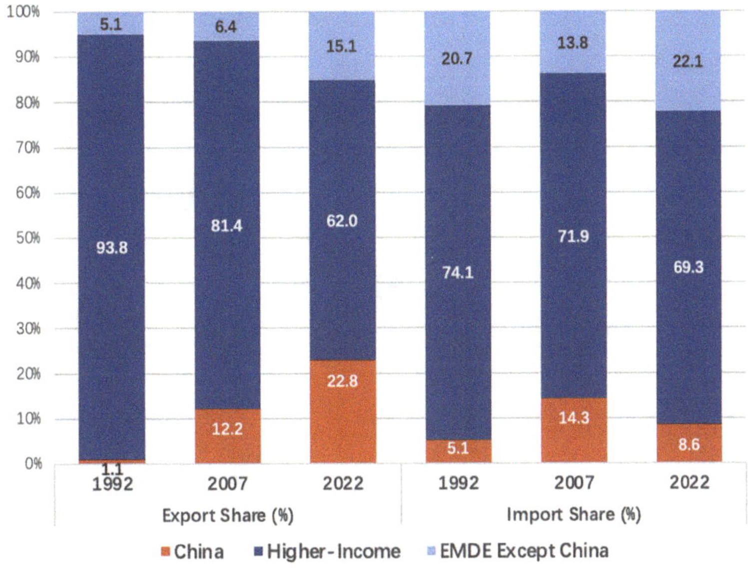


**Fig. 1** World low carbon technology trade (percentage share)

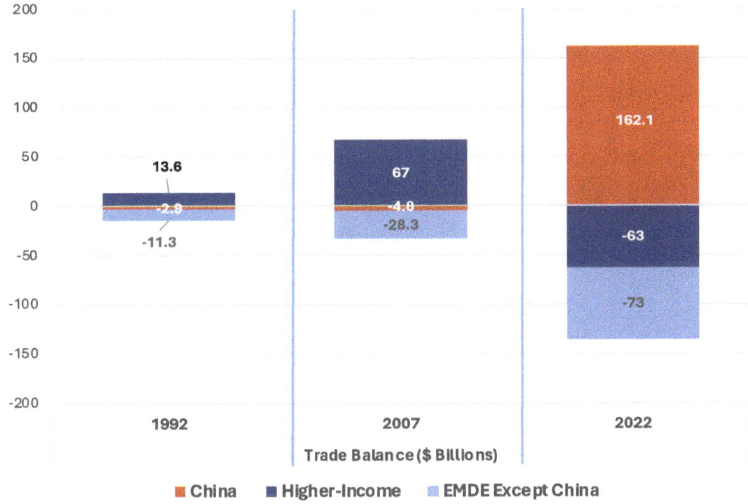


**Fig. 2** World low carbon technology trade balance (billion USD)

Another major development in global LCT export occurred in 2017 with the explosive growth in the clean transportation sector, and in particular, electric vehicles. The export of clean transportation surged dramatically, marking a significant change in the composition of the LCT trade. Its share of the total LCT export vaulted from a mere 0.02% in 2016 to 5.9% in 2017, and subsequently raised to over 20% in 2022.

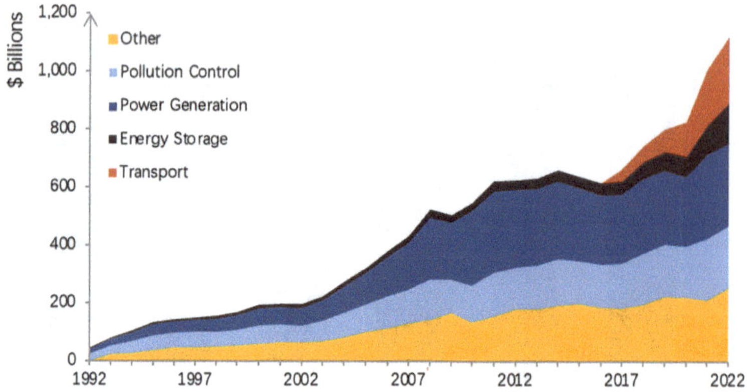


**Fig. 3** Total world exports in low carbon technologies

This surge reflects broader trends in global priorities and innovations, particularly the rapid adoption and development of electric vehicles and other sustainable transport solutions, which is consistent with global efforts to tackle urban pollution and climate change.

These trends in the global LCT export market indicate the building blocks of a clear and sustained push for decarbonization.

China's LCT exports are consistent with global trends, but with its own unique characteristics, as shown in Fig. 4. The country has focused significantly on advanced technologies for decarbonization, which collectively drive the bulk of its LCT exports (Table 1).

In terms of destination, the main consumer of Chinese LCT products are high-income countries which account for over 70% of total exports. Notably, exports to

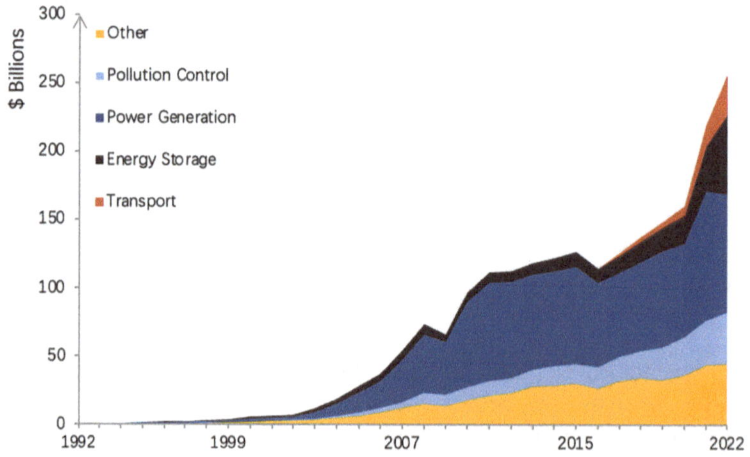


**Fig. 4** Total Chinese exports in LCTs by sector

**Table 1** Share of total Chinese exports in LCTs by trading partner income and region

| Income group | 1992 | 2007 | 2022 |
|---|---|---|---|
| High income | 63.8 | 79.6 | 70.1 |
| Upper middle income | 13.7 | 0.5 | 15.9 |
| Lower middle income | 21.5 | 7.0 | 13.6 |
| Low income | 1.0 | 2.9 | 0.5 |
| *Region* | | | |
| East Asia and Pacific | 61.4 | 50.9 | 30.8 |
| Europe and Central Asia | 10 | 24.4 | 37 |
| Latin America and Caribbean | 0.8 | 4.4 | 6.9 |
| Middle East and North Africa | 4.4 | 2.3 | 4.7 |
| North America | 6.7 | 13.2 | 13.1 |
| South Asia | 16.1 | 3.5 | 5.3 |
| Sub-Saharan Africa | 0.6 | 1.3 | 2.2 |

Europe have shown substantial growth. In contrast, exports to North America have remained relatively stable over the years.

Other income groups are also seeing growth in imports of Chinese LCTs. Exports to upper-middle and lower-middle-income countries represent 15.9% and 13.6%, respectively. However, the share of exports to low-income countries, accounting for just 0.5% of Chinese exports in 2022. These changes in export market underscores developments in LCT demand, suggesting an export orientation towards countries with more progressive climate policy.

# 3 Empirical Design

## *3.1 The Determinants of LCT Trade with China*

Trade in LCT products has been increasing substantially over the past years between China and China's trade partners. However, such growth is heterogeneous across different groups of countries. Therefore, in this part of the analysis, we explore the potential determinants of LCT trade with China for China's trade partners. In other words, we estimate the following regression model:

$$\mathrm{LCR}_{i,t} = \alpha + \beta X_{i,t-1} + \phi_i + \eta_t + \varepsilon_{i,t} \quad (1)$$

where $\mathrm{LCR}_{i,\,t}$ is China's total exports/imports of LCT to/from trade partner i in year t; $X_{i,t-1}$ includes all the variables that could affect the LCT trade with China in trade partner i in year $t-1$, which is one-period lagged. The choice of the explanatory variables follows the literature in gravity equations; $\phi_i$ is trade partner fixed effects; $\eta_t$ is year fixed effects.

## 3.2 The Determinants of Mineral Trade with China

As transition minerals play an important role in the supply chains of low carbon technology, we further explore the potential determinants of trade in transition minerals with China for China's trade partners. We establish the following regression model:

$$\text{Minerals}_{i,t} = \alpha + \beta X_{i,t-1} + \phi_i + \eta_t + \varepsilon_{i,t} \tag{2}$$

where, $\text{Minerals}_{i,t}$ is China's total exports/imports of transition minerals to/from trade partner i in year t; $X_{i,t-1}$ includes all the variables that could affect the mineral trade with China in trade partner i in year $t-1$, which is one-period lagged; The choice of the explanatory variables again follows the literature in gravity equations; $\phi_i$ is trade partner fixed effects; $\eta_t$ is year fixed effects.

## 3.3 The Environmental Effects of LCT Imports from China

To estimate the impact of China's exports and imports of low carbon technology goods on a country's economic and environmental performance, we construct the following regression model:

$$Y_{i,t} = \alpha + \beta \text{LCT}_{i,t-1} + \Gamma' X_{i,t-1} + \phi_i + \eta_t + \varepsilon_{i,t} \tag{3}$$

where $Y_{i,t}$ is environmental outcome variable in trade partner i in year t; $\text{LCT}_{i,t-1}$ is China's total exports of LCT to trade partner i in year $t-1$, which is one-period lagged given the time needed for LCT adoption; $X_{i,t-1}$ includes all the control variables in trade partner i in year $t-1$, which is also one-period lagged; $\phi_i$ is trade partner fixed effects; $\eta_t$ is year fixed effects.

## 3.4 Economic Mechanisms

The baseline regressions illustrate the overall effects of LCT imports from China on a country's environmental variables. Next, we examine the mechanisms through which the above effects take place. The regression model is as follows:

$$Y_{i,t} = \alpha + \beta_1 LCT_{i,t-1} \times CV_{i,t-1} + \beta_2 LCT_{i,t-1} + \beta_3 CV_{i,t-1} + \Gamma' X_{i,t-1} + \phi_i + \eta_t + \varepsilon_{i,t} \tag{4}$$

where the notations are the same as Eq. (3), except the newly included variable $CV_{i,t-1}$. $CV_{i,t-1}$ denotes the channel variable in trade partner i in year $t - 1$. The effect of $LCT_{i,t-1}$ on the outcome variable can be affected by the value of $CV_{i,t-1}$ through the interaction term $LCT_{i,t-1} \times CV_{i,t-1}$. According to the relevant theories in economic growth and trade as discussed in the literature review, we consider the following variables as potential channels: Human capital, political stability, foreign direct investment inflows, as well as foreign aid received. Finally, although we have lagged all the right-hand-side variables to avoid reverse causality, we recognize that the empirical design delivered in the current section may still have endogeneity problems. We plan to further address the issue in our research in the future.

# 4 Empirical Findings

## *4.1 The Determinants of LCT Trade with China*

First, economic size proxied by GDP scale suggests that countries with higher GDP tend to import more LCT products from China, which is consistent with the data and the predictions from gravity model. While, economic development level proxied by GDP per capita tends to have opposite effects on LCT imports and exports. Second, inflation and the uncertainty about the inflation (proxied by the 4-year rolling standard deviation) are negatively correlated with LCT trade with China, either imports or exports. Meanwhile, it is surprising to observe a positive correlation between the 4-year rolling standard deviation of real effective exchange rate and LCT trade with China. Finally, trade openness and political stability both stimulate LCT imports from China, while tariff seems to play a less important role in LCT trade with China.

## *4.2 The Determinants of Mineral Trade with China*

Trade openness is found to have a negative effect on mineral imports from China, partially due to the effect of trade diversification. Weak evidence has been documented that economically more developed countries tend to import more minerals from China and export less minerals to China.

## 4.3 *The Environmental Effects of LCT Imports from China*

We find that across all the $CO_2$ measures, LCT imports from China are found to significantly reduce importer countries' $CO_2$ emissions. Using more disaggregated level data, we further find that liquid fuel and fuel combustion in the transportation sector are among the fuel categories that decrease the most. One explanation for the finding could be that the liquid fuel used in vehicles has been reduced due to the increasing imports of electronic vehicles or related technology from China. Finally, we recognize that such environmental effects of LCT imports haven't been documented for NO and PM2.5.

## 4.4 *Economic Mechanisms*

### 4.4.1 Human Capital

The results suggest that the environmental benefits of LCT imports from China are amplified by the education level in the importer countries, as human capital not only is linked to returns on R&D but also plays the key role in the adoption of foreign technology such as LCT.

### 4.4.2 Political Stability

The second potential mechanism is political stability. The effective use of low carbon technology imported from China can be dependent on the institutional environment of the importer countries. Countries with greater political stability are able to plan and achieve longer-term goals, especially those environment and technology related ones. LCT imports from China have larger effects on $CO_2$ emissions in those countries with better political stability.

### 4.4.3 Foreign Direct Investment

FDI inflows is another variable that can potentially improve the adoption of LCT technology obtained from China. According to the previous research, FDI is usually associated with greater technology and human capital stocks, which are both important in the process of LCT adoption. Again, the interaction term between LCT imports and FDI inflows are included in the regressions. A certain amount of LCT imports from China can reduce more $CO_2$ emissions given a larger amount of FDI inflows. However, no such effect is found for $CO_2$ emissions per GDP.

#### 4.4.4 Foreign Aid

Development aid received from foreign countries is another factor that could facilitate the adoption of LCT technology. The majority of foreign aid is given by developed countries, which are usually with advanced technology and human capital. Such knowledge of technology and human capital could spillover to the foreign aid receipt countries. Accordingly, the environmental effects of importing LCT products from China are enhanced when the importer countries receive foreign aid.

### *4.5 Gender Analysis*

As increased gender equality unlocks economic opportunities by broadening the talent pool and driving growth, particularly in low-carbon technologies and mineral trade.

In particular, we find that gender equality is highly correlated with countries' LCT exports to China. This likely suggests that female participation and the environment of inclusion in a society contribute to the competitiveness of the economy. Therefore, the economic and environmental opportunities offered by sustainable trade and sustainable supply chains can be better seized in a country with better inclusion of women in the labor force.

## 5 Policy Recommendations

### *5.1 Boost R&D Subsidies for Renewable Energy to Innovate and Lead Globally in Energy Security*

China should maintain and potentially increase R&D subsidies for renewable energy technologies. This support should not be limited to promoting exports but also focus on enhancing domestic utilization and adoption of renewable energy solutions. By fostering innovation and technological advancements within the country, China can strengthen its position as a global leader in renewable energy while simultaneously addressing domestic energy needs and environmental challenges. Sustained R&D subsidies will drive innovation, making Chinese renewable energy technologies more competitive globally, reduce reliance on fossil fuels, lower carbon emissions, and improve energy security. Specifically, China can create a dedicated national fund to finance research and development projects in renewable energy, offer tax incentives to companies and research institutions that invest in renewable energy R&D, develop regional innovation hubs focused on renewable energy R&D to drive regional economic growth and specialization, and strengthen cooperation with the Global South on policy banks and R&D in general.

**First**, establish a national renewable energy R&D fund: Allocate a substantial annual budget to the fund, sourced from government revenues, international grants, and private sector contributions. Implement competitive grant programs to support innovative projects in solar, wind, biomass, and other renewable energy technologies. Encourage collaborations between universities, research institutions, and private companies to foster innovation and technology transfer. Establish a robust monitoring and evaluation framework to ensure transparency and effectiveness of the fund's expenditures and to measure the impact of funded projects.

**Second**, provide tax incentives for renewable energy R&D: Introduce tax credits equivalent to a percentage of the R&D expenses incurred by companies working on renewable energy technologies. Allow accelerated depreciation on capital expenditures related to renewable energy R&D equipment and facilities. Offer additional tax deductions for the salaries of scientists, engineers, and other personnel directly involved in renewable energy R&D. Provide tax reductions or exemptions for income derived from patents and intellectual property generated through renewable energy R&D activities.

**Third**, construct regional renewable energy innovation hubs: Invest in state-of-the-art research facilities and laboratories in key regions with high renewable energy potential. Provide seed funding, mentorship programs, and business incubation services to renewable energy startups located in these hubs. Facilitate partnerships between local governments, academic institutions, and private enterprises to jointly undertake large-scale R&D projects. Implement programs to attract and retain top domestic and international talent in the field of renewable energy through grants, scholarships, and attractive living conditions in innovation hubs.

**Fourth**, establish partnerships between China's policy banks and their counterparts in the Global South: Establish renewable energy project pipeline facilities and other partnerships that could lower the cost of capital for renewable energy and the transaction costs of technology transfer from China to the Global South. Strengthen the cooperation of R&D between China and the Global South in general. The partnerships should pay particular attention to the trade partners that mainly export transition minerals. The partnerships can help trade partners reduce carbon emissions through producing minerals and move up the value chain on transition minerals.

## *5.2 Stabilize Export Prices to Foster Steady Trade and Investment Relationships with Partners*

China should implement policies to maintain stable export prices, particularly with its major trading partners on low-carbon technology products and related minerals. Stable export prices can reduce uncertainty for exporters, importers and investors, promoting steady trade and investment relationships. Specifically, China can monitor export prices for predictions and warnings, expand and enhance bilateral currency

SWAP agreements with key trading partners, engage in regular macroeconomic policy coordination with key trading partners to promote export price stability.

**First**, establish a comprehensive system to monitor export prices closely, especially in key sectors such as low-carbon technology and related minerals: By tracking price trends, market demand, and global economic factors, China can identify early warning signs of price volatility and respond proactively. This monitoring system could include data analytics tools to predict potential price fluctuations, allowing exporters to adjust strategies in advance. Regular communication of these predictions and warnings to stakeholders would help mitigate risks and stabilize export prices over the long term.

**Second**, strengthen bilateral currency SWAP agreements: Negotiate and establish bilateral currency SWAP agreements with major trading partners to facilitate trade and investment without relying on third-party currencies. Set clear terms and conditions for currency SWAPs, including SWAP amounts, duration, and interest rates, to ensure mutual benefits and stability. Use currency SWAPs to provide liquidity support during periods of financial stress, reducing the risk of price volatility. Monitor and periodically review the effectiveness of currency SWAP agreements, making adjustments as necessary to optimize their impact on export price stability.

**Third**, coordinate macroeconomic policies with key trading partners: Establish formal mechanisms for regular dialogue and coordination on macroeconomic policies with major trading partners, focusing on monetary, fiscal, and trade policies. Align interest rate policies and inflation targets to reduce discrepancies that could lead to export price fluctuations. Collaborate on economic forecasts and policy responses to global economic developments to ensure synchronized and mutually supportive economic strategies. Foster transparency and data sharing between countries to build trust and facilitate informed decision-making in export price management.

## *5.3 Encourage Outward FDI in Renewable Energy and Low Carbon Technologies to Promote Global Cooperation*

China should continue to encourage and facilitate outward FDI on renewable energy and low carbon technologies to destination countries, especially those major trading partners. Outward FDI helps integrate China more deeply into the global economy, fostering stronger economic ties and collaboration. Firms can gain access to new markets, resources, and technologies, enhancing their productivity and competitiveness. Outward FDI can also diversify income sources and reduce economic vulnerability to domestic market fluctuations, especially for the renewable energy sector. Specifically, China can establish comprehensive financial incentives and support mechanisms to encourage Chinese enterprises to invest in renewable energy and low carbon technology sectors abroad, foster bilateral and multilateral partnerships to facilitate technology transfer and improve market access for Chinese renewable energy and low carbon technology firms, and pay particular attention to South-South cooperation.

**First**, provide financial incentives and support for outward FDI in renewable energy and low carbon technology: Implement tax breaks, low-interest loans, and grants for companies investing in renewable energy and low carbon technologies in foreign markets. Create dedicated funds or financing institutions to provide capital for overseas investments in these sectors. Offer risk mitigation tools such as insurance and guarantees to protect against political and economic risks in host countries. Additionally, streamline administrative processes and reduce bureaucratic hurdles for outward FDI, making it easier for companies to navigate regulatory environments.

**Second**, develop bilateral and multilateral partnerships for technology transfer and market access: Negotiate and establish agreements with host countries that promote technology exchange, joint ventures, and collaborative research and development in renewable energy and low carbon sectors. Engage in diplomatic efforts to create favorable regulatory environments for Chinese investments, including removing trade barriers and ensuring fair competition. Work with international organizations and development banks to co-finance and support large-scale renewable energy projects that involve Chinese companies. Promote China's renewable energy and low carbon technologies at international trade fairs and forums to attract potential partners and investors.

**Third**, strengthen South-South cooperation: China should actively pursue and strengthen South-South cooperation by establishing joint ventures and FDI partnerships with countries in the Global South. These initiatives should focus on promoting the transfer of renewable energy and low carbon technologies, facilitating the extraction and processing of transition minerals, and building local capacity through comprehensive technology transfer and training programs. By doing so, China can ensure that developing countries co-benefit from technological advancements and move up the value chain, leading to sustainable economic growth and reduced carbon emissions. Strengthening South-South cooperation will not only support the development of partner countries but also enhance China's leadership in global green technology initiatives. A good example would be China's EV investment in Hungary over the last years.

### *5.4 Advance Trade Liberalization to Foster Sustainable Trade Practices and Interconnected Supply Chains Globally*

China should pursue further trade liberalization for the development of sustainable trade and supply chains. Further trade liberalization can create more business opportunities for domestic firms in the renewable energy sector and increase their productivity and competitiveness. Specifically, China can proceed diversification of export destinations, expansion of import scale, specialization of service trade, multi-layered regional economic cooperation and differentiation of the Belt and Road Initiative.

**First**, diversification of export destinations: Given the current complex and ever-changing international situation, Chinese export companies should not only target mature markets in Europe and the United States but also actively explore emerging markets such as BRICS countries and Central Asia. When exporting products to different countries, Chinese companies should provide differentiated products based on the income levels and real situations of the destination to better meet local demands while enhancing competitiveness.

**Second**, expansion of import scale: From a macro policy perspective, imports should be continuously expanded, tariffs reduced, and trade costs minimized. This would allow more products to enter the domestic market, enhancing consumer satisfaction and sense of gain. From a business perspective, if imported goods are intermediate products, cost reductions will benefit sustainable profit growth and local fiscal revenue. If imported goods are final products, reduced tariffs might bring short-term competition, but in the long run, increased industry productivity will benefit industry development.

**Third**, specialization of service trade: China is already the largest goods trading nation globally, but there is still significant room for growth in service trade. Efforts should be made to expand the total volume, adjust the structure, and establish unique features. According to the recently released "China Service Trade Development Report 2022," the total volume of China's service trade reached $889.1 billion, which is still behind the total volume of the United States. Structurally, China has a large deficit in key industries, such as education services, one of the largest deficit industries. Measures should be taken to expand the service trade surplus. As China opens its doors, its rich tourism resources should attract more foreign tourists, but soft infrastructure needs improvement to enhance travel service quality. China should identify comparative advantages in its service trade industries to establish unique features, such as the traditional Chinese medicine industry. Clean energy services that involve the integration of renewable energy solutions with products to enhance sustainability and reduce carbon footprint could be another source of growth.

**Fourth**, multi-layered regional economic and trade cooperation: China is currently advancing its process to join the CPTPP, but due to the many member countries and relative complexity, achieving this in the short term is challenging. It is also crucial to promote the China-Japan-Korea Free Trade Area from factor openness to institutional openness. The geographic proximity and complementary advantages of China, Japan, and Korea mean that expanding economic and trade cooperation among these three countries will help promote regional economic integration and strengthen regional industrial chain cooperation.

**Fifth**, differentiation of the Belt and Road Initiative: As the Belt and Road Initiative enters its next decade, the focus for the future should be on strengthening economic and trade cooperation with Russia's Far East on the land-based Silk Road. The maritime Silk Road should adopt a "dual-track" approach: promoting the China-Japan-Korea Free Trade Area to the north and enhancing economic and trade cooperation with the Middle East, West Asia, and North Africa to the south, particularly advancing free trade agreements with Iran, Saudi Arabia, and Egypt. By developing the land-based Silk Road to the east and the maritime Silk Road to the north, a large

Northeast sea-land corridor can be established, reshaping China's strategic landscape for external openness.

## 5.5 Improve External Communication to Eliminate Misunderstandings about China's Overcapacity

China's overcapacity is often misinterpreted, particularly in international contexts where it is wrongly linked to product dumping in developed countries like the United States. To effectively counter these misconceptions, China should enhance its external communication strategies. This includes ensuring consistency in messaging between domestic and international platforms, and articulating China's position clearly and confidently in global forums. Specifically, China should focus on the concept of overcapacity, the causes of overcapacity, the interpretation of industrial policy and the fallacies in the U.S. anti-dumping measures.

**First**, it is crucial to grasp the concept of overcapacity. Whether China experiences overcapacity depends on how it is measured. If overcapacity is defined as the difference between potential production and actual production, then China indeed shows some signs of overcapacity to a certain extent. However, there is a fundamental difference between the overcapacity China refers to and the overcapacity-induced dumping claimed by the United States. From this perspective, overcapacity is a misleading concept.

**Second**, the causes of overcapacity need to be analyzed. Overcapacity is a common issue in global economic development. According to our calculations, China's capacity utilization rate, similar to that of the European Union, the United States, and Brazil, falls within a reasonable range of overcapacity. Since this is a common issue, the root causes of overcapacity need further exploration. The causes of overcapacity can be attributed to four main factors:

1. **Market Factors**: This mainly manifests as insufficient effective demand, with the current overcapacity primarily stemming from weakened demand following the 2008 global financial crisis.
2. **Corporate Factors**: The rush towards new industries, leading to phenomena such as "herd behavior."
3. **Industry Factors**: Certain industries have not deeply explored new productive capacities, with investment methods remaining crude and entry barriers low.
4. **Local Government Factors**: Fiscal investment has become a significant driving force behind overcapacity.

Among these factors, insufficient effective demand is the primary cause of overcapacity. The root cause of China's current overcapacity lies mainly in the global market's insufficient effective demand. To address this issue, China should actively work to build a unified national market, aiming to alleviate the pressure caused by insufficient global demand.

**Third**, China's industrial subsidy policy needs to be correctly interpreted. There is a logical error in the assertion that "industrial subsidies lead to overcapacity, which then leads to product dumping." When evaluating whether industrial subsidies impact exports, the standard should be whether China's subsidy policies violate the relevant rules established by the WTO. Currently, China's industrial subsidies are primarily focused on R&D to promote technological innovation, and do not violate the WTO's red subsidy provisions. Furthermore, the beneficiaries of these subsidies include a variety of ownership structures, such as state-owned enterprises, private enterprises, and foreign-funded enterprises.

**Fourth**, the fallacies in the U.S. anti-dumping measures against China should be clarified.

1. **Insufficient Evidence**: Chinese imports of automobiles have not caused substantial harm to the U.S. domestic automotive industry. China's exports of new energy vehicles to the U.S. account for only 1% of total U.S. auto sales, which is insufficient to cause significant damage to the U.S. domestic auto industry. In reality, the challenges faced by the U.S. automotive industry primarily stem from its own industrial hollowing-out and incomplete value chains, rather than competition from Chinese imports.
2. **Excessive Severity**: The average U.S. tariff rate is around 37%, and China is designated as a non-market economy, subject to anti-dumping duties as high as 100% on new energy vehicles. This approach is evidently unreasonable. The establishment of a socialist market economy is a great achievement of socialism with Chinese characteristics. While China's economy does experience some imbalances and underdevelopment, it is undoubtedly a market economy, with the market playing a decisive role in resource allocation.

## Reference

1. Arslanalp, S., Kostial, K., & Quirós-Romero, G. (2023). *Data for a greener world: A guide for practitioners and policymakers*. IMF. https://www.imf.org/en/Publications/Books/Issues/2023/04/04/Data-for-a-Greener-World-A-Guide-for-Practitioners-and-Policymakers-522462

# Chapter 13
# Green Opening-up and South-South Cooperation

## 1 Introduction

As a major developing country, China has consistently been a staunch supporter and active participant in South-South cooperation, advocating for the sharing of development experiences and the promotion of economic and trade development among developing countries. For decades, China has actively facilitated the exchange of knowledge, experience, policies, technologies, and resources among southern countries. This cooperation spans various fields, including climate change, agriculture, health, and infrastructure [1].

Green development, as a critical aspect of global sustainable development, requires collaborative efforts from all countries. The achievement of significant international agreements such as the Kyoto Protocol and the Paris Agreement, along with global support for the United Nations 2030 Agenda for Sustainable Development, underscores the collective responsibility and obligation to advance green governance. This cooperation is essential in both North-South and South-South contexts to foster international cooperation for green development [2].

---

The content of this chapter is extracted from the main findings of the 2023-2024 CCICED Special Policy Study (SPS) on "Green Opening-up and South-South Cooperation".





*Adhering to Green and Open Cooperation to Advance Modernization*,
https://doi.org/10.1007/978-981-95-5547-5_13

# 2 Conceptual Framework of South-South Cooperation

## *2.1 The Importance of Green Opening-up and South-South Cooperation for China in the Current Context*

### 2.1.1 Common Development Challenges Facing China and Other Developing Countries

In the face of rapid global environmental changes and evolving economic landscapes, China's role in the global South, particularly within the realms of green development and sustainability, has become more and more important.

As a developing country, China and other countries in the global South share a range of development challenges that are exacerbated by environmental and climate changes. In response to climate change, global demand for low carbon technologies is at an all-time-high. At the same time, the rapid changes in the global environment and climate may also exacerbate gender inequality. The resulting challenges, such as food and water shortages, frequent extreme disasters, and intensified refugee and migration crises, have a disproportionate impact on women's rights.

### 2.1.2 Opportunities for Developing Countries

The global push for a green transition represents a significant economic and environmental prospect for the Global South. Moreover, South-South Cooperation provides a pathway to bolster the voices and capacities of vulnerable populations. At the same time, Green Opening-up and South-South cooperation provide a platform for women to speak out.

### 2.1.3 Relationship Between China and Other Developing Countries

China's trajectory as a developing nation that has emerged as a global economic powerhouse offers insightful lessons and a model for sustainable development through green policies. Its major initiatives exemplify strategies that can be adapted and adopted by other developing countries. In addition, China has played a pivotal role in offering liquidity and development finance, thus pioneering the creation of Southern-led financial institutions. These entities not only expand the influence of the Global South in international economic governance but also provide vital new and alternative sources of funding. These initiatives facilitate greater autonomy and agency for developing countries in shaping their development trajectories.

## 2.2 *Existing Framework for China's Green Opening-up and South-South Cooperation*

Green opening-up, being a non-confrontational issue necessitating close collaboration between developed and developing nations, could serve as a pivotal starting point in addressing these complexities. China not only possesses a cultural and historical tradition of harmonious coexistence with nature but also boasts a range of government departments overseeing South-South cooperation from a multidisciplinary perspective. It has initiated a suite of supportive and directive policies, leveraging the strengths of the private sector, civil society, and international institutions to jointly propel green opening-up and South-South cooperation forward.

### 2.2.1 Culture and Values

Green development in China is deeply rooted in traditional values and has evolved alongside international norms. Stemming from the ancient concept of "the unity of man and heaven," traditional Chinese values have embraced ecological principles and the laws of nature, emphasizing the harmonious coexistence of humans and the natural world. Notions such as "green mountains are golden mountains" and "harmonious coexistence of humans and nature" reflect China's rich traditional ecological culture and imbue contemporary green development with new significance.

### 2.2.2 Institutional Set-up and Division of Labor

South-South cooperation covers multilevel and multidisciplinary international cooperation among developing countries, including bilateral and multilateral and regional cooperation, as well as shared exchanges in the fields of finance, trade, environment, agriculture, health and other areas, and at the political, economic and diplomatic levels. Foreign aid, as an important manifestation of South-South cooperation, falls under the responsibility of the central government, with ministries and departments responsible for policy formulation and fund allocation in relevant areas according to their respective responsibilities, and for certain coordination and cooperation.

### 2.2.3 Policy System

China has a long history of policy development and top-level design for green opening-up and South-South cooperation, having formulated numerous policies, plans, guidelines, and action programs for green South-South cooperation.

At the macro-planning level, since the 12th Five-Year Plan, China has included the reduction of carbon dioxide emissions per unit of GDP (carbon intensity) as a binding indicator in the national economic and social development plan. At the specific policy level, policies related to green opening-up and South-South cooperation are primarily normative and guiding documents. These policies address key issues such as Chinese enterprises going global, the construction of a green financial system, the development of a green Belt and Road Initiative, global climate governance, and alignment with international green standards.

In addition to providing guidance and working guidelines, China has established a series of management methods and guidelines for the South-South Cooperation Fund on Climate Change, the Global Development and South-South Cooperation Fund, the Global Development Initiative Project Pool, and the Global Development Initiative Fund Pool. These guidelines clarify the allocation of authority and responsibility, and outline the priorities for the processes of project preparation, implementation, and evaluation.

## *2.3 Financial Mechanism*

### 2.3.1 Wide Coverage and High Total Amount of Public Funds

From 2011 to 2014, China provided a cumulative total of 270 million yuan (approximately US$44 million) to promote South-South cooperation on climate change [3]. Established in 2015 with an initial $2 billion from China to support developing countries in implementing the 2030 Agenda for Sustainable Development, the South-South Cooperation Assistance Fund was upgraded to the Global Development and South-South Cooperation Fund in 2022, with a capital increase to $4 billion [4]. In 2015, China allocated 20 billion yuan to establish the South-South Cooperation Fund on Climate Change and introduced the "Ten Hundred Thousand" Initiative for South-South Cooperation along with the "Belt and Road" South-South Cooperation Projects on Climate Change. These initiatives aim to support developing countries in coping with climate change through financing, capacity-building, and infrastructure construction, helping them transition to green and low-carbon development [5].

### 2.3.2 Continuing to Leverage Public Funds to Promote High-Quality Chinese Enterprises Going Global

Currently, there is a significant gap between the availability of public funds for green opening-up and the demand for South-South cooperation, highlighting the need for innovative participation from the private sector. The 2011 National Environmental Protection Twelfth Five-Year Plan explicitly encourages the establishment of environmental protection industry development funds through various channels. It guides venture capital enterprises, equity investment firms, and other businesses to develop and expand the environmental protection industry. Additionally, it encourages increased investment in environmental protection from venture capital enterprises, equity investment firms, social donations, and international aid funds [6].

China has made significant progress in mobilizing private-sector funds. Within the framework of the Global Development Initiative, CIDCA mobilized $12 billion in special funds from domestic and foreign financial institutions for the first time in 2023.

# 3 China-ASEAN Cooperation

## 3.1 *The Landscape of Southeast Asia Energy and Economic Development*

### 3.1.1 Navigating Energy Security and Climate Change Impacts in Southeast Asia's Low Carbon Transition

Southeast Asia is projected to be the world's fourth largest economy by 2030, with the combined GDP of its member countries reaching USD 8 trillion. Energy consumption will also increase along with the economic development. Regional energy demand is projected to triple by 2050 from 2020 baseline (Fig. 1).

ASEAN's dependence on fossil fuels will result in 2.70 GtCOe2 of emissions by 2050 if no significant policy improvements are made. These emissions can lead to heightened climate change impacts in the region and incur both economic and non-economic losses. ASEAN member countries heavily depend on importing fossil fuels through both intra-ASEAN and global trade to meet their domestic energy needs. The region's energy security is jeopardized by its reliance on fossil fuels, which is exacerbated by the volatile nature of the global supply chain. The presence of

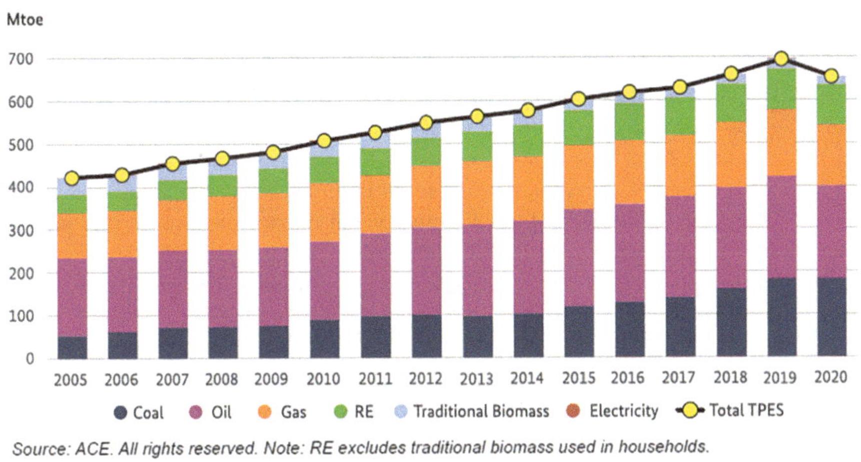


**Fig. 1** ASEAN total primary energy supply by fuel, 2005–2020

inflation and price volatility in the fossil fuel market poses a significant risk to the affordability of energy prices.

Southeast Asia also faces the challenge of worsening impacts of climate change. These climate change impacts are highly affecting the productive economic sectors in the AMS. ASEAN's aspiration to be the 'epicentrum of growth' by 2030 is threatened by the barrage of climate-related impacts in the region. The latent economic loss incurred from climate change impacts pose multiple challenges to regional and national efforts for poverty reduction and the attainment of Sustainable Development Goals (SDGs).

### 3.1.2 The Challenges and Opportunities towards a Decarbonized ASEAN

***ASEAN Decarbonization Pathways***

According to the World Bank (2023), Southeast Asia as a whole has consistently experienced economic growth rates ranging from 1% to 1.3% between 2006 and 2023. This sustained stability presents a unique opportunity for the region to proactively address global greenhouse gas emissions by embarking on a deliberate path toward decarbonization. Luciani (2020) contends that this transition, both as a narrative and a movement, holds the potential to invigorate economic growth but also foster employment opportunities. However, it will also introduce risks to existing economic norms heavily reliant on fossil fuels. To strike a balance between this opportunity and threat, embracing the transition becomes imperative for ASEAN to strategically pursue green development, ensuring a prosperous future while safeguarding environmental sustainability.

***Cooperatives and Aspirations in the Energy Sector among AMS.***

The 'ASEAN Way' of diplomacy is characterized by non-intrusiveness, signifying that member countries acknowledge the need to handle domestic issues internally. Despite frequent engagement in meetings and the adoption of various mechanisms, ASEAN—an international organization committed to regional integration— rarely achieves unified coordination in its foreign policies. However, through this study, we believe that, as global norms for climate mitigation, the abundance of renewable energy resources, and strategic green global value chain transitioning the energy towards low-carbon energy, it's really visible for ASEAN to become a role model for the other emerging and fossill intensive countries in green development

***Scenarios and Implications for Renewable Energy Growth***

Deploying renewable energy is the most accessible action to reduce carbon emissions in ASEAN. This is because the ASEAN member states have already expressed cooperative aspirations and goals, and the region has geographical potential with

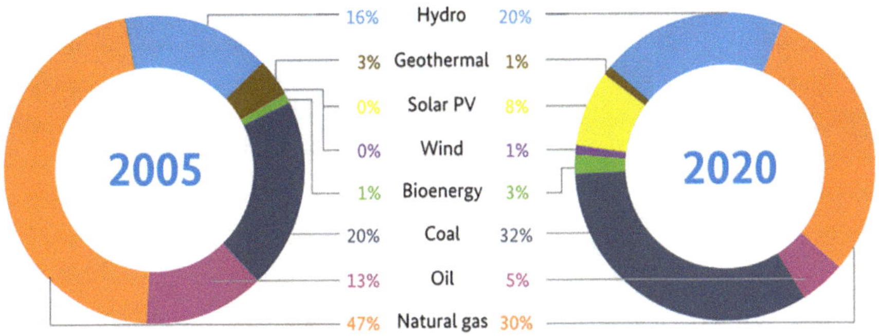


**Fig. 2** ASEAN installed capacity by Fuel 2005 versus 2020. *Source* The ASEAN Energy Outlook 2022 (ACE, 2022)

renewable energy resources. Additionally, there is a global value chain in manufacturing renewable energy. Most importantly, ASEAN's power and influence can become the backbone of Asia in energy transition.

***Renewable Energy Development in the ASEAN Member States.***

AMS has been actively involved in the development of renewable energy since 2005 as a means of addressing climate change. This effort has been furthered by implementing hydro power generation, geothermal energy, and bioenergy (Fig. 2).

Variable renewable energy (VRE) sources, which have no implications of significant changes due to natural changes and resource exhaustion, accounted for less than 30% of the total renewable energy in 2020. The quantity must be augmented as this particular form of renewable energy clearly has the potential to serve as the foundation for the ASEAN energy mix, owing to its distinct benefits over other forms of renewable energy.

***Scaling up Green Industry and Renewable Energy Technologies Manufacture***

- **Green Industry**

As of 2020, electricity consumption in the ASEAN region is the second greatest share, amounting to 22.7%. The industrial and transportation sectors have consistently maintained their positions as the leading consumers of electricity, accounting for 38.4% and 35.5%, respectively (ACE, 2022). Given this ambition, the industry sector can serve as a pivotal point for AMS to expedite its efforts in environmental sustainability (Fig. 3).

China's overseas industrial parks are crucial in promoting the greening of the industry sector, as they are closely linked to the country's expansion in the global market and the establishment of a global value chain. According to the Ministry of Commerce PRC (2019), China had a total of 113 foreign economic and trade zones in 2019. These zones were spread over 46 different nations, including countries in Southeast Asia. As a result of the greening initiative, the Ministry of Ecology and

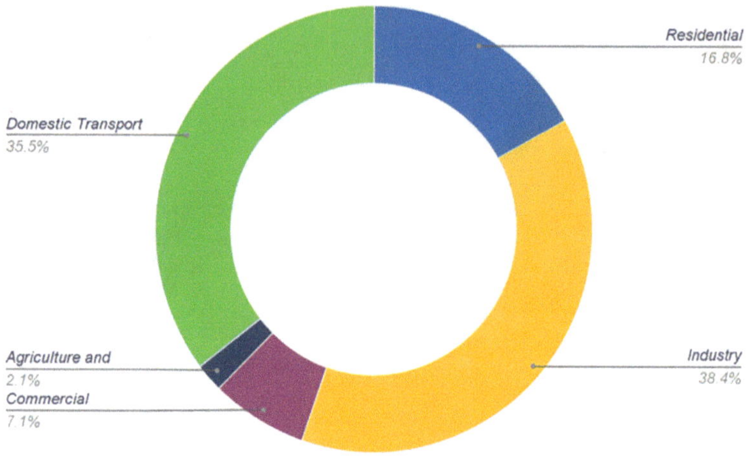


**Fig. 3** ASEAN total energy consumption by sector in 2020. *Source* Curated by Author from The ASEAN Energy Outlook 2022 (ACE, 2022)

Environment of PRC has released guidelines for promoting environmentally friendly foreign investment cooperation. These rules encourage enterprises to establish, construct, and enhance the growth of green industries.

- **Renewable energy Technologies and manufacturing**

Given that variable renewable energy is anticipated to become the global dominant energy source by 70% in the future, it is imperative to acquire a regional position in order to expand the manufacturing of these renewable energy sources to manage the global supply. The renewable energy technologies industry is a strategic prospect for ASEAN as a climate leader in the world, as it is geographically situated in the vicinity of critical mineral sources that are a resource for renewable energy technology manufacturing. Additionally, the industry has a strong manufacturing foundation in the global value chain.

Currently, Southeast Asia is the manufacturer of 9% to 10% of the world's solar photovoltaic (PV) cells and modules, as well as 6% to 10% of all electric two-wheelers. This lays a good foundation for strengthening ASEAN's renewable energy technology status and expanding green technology and manufacturing industry cooperation between China and ASEAN.

## 3.2 *Achieving a Decarbonized Southeast Asia through International Collaboration: ASEAN-China Cooperation for Renewable Energy Development*

### 3.2.1 Jointly Building the BRI Provides Broad Space for Green Energy Cooperation Between China and ASEAN Countries

In jointly building the Green BRI, China and ASEAN countries have actively promoted green policy coordination, green infrastructure connectivity, unimpeded green trade, and green financial integration. China has taken the initiative to align the development strategies of participating countries, formulated the China-ASEAN Action Plan on Environmental Cooperation, the Five-Year Plan of Action on Lancang-Mekong Cooperation (2023–2027), and other development plans, and actively participated in the green industry development, clean energy transition, green technology innovation, climate governance, and ecological city construction of participating countries.

### 3.2.2 Remarkable Results Have Been Achieved in the Cooperation Between China and ASEAN Countries for Promoting Innovation and Investment in Clean Energy Technology

With the guidance of the concept of green development, China and ASEAN countries have increasingly deepened their cooperation in clean energy. Addressing global climate change and achieving national energy security are common challenges for China and ASEAN countries. During the process, China and ASEAN countries have established a mature cooperation mechanism with the vision of jointly building the Green BRI. The remarkable cooperation achievements being achieved, the cooperation foundation and the experiences established by both sides will support the in-depth and extensive cooperation in the future. They will further provide impetus for ASEAN countries to achieve the Net-Zero target early.

## 3.3 *Best Practices of ASEAN-China Renewable Energy Cooperation*

### 3.3.1 Cirata Floating Solar PV Power Project in Indonesia

Cirata Floating Solar Photovoltaic Power Project in Indonesia is the largest floating solar PV power project in the Southeast Asia region, with a total installed capacity of 192 MWDC. The project is jointly invested and developed by Indonesian state power firm and Abu Dhabi Future Energy Company and constructed by POWERCHINA in an EPC model. This project has been selected as a national key strategic policy project in Indonesia, which is of significant milestone significance for Indonesia's development of clean energy and achievement of energy transition.

### 3.3.2 Solar PV Electricity Export from Indonesia to Singapore

In October 2021, Singapore's Electricity Market Authority (EMA) announced a plan to import 4GW of electrical power from overseas, attracting the attention of many local and international electricity investors in Singapore. Indonesia is adjacent to Singapore, and the straight-line distance between Batam Island in Indonesia and Singapore does not exceed 20 km. Indonesia has abundant solar energy resources and a large amount of land suitable for solar energy project development. Developing and constructing solar photovoltaic power generation and storage projects on islands near Singapore and sending green electricity to Singapore through submarine cables is the best option to achieve complementary advantages for both sides. In April 2023, at the sixth Singapore Indonesia Leaders' Informal Meeting, the two countries signed a memorandum of understanding on renewable energy cooperation, aiming further to promote cooperation in the field of renewable energy, expand green power trading between the two countries, and improve the cooperation framework and infrastructure for power transmission.

### 3.3.3 Laos Wind Power Exports to Vietnam

The Monsoon 600 MW wind power project in Laos was born which is located in the low mountain areas of the Sekong and Attapu provinces in central and southern Laos. It is the first wind power project constructed in Laos and currently the largest single wind power project invested in Southeast Asia. The entire project site covers an area of 1100 ha, including 133 wind turbines with a single unit capacity of 4.51 MW. The wind farm will also have 5 booster stations and nearly two kilometers of 500 kV transmission lines. Impact Energy Asia Development Limited (IEAD) from Thailand invested in and developed this project. The Asian Development Bank (ADB) and several commercial banks have jointly provided financing plans for the project, the largest syndicated renewable energy project financing transaction among ASEAN countries.

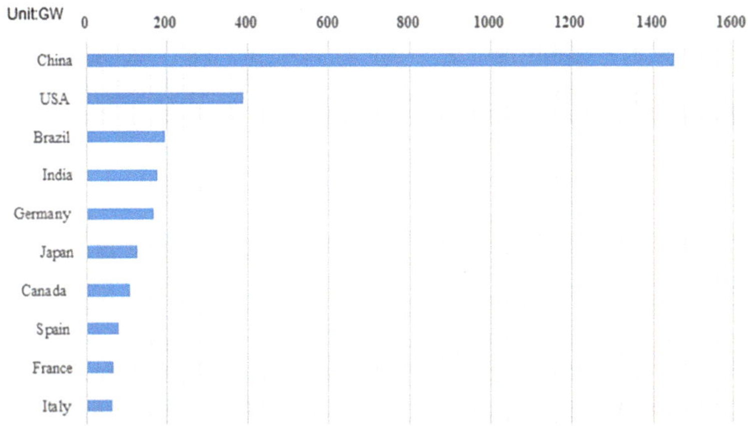


**Fig. 4** Distribution of total installed capacity of renewable energy in major countries

## *3.4 Potential to Scale up ASEAN-China Cooperation for Green Development*

### 3.4.1 China Is the world's Largest Clean Energy Market

After nearly a decade of rapid development, China has become the world's largest producer and consumer of green and clean energy, and its overall market size has remained the world's largest for many years. According to data released by IRENA, as of the end of 2023, China's total installed renewable energy capacity was 1453 GW, ranking first in the world. The specific capacity distribution of major renewable energy-installed countries in the world is shown in Fig. 4.

China has a vast territory, significant regional differences, and significant differences in resource endowments, which have led to the formation of various development and application scenarios in the process of renewable energy development, including cascade development and scheduling models for rivers, large-scale development modes for wind and solar power, complementary development modes for water, wind, and solar energy, as well as agricultural and solar PV complementary modes, fishery and solar PV complementary modes, etc. These multi-scenario development modes also provide useful references for developing renewable energy in other countries and regions.

### 3.4.2 China's Leading Advantages for Renewable Energy Technology

China's rapid progress in renewable energy cannot be separated from its continuous innovation and progress in renewable energy technology. In the field of hydropower, Chinese companies are leading the way in surveying, designing, and constructing large-scale hydropower stations. In wind power, Chinese companies are also constantly creating new records in important indicators such as wind turbine unit capacity and reliability. As for solar power generation, Chinese companies have maintained a leading position worldwide for many years in areas that reflect their core competitiveness, such as PV cell conversion efficiency and cost control.

### 3.4.3 Comprehensive System of Renewable Energy Technology Standard

With the development and technological progress of the renewable energy industry, China has also made tremendous achievements in establishing the renewable energy technology standard system. China has established a complete renewable energy standardization management mechanism, from the national standardization management department to the competent department of the renewable energy industry, from the National Standardization Technical Committee to the Renewable Energy Industry Standardization Technical Committee. At the same time, China has preliminarily established an institutional system that connects and complements renewable energy technology standards at different levels. Regarding standard system construction, China has established relatively complete technical standard systems in many fields represented by new and renewable energy, such as hydropower, wind power, and solar energy.

### 3.4.4 Strong Production Capacity Advantage in the Field of Renewable Energy Development

China's rapid development in the field of renewable energy cannot be separated from the support of strong production capacity advantages. In the arduous development process of the past few decades, China has established a large and mature renewable energy market system, formed a sound technical standard system and scientific research and innovation capabilities, and built the world's most complete renewable energy equipment research and development, manufacturing industry chain. The capabilities and advantages of Chinese companies in the field of renewable energy have laid a solid material foundation for future cooperation between China and countries around the world, especially ASEAN countries.

## 4 China–Africa Cooperation

In the context of global climate change, Africa has to deal with dual challenges of development and climate change. Over 90% of African governments have submitted Nationally Determined Contributions (NDC) to the United Nations Framework Convention on Climate Change (UNFCCC), 70% of African governments have very specific renewable energy targets, and over 25% of African governments have committed to net-zero goals.

Sino-African cooperation through the Forum on China–Africa Cooperation (FOCAC) and the Belt and Road Initiative (BRI) will be crucial for Africa to meet the dual challenges. This research is designed to assess the progress of the Sino-Africa cooperation in the field of green development and explore issues and further opportunities.

## 4.1 Overview of Africa's Climate Action and Green Development

African countries are disproportionally affected by climate change. Over a decade ago developed countries committed to providing $100 billion each year to developing countries in their fight against climate change [7]. However, until 2022, this funding had not materialised, leaving more vulnerable countries, including African countries, struggling to address climate change [8].

Yet despite high levels of vulnerability, African countries have enormous potential for renewable energy production in the forms of solar, wind, hydropower, geothermal, and biomass energy. This potential for cleaner energy production is matched by high levels of climate ambition across the continent. Africa has developed strategies for climate action and green development, which operate through distinct policy frameworks at both the continental, regional and national levels.

### 4.1.1 African Union's Climate-Related Policies

Africa's commitment to addressing climate change is guided by the African Union Agenda 2063, the UN Sustainable Development Goals (SDGs) and the Paris Agreement. Agenda 2063, adopted in 2015, is Africa's blueprint for sustainable development for a 50-year period, with the goal of achieving 'an integrated, prosperous and peaceful Africa, driven by its own citizens, representing a dynamic force in the international arena' [9]. Agenda 2063 Goal 7 is closely linked with Sustainable Development Goal 13, which calls for immediate action to fight climate change and its impacts [10], and thus shows how determined African countries are to address climate change. More strategic documents on climate change have been issued since 2021, including a Green Recovery Action Plan in 2021 and a new action plan for climate change and resilience development strategy in 2022.

The first Africa Climate Summit held in September 2023 in Nairobi, Kenya shows the strong determination of African countries to find solutions to global climate change. The Nairobi Declaration on Climate Change was the main outcome of the Summit to unify African states' stance on climate action. The Declaration calls for reforms of the global financial system to facilitate climate financing and a global tax to fund climate action. It also set an ambitious target to increase Africa's renewable generation capacity to at least 300 GW by 2030.

### 4.1.2 African Countries' Renewable Energy Policies

To boost renewable energy development, 30 African countries have established specific policy frameworks or renewable energy targets. Starting in 2008, several countries have set up targets on the percentage of renewable energy contribution to total energy generation. The targets are mainly assigned in the National Energy (Efficiency) Strategy or as part of their Climate and Sustainable Development Policy. The target percentage is normally below 50% while the detailed roadmap or action plans are mostly missing.

### 4.1.3 Climate Ambition in African Nationally Determined Contributions (NDCs) and Financing Needs

Fifty three African countries have now submitted their Nationally Determined Contributions (NDCs) to the UN Framework Climate Change Convention (UNFCCC). Of these 53 submitted NDCs, 43 are updated NDCs (also known as Enhanced NDCs (ENDCs)). The only countries that have not submitted NDCs are Libya and Saharawi Arab Democratic Republic. The large number of updated NDCs highlights the continent's strong commitment to climate change mitigation and sustainable development (Fig. 5).

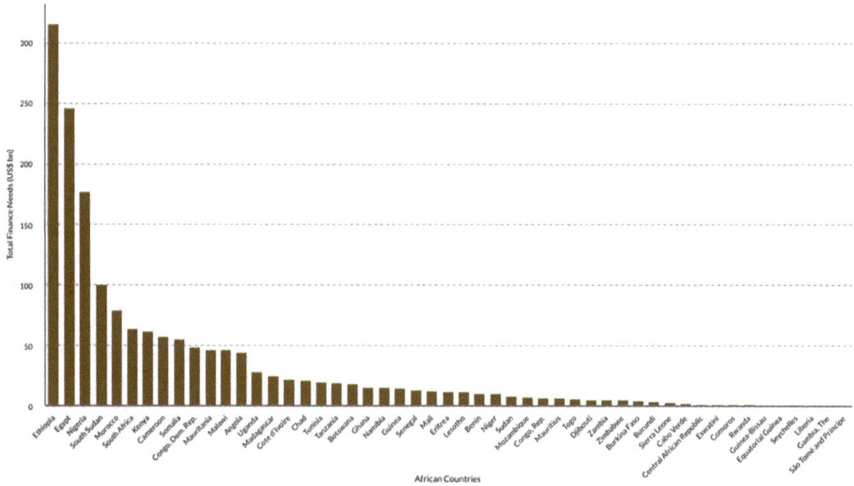


**Fig. 5** Africa's NDC financing costs

**Table 1** Installed capacity of renewable energy generation (GW) in 2022

| | Renewable energy | Hydro | Solar PV | Wind |
|---|---|---|---|---|
| Africa | 58.7 (1.7%) | 39.0 (2.8%) | 11.4 (1.1%) | 7.7 (0.8%) |
| China | 1160.8 (34.3%) | 413.5 (29.7%) | 392.4 (37.1%) | 366.0 (40.7%) |
| World | 3381.8 (100%) | 1392.5 (100%) | 1055.0 (100%) | 898.8 (100%) |

Forty-nine African countries indicate both mitigation and adaptation needs, while São Tomé and Príncipe as well as Zimbabwe only provide mitigation cost estimates. Notably, inconsistencies in the extent of country cost breakdowns limit the scope of analysis that can be done at continental level. However, data indicates that Africa's adaptation costs of about US$438 billion are about half the continent's climate mitigation finance needs of approximately US$870 billion.

### 4.1.4 Africa's Needs and Potential for Green Development

Compared to other regions in the world, the renewable energy development in Africa has been very slow (Table 1).

African countries boast abundant sunlight, wind, hydro, and geothermal potential, making them prime candidates for clean energy development. Currently, renewable energy accounts for over half of the total electricity capacity in 17 African countries. At the same time, the development of renewable energy is also of great significance to gender equality and women's development. First, by providing a stable supply of clean electricity, the quality of life of local residents has been improved, and the nighttime learning and living conditions of women and children have been improved. It also creates employment opportunities for African countries, especially for women (Fig. 6).

## *4.2 Status and Challenges of Sino-African Cooperation in Green Development in Africa*

### 4.2.1 Progress of Sino-African Cooperation in Renewable Energy Development

According to China's Global Energy Finance (CGEF) Database [11], China Development Bank (CDB) and Export-Import Bank of China (Exim Bank of China) provided US$49 billion in loans to African countries to fund energy projects between 2000 and 2022. Of the total, $13 billion was invested in hydropower, $611 million was committed to wind, and $480 million to geothermal and solar financing accounted for $367 million.

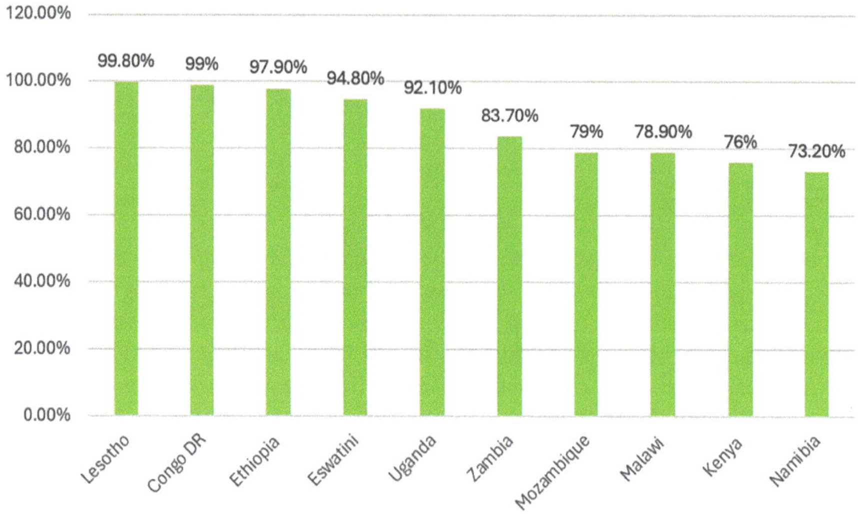


**Fig. 6** Top 10 African countries with the highest renewable energy share of electricity capacity in 2021 (Generated by Development Reimagined. Data sources from IRENA, "Renewable Energy Capacity Statistics 2022")

Among the recipients of Chinese support in renewable energy development, Ethiopia has been one of the biggest. Over the years Chinese enterprises have funded several energy generation projects in Ethiopia worth over $4 billion, accounting for over 50% of new power generation capacity. Kenya is another primary recipient of Chinese investments in renewables. China has financed and built large solar and wind farms across Kenya, helping expand renewable energy access, particularly in rural areas. Most notably, the 310 MW Lake Turkana Wind Power project, the largest wind power farm in Africa, was constructed by a Chinese consortium. This project came online in 2018, providing about 15% of Kenya's electricity capacity. These projects not only provide clean energy, but also bring substantial benefits to the development of women and children in Africa by creating jobs and improving education and living conditions.

### 4.2.2 Challenges Faced by Africa in Green Development and China's Role

Although African countries have great ambitions to address climate change and possess abundant renewable energy resources, the lack of adequate climate finance is a barrier to Africa's green development.

China has played a constructive role so far in availing climate finance to help Africa develop its clean energy sector, and it is important for China to continue to issue innovative finance instruments. Since 2000, it is estimated that China has provided US$25 billion in loan financing in Africa's clean energy sector [12]. Since FOCAC 2021,

Development Reimagined has found that 95 climate action projects with contributions from China have been initiated, covering renewable energy, climate-friendly infrastructure, clean technology and transportation areas, also capacity building to Africa [13]. During the Belt and Road Forum last year, China launched a new Green Investment and Finance Partnership (GIFP) with more than US$100 billion in financing for energy transition in the BRI countries. Certainly, based on the evidence outlined in this paper so far, targeting a high proportion of this new funding to unleash Africa's green development potential will be an excellent strategic move by China [14].

### 4.2.3 Africa's Potential in Green Manufacturing

The development of manufacturing capacity is in line with Africa's push for industrialization and localized manufacturing. According to SE4All [15], investment in energy transition technologies creates 3 times as many jobs as fossil fuels per $1million invested. Through 2030, between eight million and 14 million energy transition jobs could be created in Africa, potentially driven by government spending on public services and investment in transition technologies.

Nevertheless, Africa is at the bottom of the global value chain with its share of global manufacturing at around only 1.9% [16]. According to International Energy Agency, under the net zero scenario, the electricity generation from solar PV will be 1859 TWh, and from wind will be 593 TWh in Africa in 2050, i.e. about 930 GW of installed capacity for solar PV and 200 GW of wind power. The renewable energy market of this scale cannot and should not all be imported from outside the region. Instead, local manufacturing should be vigorously developed, African countries should develop local manufacturing capacity for solar PV and wind power capacity to meet its energy needs and sustainable development goals. Moreover, Africa is resource rich in several raw materials critical for renewable energy products. The continent has high availability of raw materials with large reserves of inputs such as cobalt, manganese, and lithium. These resources provide an attractive option for companies looking to install renewable manufacturing facilities closer to resource supply chains.

China leads the manufacture of renewable energy products. In the case of solar PV, as of the end of 2023, the production capacity of each link of photovoltaic equipment was about 1000 GW while the global market demand for that year was only 400 GW. Due to the tariff barriers faced by the United States and Europe in recent years, some of China's solar industry's production capacity has shifted to Southeast Asia.

However, so far there has been so far only one significant Chinese investment in renewable energy Manufacturing in Africa Overall, while China Africa cooperation on green development has increased and improved, it has significant further potential for growth due to the combination of high African potential and ambitions with China's existing capacity and experience in cross-country connectivity and industrialization.

## 5 Policy Recommendations on Enhancing Green Opening-up and South-South Cooperation

### *5.1 Formulate a Roadmap for Promoting Green Opening-up and South-South Cooperation*

#### 5.1.1 Short-Term Initiatives

1. Improve the High-Level Coordination Mechanism for Green Opening-Up and South-South Cooperation: Based on the existing sectoral management foundations for South-South cooperation, climate change cooperation, and green energy transition cooperation, invite CIDCA or its subordinate agencies to join the China Council for International Cooperation on Environment and Development (CCICED) as a steering committee. This integration will facilitate better interactions with the international community on issues related to green opening-up and South-South cooperation.
2. Incorporate Green Opening-Up and South-South Cooperation into international cooperation frameworks, including setting international cooperation goals, policy planning, management regulations, supervision, and evaluation. Simultaneously, expand the consultation of experts in the fields of environment and climate change during policy formulation and expert consultation processes. This will ensure that policies are well-informed and aligned with international standards and practices.
3. Expand South-South Cooperation Financing Modalities: maximize the use of existing funds, especially specialized funds in the fields of green development and climate change response, such as the South-South Cooperation Fund on Climate Change and China's contributions to the Global Environment Facility. Simultaneously, expand financing channels for South-South cooperation by attracting private sector investment. Explore innovative financial support modalities, such as public-private partnerships, green bonds, and other sustainable finance instruments to enhance resource mobilization for green projects.
4. Incorporate pre-feasibility study as a requirement for future projects. To minimize potential environmental and socio-economic risks, the Chinese government and funding recipient country governments should implement a pre-feasibility study requirement for all projects. Project financiers and developers should be required to conduct a prefeasibility assessment of the project. Financers and developers should undertake a thorough prefeasibility assessment, encompassing three critical stages: 1. Securing Access to the Project Site: This involves obtaining the rights to buy or lease the land necessary for the project, a crucial step as funders typically require secured land access before providing feasibility funding. 2. Conducting Comprehensive Assessments: This stage should include evaluations of the basic site characteristics, environmental impacts, grid connection possibilities, as well as solar and wind resource assessments. Additionally, an analysis of plant capacity and technical options should be carried out. 3. Financial

and Cost Analysis: Developers should develop a high-level financial model that includes indicative electricity tariffs based on the assessments. This structured approach ensures that all potential risks are adequately addressed before project development proceeds.

### 5.1.2 Medium-Term Strategies

1. Formulate a policy framework and development strategy that aligns green opening-up with the promotion of South-South cooperation projects. This framework should consider the needs of developing countries and China's comparative advantages and output capacity in the green sector. Start with the formulation of cooperation plans in priority areas, improving specific cooperation programs for key countries and regions. This should be supported by a robust disciplinary mechanism and regulatory system to ensure effective implementation and oversight.
2. Enhance practical guidance for green opening-up and South-South cooperation across industries, utilize advanced technologies and resources from the trade, investment, industry, information, agriculture, and climate sectors to provide positive guidance for South-South cooperation at the project level.
3. Engage a wide range of partners, including government departments, enterprises, international organizations, NGOs, and social groups. Actively involve Chinese local governments, especially those in border provinces, as well as "going out" enterprises and NGOs. Explore tripartite cooperation with other donors to broaden the participation of various actors in green opening-up and South-South cooperation.
4. Strengthen talent development and capacity building. Enhance the capacity building of talents related to green opening-up and South-South cooperation, especially of staff in relevant domestic departments. Strengthen the reserve of industry experts in green development and international exchanges. Improve the theoretical foundation for policymaking in green opening-up and South-South cooperation while fostering communication and cooperation with international organizations. Support the development of a robust discourse system for international green development cooperation to enhance China's influence and participation in global environmental governance.
5. Adopt an Issue Oriented Model. Previously, China's approach to South-South Cooperation was primarily country-focused. Moving forward, it would be beneficial for China to place climate change and sustainable development at the forefront of its South-South Cooperation efforts, emphasizing global issues such as the climate crisis rather than concentrating solely on individual countries. Adopting this systematic model of cooperation would enable China and the Global South to address global systemic challenges as a unified community, enhancing the collective ability to implement effective solutions.

### 5.1.3 Long-Term Vision

Improve the whole-process management of green opening-up and south-south cooperation. First, include comprehensive environmental, social, and governance (ESG) indicators in the evaluation system of South-South cooperation projects. Based on the early-stage demand research and country-specific programs, attract a wider range of green development projects to consider South-South cooperation. Second, establish a complete legal system, policy planning, and industry guidelines to provide protection and guidance for project implementation. This framework will safeguard the rights and interests of relevant organizations or enterprises, promote China's advantageous industries globally, and enhance the efficiency of project implementation. Third, enhance the external communication of green opening-up and South-South cooperation during the project implementation period. Clearly explain China's green opening-up initiatives and the progress of South-South cooperation projects to the international community. Fourth, the construction of the post-evaluation mechanism after the completion of the project needs to combine existing domestic practices and international experience, formulate evaluation indicators based on different industry fields, and use green openness-related indicators as important evaluation bases, including the ecological and environmental impacts brought by the project, the impact of the degree of technology transfer on the local green development transformation, etc. At the same time, it is possible to consider incorporating gender equality-related standards into the whole process management of project design, implementation, and evaluation. Gender mainstreaming, as an important social development issue, is also an important part of "soft aid". It can not only effectively connect with international standards and enhance the international recognition of green South-South cooperation projects, but also focus on vulnerable groups and achieve the "soft and hard combination" of South-South cooperation projects.

## *5.2 Promote the Acceleration of China-ASEAN Renewable Energy Cooperation*

### 5.2.1 Enhance a Regional-Level Comprehensive Strategic Partnership Between ASEAN and China for Clean and Green Energy Development and Energy Infrastructure Connectivity

Expanding the ASEAN-China partnership to develop clean energy and decarbonize the region's industry and energy infrastructure is crucial. China's support is vital for accelerating renewable energy adoption. Broadening the existing partnership agreement to tackle climate concerns could lead to increased economic prosperity and energy security. China's commitment to global climate efforts can be extended to include supporting energy transition in collaboration with the Association of Southeast Asian Nations (ASEAN). This partnership can kickstart renewable energy projects in the region and enhance energy infrastructure connectivity, supporting the ASEAN connectivity agenda by fostering the realization of the ASEAN Power Grid program.

### 5.2.2 Support the Green Manufacturing of Goods and Materials for Clean Energy Technology through Industrial Decarbonization Measures Using Green Electricity and Green Processing in ASEAN

The ASEAN region is expected to experience an increase in carbon emissions due to industrial development. To tackle this issue, it is crucial to encourage Chinese companies to promote low-carbon technology and manufacturing processes in this region. This can be achieved through technical assistance, technology transfer, and partnerships, all of which are important steps in making the ASEAN industry less carbon-intensive. One potential area for collaboration is the utilization of captive coal-fired power plants in nickel facilities in Indonesia as a pilot site for China-ASEAN cooperation on low-carbon industrialization. Furthermore, directing ASEAN's industrial transformation towards export-oriented clean energy technology can contribute to regional green development.

### 5.2.3 Enhancing the Role of ASEAN-China Clean Energy Cooperation Center as a Platform for Knowledge Sharing and Capacity Building

The ASEAN-China Clean Energy Cooperation Center is a hub for fostering innovative research and development of state-of-the-art clean energy technologies. The expertise cultivated within this center plays a crucial role in helping ASEAN address the technical obstacles associated with implementing renewable energy. Additionally, the center collaborates with private sector enterprises from China operating in ASEAN to support clean energy investments through technical analysis, feasibility studies for strategic projects, and recommendations to facilitate increased Chinese investments in the region.

### 5.2.4 Develop Regional-Level Guidelines for Responsible Practices in Managing Transition Minerals Processing and Greening Supply Chain for Clean Energy Technologies

Amidst the growing focus on mineral exploration and processing for sustainable development in the ASEAN-China region, it is essential to prioritize cooperation to ensure Environmental, Social, and Governance (ESG) standards. This alignment is crucial to ensure that the processing of transition minerals into materials for clean energy technology does not result in increased carbon emissions or adverse effects on the environment. ASEAN has introduced the Taxonomy for Sustainable Finance to provide guidance to investors interested in supporting environmentally friendly or green projects. Similarly, China has formulated the Green Development Guidance for its overseas investments, focusing on aligning with climate goals and promoting sustainable development. These initiatives can work hand in hand to encourage and guide sustainable investments in the region, including offering specific investment guidelines for transition minerals.

#### 5.2.5 Establish a Project Pipeline Facility to Support Rapid Utility-Scale Renewable Energy Deployment both on-Grid and off-Grid

One of the causes of lagging renewable energy development in ASEAN is a lack of bankable projects. Due to unstable policies, investor risk perceptions towards renewable energy project development are high. The aim of this pipeline facility establishment is to provide a channel for ASEAN countries to access capital from China for renewable energy projects. The facility can also serve as an incubator for the early stages of project development by identifying technically and financially sound projects to be presented to investors. Through this measure, the facility can also identify which projects need leverage to get the necessary financing by providing de-risking instruments such as credit guarantees or equity. To ensure the operation of this project pipeline facility, it is recommended that the facility become one of the deliverables under the regional comprehensive strategic partnership. Engagement with national-level energy sector stakeholders is also crucial to assist them in preparing projects that will attract foreign capital.

### *5.3 Unlock the Potential of China-Africa Renewable Energy Cooperation*

#### 5.3.1 Set more Pragmatic Goals for Climate Cooperation

As part of the global response to climate change, Chinese and African governments have made strong policy commitments on climate action and green development cooperation in Africa, which have been translated into the implementation of clean energy projects. The Forum on China-Africa Cooperation (FOCAC), as the main platform for promoting China-Africa partnership, has shifted its focus to climate and environmental cooperation. As a staunch supporter of sustainable development in Africa, China has launched more than 100 clean energy and green development projects under the framework of the FOCAC to help African countries better utilize solar, hydropower, wind, biogas and other renewable energy sources. In the next step, we should continue to make good use of high-level dialogue opportunities such as the FOCAC, give play to the role of platforms such as the Belt and Road International Alliance for Green Development and the China-Africa Environmental Cooperation Center, and expand cooperation on renewable energy such as solar photovoltaic and wind power. Explore the formulation of more pragmatic action plans and goals in climate investment and financing and capacity building, such as setting overseas renewable energy installation targets and investment targets for the energy industry, and setting targets related to value-added investment and key mineral infrastructure in the trade field. More specific and pragmatic goals can ensure that all parties have enough confidence and motivation to try innovative models such as PPP and enable all participants to work towards common goals.

### 5.3.2 Provide Low-Cost, Patient and Efficient Climate Investment and Financing for Africa

Further expand and enrich the financing model in the field of China-Africa green development, continue to provide low-cost, patient and efficient climate investment and financing for Africa, and increase investment in "small and beautiful" projects. Africa faces a huge funding gap in achieving its climate ambitions. Therefore, China should focus on advantageous industries such as hydropower and continue to provide loans, especially long-term loans and concessional loans. Supporting regional green infrastructure construction is the key to promoting the green economic transformation of African countries. In addition, more Chinese private enterprises participating in solar energy, wind energy, environmental product manufacturing and other environmentally friendly "small and beautiful" project investments will inject new vitality into Africa's green development. Make full use of existing special funds in the field of China-Africa green cooperation, such as the China South-South Cooperation Fund on Climate Change and China's donation to the Global Environment Fund, to improve the efficiency of fund use. Use cooperation projects such as the "Green Development Investment and Financing Partnership" to develop innovative financial instruments and hybrid financing paradigms, explore innovative fiscal support methods, leverage private sector participation, encourage multiple entities to participate in hybrid financing, create a North-South-South trilateral cooperation paradigm and project pilot, and form complementary resources and advantages. Coordinate with international financing institutions and bilateral development partners, encourage third-party market cooperation, and invite international investors to jointly develop renewable energy industrial parks with Chinese investors.

### 5.3.3 Support Africa to Establish a Business Environment Conducive to Green and Low-Carbon Cooperation

Support Africa to establish a business environment conducive to green and low-carbon cooperation, and help Africa integrate into the global environmental product supply chain, including improving fiscal incentives, establishing joint ventures, cultivating African green manufacturers and other measures to promote the localization of the industrial chain. Strengthen policy dialogues in green areas such as renewable energy, empower African governments to formulate top-level designs for green and low-carbon development, and create a good business environment for Chinese companies to invest and cooperate in Africa. Encourage the establishment of joint ventures with African companies to cultivate a local green industrial environment in Africa. Relying on cooperation initiatives such as the "Green Silk Road Envoy Program" and the "Belt and Road" Ecological Talent and Environmental Protection Exchange Program, strengthen the skills training of engineering and technical personnel in related industries in Africa, and promote local employment and professional development of technical personnel. Support the construction of regional green infrastructure and improve regional logistics networks. Help Africa integrate into the global supply chain of environmental products, encourage Chinese companies to invest in the manufacturing industry in African countries, establish joint ventures to promote the localization of environmental product manufacturing, promote employment and reduce dependence on imports.

### 5.3.4 Build Renewable Energy Industrial Parks with the "Renewables-Driven-Renewables" Model

Based on policy commitments, African governments should focus on two main areas with China's support: i) expanding renewable energy deployment, especially solar photovoltaic and wind power; ii) developing local renewable energy manufacturing capabilities by establishing renewable energy industrial parks. In fact, such renewable energy industrial parks can be designed as a "renewables-driven-renewables" model, that is, the power supply of the industrial park should come from renewable energy such as solar photovoltaic power stations and wind farms as much as possible; at the same time, the industrial park can assemble and manufacture the parts required for renewable energy equipment such as solar panels and wind power equipment to expand the deployment of these equipment in African countries. Considering that the domestic green product market in some African countries is relatively small at this stage, it is possible to consider establishing green industrial parks at hub-type city nodes, which can be jointly developed by relevant countries and supply renewable energy equipment and parts to neighboring countries. Through this model, the two-way promotion of renewable energy supply in industrial parks and the development of new energy equipment manufacturing industry can be promoted.

### 5.3.5 Support the Ability to Enhance Renewable Energy Development in Africa

Support the ability of African governments to formulate correct investment policies to attract investors in the field of renewable energy manufacturing. Rely on the "Green

Silk Road Envoy" program and the "Belt and Road" Ecological Talent and Environmental Protection Exchange Program to strengthen talent training in Africa's solar photovoltaic and related key industries. Fund the training, employment and career development of African engineers and technicians, focusing on solar photovoltaic and related key industries; share best practices to meet the needs of industrial chain localization, technology application and project development.

## References


1. Xinhua Net. Setting a Good Example for South-South Cooperation. The State Council of the People's Republic of China. (EB/OL). 12 Sept 2023, 24 June 2024. https://www.gov.cn/yaowen/tupian/202309/content_6903494.htm#1
2. Li, Y. The history and achievements of China's south-south cooperation in addressing climate change. *World Environment, 06*(2020), 77–78.
3. Qu, C. Observing China|How Is China 'Doing Its Best Within Its Capabilities' in South-South Climate Cooperation?
4. China and South-South Cooperation. The State Council of the People's Republic of China, 27 Sept 2015. https://www.gov.cn/xinwen/2015-09/27/content_2939505.htm
5. China Climate Change South-South Cooperation Fund. Belt and Road Energy Cooperation Network, 22 Jan 2018. http://obor.nea.gov.cn/v_finance/toFinancialDetails.html?countryId=215&status=2
6. Notice of the State Council on Issuing the 12th Five-Year Plan for Environmental Protection. The State Council of the People's Republic of China, 15 Dec 2011. https://www.gov.cn/gongbao/content/2012/content_2034724.htm
7. UN. Finance and Justice. https://www.un.org/en/climatechange/raising-ambition/climate-finance
8. Carbon Brief. (2021). Rich countries fall $10bn short in climate finance pledges. https://www.carbonbrief.org/daily-brief/rich-countries-fall-10bn-short-in-climate-finance-pledges/
9. African Union. 2021. Agenda 2063: The Africa We Want. Retrieved from https://au.int/en/agenda2063/overview
10. UN. (2022). Sustainable Development Goals: Goal 13 https://www.un.org/sustainabledevelopment/climate-change/
11. Global Development Policy Center. China's Global Energy Finance Database (EB/OL). 26 June 2024. https://www.bu.edu/cgef/#/intro
12. Development Reimagined. (2022). Bolstering Africa's strategic climate change engagement with China. Development Reimagined. https://developmentreimagined.com/2022-report-bolstering-africas-strategic-climate-engagement-with-africa/
13. Development Reimagined. Database: Understanding the progress of China-Africa climate actions since FOCAC 8 (EB/OL). 26 June 2024. https://developmentreimagined.com/tracking-progress-in-china-africa-climate-declarations-since-focac-8-2021/
14. The Third Belt and Road Forum for International Cooperation. Building an Open, Inclusive and Interconnected World for Common Development (EB/OL) 18 Oct 2023, 02 July 2024. http://www.beltandroadforum.org/english/n101/2023/1018/c124-1175.html
15. UNCTAD. The Potential of Africa to Capture Technology-Intensive Global Supply Chains (EB/OL). 16 Aug 2023, 02 July 2024. https://news.un.org/en/story/2023/08/1139782#:~:text=Grynspan%20noted%20that%20Africa's%20tech,chance%20to%20capture%20technology%2Dintensive
16. IEA. (2023). World energy outlook 2023. IEA. https://iea.blob.core.windows.net/assets/86ede39e-4436-42d7-ba2a-edf61467e070/WorldEnergyOutlook2023.pdf

# Chapter 14
# Low-Carbon Transition of Traditional Energy Regions of China

## 1 Introduction

"Research Project on Low-carbon Transition of Traditional Energy Regions of China" focuses on the green and low-carbon transition in the Coal Triangle Area, while also considering issues of fairness, justice, and energy security during the transition process.

## 2 Research Background and Significance

### *2.1 Research Background*

#### 2.1.1 The 28th United Nations Climate Change Conference (COP28) Explicitly Stated that Countries around the World Should Gradually Phase out the Use of Fossil Fuels, and a Series of Initiatives and Alliances on Coal Phase-out Have Been Reached Internationally

For the first time, the "United Arab Emirates Consensus" reached at COP28 explicitly set a critical goal: countries must transition away from fossil fuels in an equitable, orderly, and fair manner within the energy system [1].

---

The content of this chapter is extracted from the main findings of the 2023-2024 CCICED Scoping Study on "Low-Carbon Transition of Traditional Energy Regions of China".



*Adhering to Green and Open Cooperation to Advance Modernization*,
https://doi.org/10.1007/978-981-95-5547-5_14

#### 2.1.2 The Party Central Committee Attaches Great Importance to the Comprehensive Green and Low-Carbon Transition of the Economy under the Dual-Carbon Goals, but Faces the Risk of High-Carbon Lock-in in Traditional Industries

According to relevant research by the project team, in 2020, China's key industries (including power, steel, cement, and petrochemical industries) had carbon locked in at 196 billion tons, the highest among all countries globally, accounting for 65% of China's carbon budget, posing a significant challenge to the dual-carbon goals [2 , 3].

#### 2.1.3 The Chinese Coal Triangle Area (Comprising the Provinces of Inner Mongolia, Ningxia, Shaanxi, and Shanxi) Is the world's Largest Coal Energy Concentration Area and Faces Severe Transition Challenges

There are significant differences in the geographical distribution of fossil and non-fossil energy resources in China. Achieving goals such as the clean utilization of fossil energy, the large-scale development of clean energy, the comprehensive utilization of various energy sources, the electrification of end-use energy, and the integration of industrial energy is essential for the sustainable development of energy production [4, 5].

### *2.2 Overview of the Study Area*

The Coal Triangle Area occupies 16.8% of the national land area, accounts for 7.48% of the national population (as of 2023) and generated 7.1% of the national GDP (in 2023). In 2023, while the per capita GDP in the Coal Triangle region, except for Inner Mongolia, were below the national average.

## 3 International and Domestic Experiences and Insights on Energy Transition

### *3.1 Case Analysis of International Energy Transition*

The EU has the most comprehensive regulatory framework for the green transition. The performance of the EU in terms of greenhouse gas emission reductions is convincing, it has achieved an absolute decoupling of emissions from GDP growth.

Column 3–1 EU energy transition

Column 3–1 EU energy transition [6, 7]

**Question: (1) fiscal policy**: Given the current level of investment, the yearly European climate investment deficit is estimated at 406 billion euros. **(2) slow progress in expanding renewable energy generation**: The relatively slow progress in renewables development in the EU is shown by IEA data on renewables capacity additions and energy composition. **(3) fragmentation, complex governance mechanism between the EU and national level**: EU level policies and objectives are implemented through complicated co-ordination mechanisms. There is a significant diversity among member state
**Solution**: According to the EGD and the 'fit for 55' legislative package, by 2030 greenhouse gas emissions across the EU27 should be reduced by 55% from 1990 levels: This would enable climate neutrality to be achieved by 2050, as laid down in the European climate law
**Effect**: Across the EU27, 2021 net emissions had fallen by 30.4% from 1990 levels. However, the data reveal significant variation among member states. The three best performers are Sweden, Romania and Lithuania. Among main world regions, the EU has managed to cut GHG emissions most and delivered also in absolute decoupling of emissions from GDP growth. EU27 GDP grew by 50% over this 26-year period, while GHG emissions fell by 23.6%

## *3.2 Case Analysis of International Just Transition*

Column 3–2 international just transition [7, 8]

**Question**:
The EU approach of just transition has a focus on addressing the effects of decarbonisation related labour market change. A second focus is on dealing with the distributional effects of climate policies including energy and transport poverty, but trying also to make low carbon technologies accessible and affordable to all
**Solution**: The European Commission originally designed the just transition fund (JTF) to provide social support for workers dismissed when mines or related fossil-fuel-based power plants closed, but it was then extended to meet industrial and regional policy objectives. In the face of a new geopolitical configuration, the EU has ramped up its industrial policy efforts, launching the green Deal industrial plan and the net-zero industry act
**Effect**: The main beneficiaries are mostly CEE member states that have relatively low GDP per capita levels, higher carbon intensity and a higher concentration of affected regions

# 4 Analysis of the Current Situation of the Coal Triangle Area and the Path to Peak Carbon Emissions

## *4.1 Current Situation Analysis*

### 4.1.1 Carbon Dioxide Emissions

Energy activities account for 93% of total carbon dioxide emissions, with industrial processes contributing 7%. Within energy activities, the power industry has the largest contribution, accounting for 58%, followed by industrial processes at 30%. In addition, the transportation and construction sectors contribute 3% and 2% respectively to the carbon emissions from energy activities [9–11].

### 4.1.2 Coal Production

From 2010 to 2023, coal production in the Coal Triangle Area has generally shown an upward trend. By 2023, the production in the Coal Triangle Area had reached 3.45 billion tons, making up 74% of the national coal production.

### 4.1.3 New Energy Development

In 2023, the national solar power generation in 2023 was 294 billion kilowatt-hours, and the solar power generation in the Coal Triangle Area was 82.3 billion kilowatt-hours. In 2023, the national wind power generation in 2023 was 809 billion kilowatt-hours, and the wind power generation in the Coal Triangle Area was 219.8 billion kilowatt-hours. The coal triangle generated 27% of the wind and solar electricity used in China this year (Fig. 1; Tables 1 and 2).

### 4.1.4 Energy Flow Chart

According to the data, in 2023, the coal production in the Coal Triangle Area amounted to 2.43 billion tons of standard coal. Out of this production, 1.68 billion tons of standard coal were allocated for distribution, with a distribution ratio reaching as high as 69% [12, 13].

### 4.1.5 Emission Drive Analysis

We used the Logarithmic Mean Divisia Index (LMDI) method to decompose the changes in carbon emissions from the Coal Triangle Area and from the provinces of Inner Mongolia, Ningxia, Shaanxi, and Shanxi from 2010 to 2023. The specific

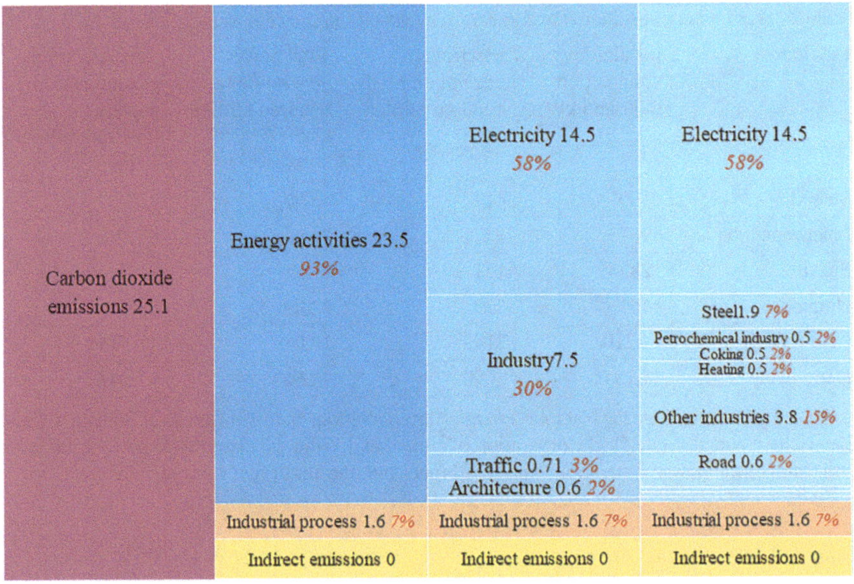


**Fig. 1** 2023 Carbon dioxide emissions inventory for the coal triangle

**Table 1** Solar power capacity and generation in the coal triangle area in 2023

| Province | Installation volume (ten thousand kW) | Electricity generation (a hundred million kWh) | Total power generation (a hundred million kWh) | Total electricity consumption by society (a hundred million kWh) |
|---|---|---|---|---|
| Shanxi | 2491 | 275 | 4376 | 2885 |
| Inner mongolia | 2296 | 205 | 7451 | 4823 |
| Shaanxi | 2292 | 111 | 2946 | 2450 |
| Ningxia | 2137 | 232 | 2246 | 1387 |
| Coal triangle area | 9216 | 823 | 17,019 | 11,545 |
| Nationwide | 60,949 | 2940 | 94,564 | 92,241 |

results are shown in Table 3, where the orange-yellow represents a positive effect, green represents a negative effect, and the deeper the color, the greater the effect (Fig. 2) [14].

An analysis reveals that economic development is the decisive factor for the growth in carbon emissions across the region and the four provinces, while the intensity of energy consumption is the primary factor for the reduction in carbon emissions. The effect of industrial structure adjustment is secondary, and the impact of population and energy consumption structure is relatively small. (1) Economic development is the decisive factor for the increase in carbon emissions. Over the 14-year period from

**Table 2** Wind power installation and generation capacity in the coal triangle area in 2023

| Province | Installation volume (ten thousand kW) | Electricity generation (a hundred million kWh) | Total power generation (a hundred million kWh) | Total electricity consumption by society (a hundred million kWh) |
|---|---|---|---|---|
| Shanxi | 2500 | 477 | 4376 | 2885 |
| Inner mongolia | 6961 | 1271 | 7451 | 4823 |
| Shaanxi | 1285 | 171 | 2946 | 2450 |
| Ningxia | 1464 | 279 | 2246 | 1387 |
| Coal triangle area | 12,210 | 2198 | 17,019 | 11,545 |
| Nationwide | 44,134 | 8090 | 94,564 | 92,241 |

*Note* Data sources include the National Energy Administration, National Bureau of Statistics, China Electricity Council, State Grid Corporation of China, and Wind database. Total power generation is the sum of the power generation from fossil fuels and non-fossil fuel sources in the province

**Table 3** LMDI decomposition results of carbon emission drivers in the coal triangle area and four provinces from 2010 to 2023.

| LMDI decompose | Coal Triangle Area | Shanxi | Shaanxi | Inner Mongolia | Ningxia |
|---|---|---|---|---|---|
| Carbon emissions | 12.64 | 4.20 | 2.27 | 5.16 | 1.01 |
| Population | 0.21 | -0.16 | 0.16 | -0.23 | 0.25 |
| Per capita GDP | 14.83 | 4.25 | 2.58 | 6.47 | 1.39 |
| Industrial structure | -0.60 | -0.59 | -0.14 | 0.79 | 0.01 |
| Energy intensity | -3.55 | -1.25 | -1.16 | -1.59 | 0.13 |
| Energy structure | -0.17 | -0.09 | -0.03 | -0.05 | 0.02 |
| | Factors contributing to the increase in carbon emissions | | | | Factors inhibiting the growth of carbon emissions |

2010 to 2023, for every 1% increase in GDP, emissions increased by 2.69 million tons. (2) Energy consumption intensity is the most significant factor in reducing carbon emissions in the Coal Triangle Area, contributing to a total reduction of 355 million tons. (3) Adjustments in industrial structure have significantly reduced carbon emissions, with the industrial structure effect contributing a total of 60 million tons of emission reductions. (4) The actual reduction in carbon emissions due to adjustments in energy structure is relatively low (Figs. 3 and 4).

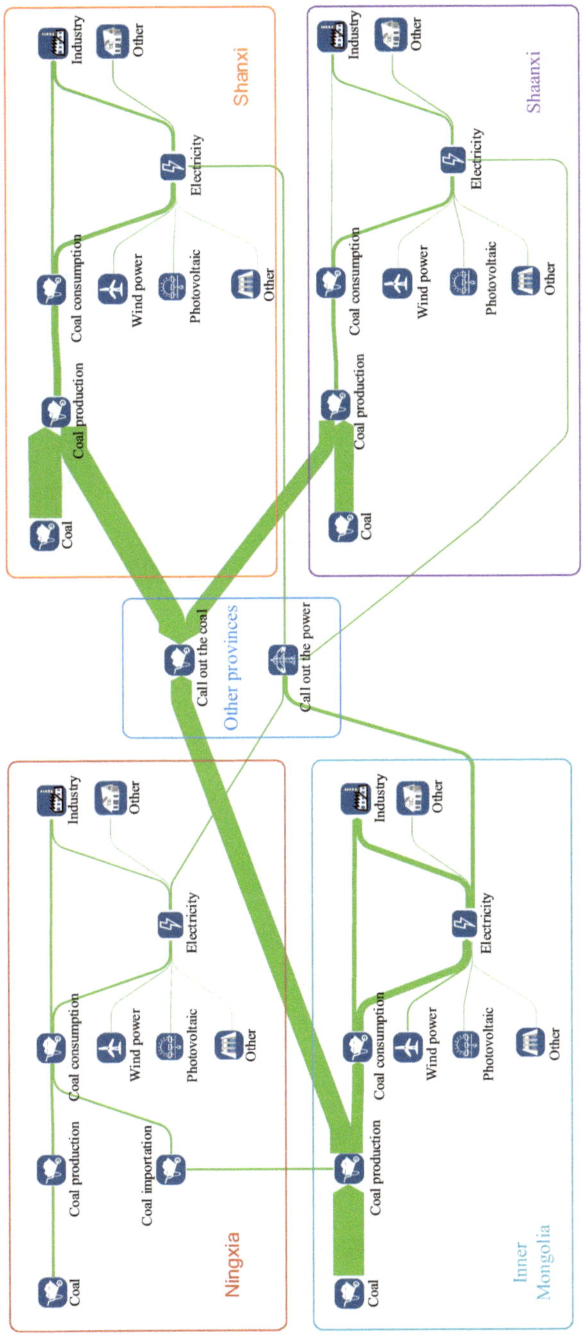


**Fig. 2** Coal and new energy production and circulation diagram of the coal triangle area in 2023

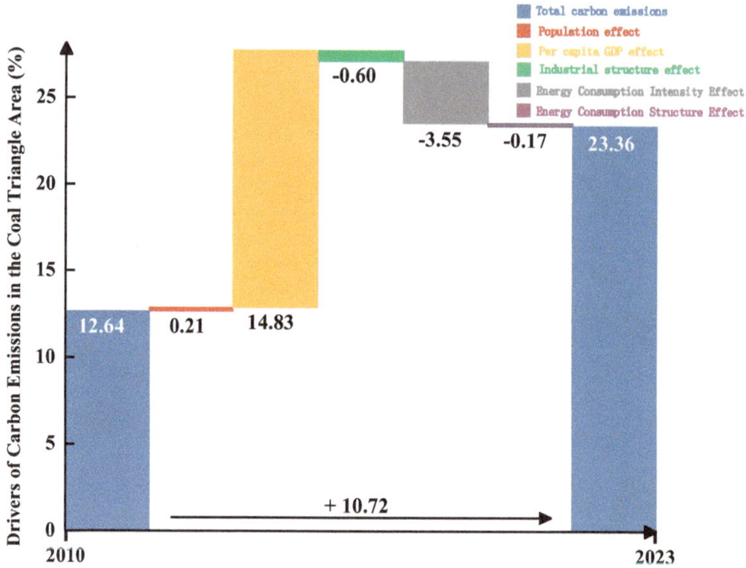


**Fig. 3** Decomposition of driving factors of carbon emissions in the coal triangle area from 2010 to 2023 [15]

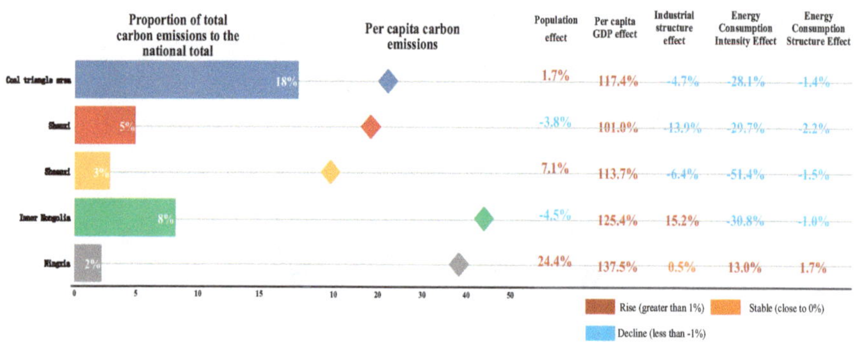


**Fig. 4** Characteristics of carbon emissions and driving factors in the coal triangle area from 2010 to 2023

## 4.2 *Historical Transition Effect Assessment of the Coal Triangle Area 2010–2023*

From 2010 to 2023, the annual average growth rates of thermal power capacity and non-fossil energy capacity in the Coal Triangle Area were both higher than the national average. However, the proportion of non-fossil energy consumption to total primary energy consumption and the proportion of non-fossil energy generation to total electricity generation in the Coal Triangle region were both lower than the national average, although the annual average growth rate of non-fossil energy consumption was higher than the national average (Fig. 5) [16].

**From 2010 to 2023, the carbon emission volume, per capita carbon emissions, and the annual average growth rate of coal production in the Coal Triangle Area were all higher than the national average, while the annual average rate of decline in carbon emission intensity was lower than the national average.** The annual average growth rate of carbon emissions in the Coal Triangle Area from 2010 to 2023 was 4.7%, which is 1.6 times the national level, and the annual average growth rate of per capita carbon emissions was 4.6%, which is 1.7 times the national level. The annual average growth rate of coal production was 4.4%, which is 1.8 times

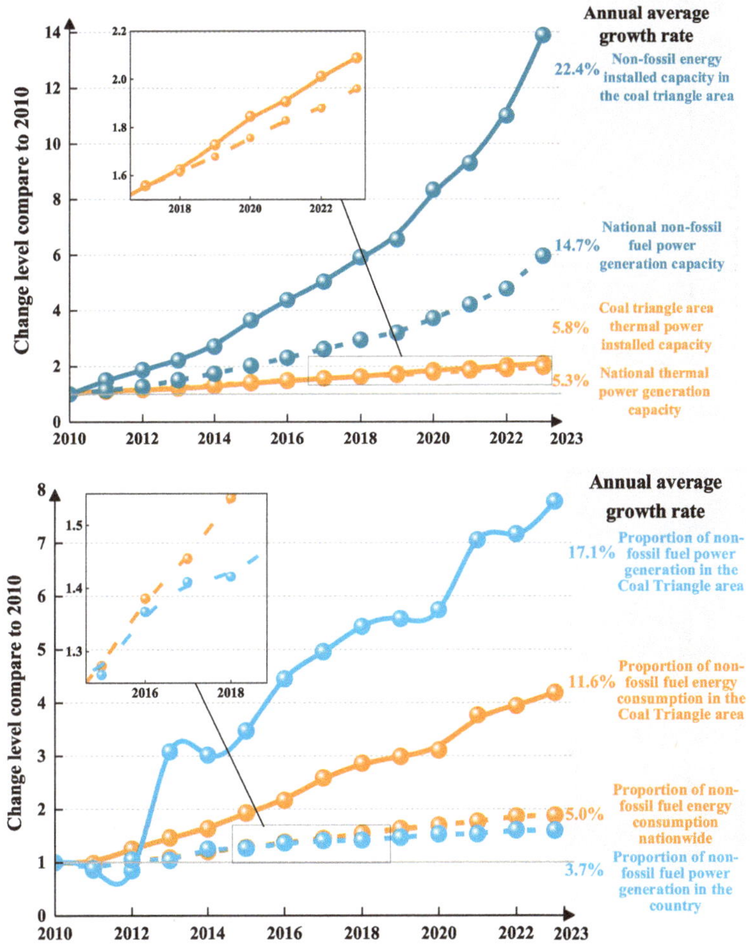


**Fig. 5** 2010–2023 coal triangle area and national energy structure changes

the national level. However, the annual average rate of decline in carbon emission intensity was 2.1%, which is lower than the national level of 3.3%.

**From 2010 to 2023, the GDP growth rate in the Coal Triangle Area showed the same trend as the national GDP growth rate, and the proportion of coal-related industries in GDP showed a downward trend but was still much higher than the national level.** In 2010, the proportion of coal-related industries in GDP in the Coal Triangle Area was 36.78%, which was 1.3 times the national level, and in 2023, it was 25.02%, which was 1.4 times the national level. The average annual rate of decline was 0.6%, which was much lower than the national level of 3.7% (Fig. 6).

Note: This project comprehensively applies the M3C-CGE module in CAEP-CP 2.0 to calculate the proportion of value-added of coal-related industries in the national and the Coal Triangle Area to GDP for different years. The coal-related industries

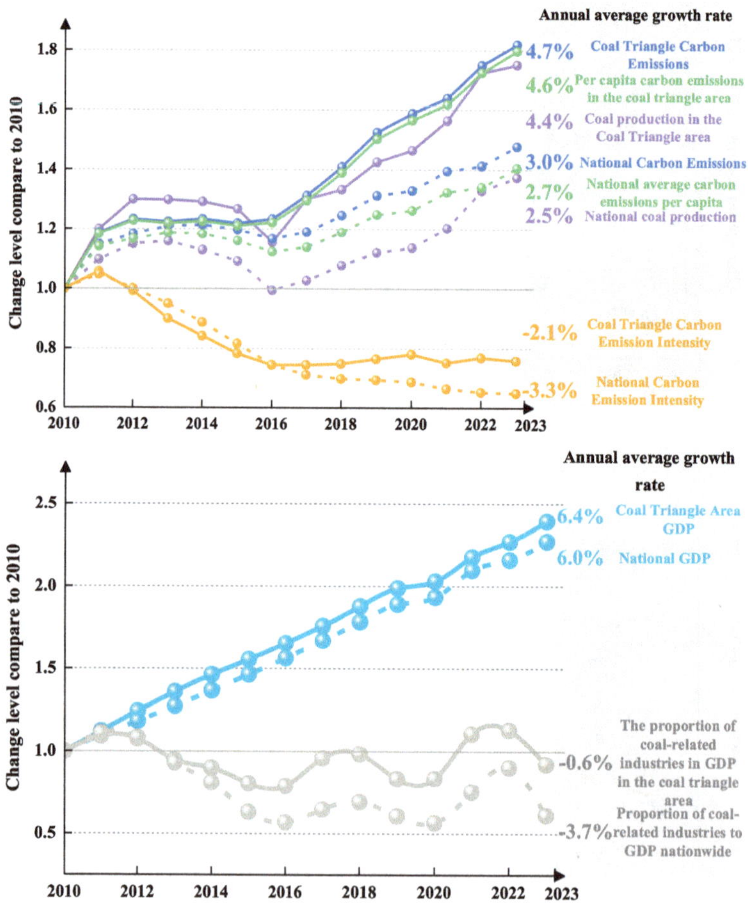


**Fig. 6** Coal triangle area and national carbon emissions (left) and the proportion of coal-related industries in GDP in the coal triangle area and nationwide (right) from 2010 to 2023

specifically include the production and supply of electricity and heat, the processing of petroleum, coking products, and nuclear fuel, as well as the manufacturing of chemical products, among other sectors.

## 5 Challenges and Issues Faced in the Transition of the Coal Triangle Area

### *5.1 The Coal Triangle Has Been Lacking in Momentum over the Past 15 Years of Energy Transition and Has Failed to Achieve Significant Overall Results*

The main reasons for the insignificant results of the energy transition in the Coal Triangle Area are the single energy structure, dependence on coal and thermal power, high and fast-growing carbon emissions, and insufficient efforts to reduce emissions [17].

### *5.2 The Coal Triangle Area Is over-Dependent on Coal-Related Industries and Has Formed a Path Dependence, Facing the Risk of Carbon Lock-in in the Future*

Energy systems based on carbon that benefit from long-term increasing returns may produce a "lock-in effect," hindering the innovation of low-carbon and renewable energy technologies. At the same time, participants benefiting from the existing system will attempt to maintain that system, further strengthening the lock-in of the existing technological system [18, 19].

### *5.3 Breakthroughs in the Transition of Key Areas Have Given Rise to a Batch of Emerging Industries*

As an important area for China's economic transformation, the Coal Triangle Area has achieved significant breakthroughs in the transition of key areas in recent years. The energy transition in the Coal Triangle Area continuously explores new industrial models. Furthermore, the combination of energy transition and rural revitalization has benchmark projects, achieving a win-win situation for "enterprise-village collective-zero-carbon energy."

# 6 Transition Path Analysis

The development objectives of the energy transition path for the triangle area are shown in Table 4.

**Table 4** Energy transition path development goals for the coal triangle area

| Development Goals | | 2023 | 2035 | 2060 |
|---|---|---|---|---|
| | | **Black energy** | **Blue energy** | **Green energy** |
| Energy production | Fossil / Non-fossil | 4% / 96% | 20% / 80% | 20% / 80% |
| | Hydrogen energy | A small test | Grey hydrogen: Green hydrogen 1.2: 1 | Green hydrogen |
| Energy consumption | Fossil: Non-fossil | 8% / 92% | 40% / 60% | 90% / 10% |
| | Hydrogen | Minority gray hydrogen | Hydrogen energy accounts for a proportion of energy consumption 12% | Green hydrogen |
| Carbon removal | Remove quantity | <One million tons | >10 Million tons | 500-800 million tons |
| Industry | The contribution of the coal industry to GDP | 45% | 32% | 5% |
| | The contribution of the new energy industry to GDP | 0.50% | 2.00% | 5.00% |
| | The contribution of the coal industry to employment | 29% | 21% | 3% |
| | The contribution of the new energy industry to employment | 0.10% | 0.20% | 0.60% |

## 6.1 The Coal Triangle Area Emission Pathways

The basic parameters for the social and economic development of the Coal Triangle Area are shown in Table 5 [20].

In the $CO_2$ emission policy scenario for the Coal Triangle Area, the $CO_2$ emissions from coal reached 2.3 billion tons in 2023, reflecting the current state of carbon emissions. From 2023 to 2030, the trend of carbon emissions is on the rise, with $CO_2$ emissions from the Coal Triangle peaking at 2.6 billion tons in 2030. After reaching this peak, from 2030 to 2035, the emissions followed a path of gradual decline, with $CO_2$ emissions from coal expected to reach 2.49 billion tons in 2035, a 5% decrease from the 2030 level. In the enhanced $CO_2$ emission scenario for the Coal Triangle Area, the peak emissions occurred in 2027 at 2.6 billion tons, which was three years earlier than in the 2030 scenario. From 2023 to 2027, the emissions followed a path of moderate growth. After reaching the peak, the emissions decreased rapidly from 2027 to 2035, with $CO_2$ emissions from coal expected to reach 2 billion tons in 2035, a 23% decrease from the peak in 2027 (Fig. 7).

## 6.2 Energy Transition Pathways

### 6.2.1 Energy Transition

Based on the China's medium and long-term emission pathway model (CAEP-CP 2.0) [21, 22], it is expected that by 2035, the total energy production will be 2.2 billion tons, of which fossil fuels will account for 77%, non-fossil fuels 18%, and hydrogen energy 5%, with the ratio of gray hydrogen to green hydrogen in hydrogen production being 1.2:1. By 2035, the coal production in the Coal Triangle Area will see a significant decrease compared to 2023. By 2035, the proportion of fossil fuels in energy consumption in the Coal Triangle Area will decrease to 40%, with hydrogen energy accounting for 12%, and the amount of carbon removal exceeding ten million tons. By 2060, the transition to green energy will be achieved, with non-fossil fuels accounting for 80% and 90% of energy production and consumption, respectively, and all hydrogen energy production and consumption being green hydrogen, with the amount of carbon removal being approximately 500–800 million tons. In terms of coal dispatching, the Coal Triangle Area also shows distinct characteristics. It is expected that by 2035, the area will dispatch out 1.11 billion tons of standard coal, with a dispatching ratio of 65%. In addition to coal dispatching, in terms of electricity dispatching, the Inner Mongolia Autonomous Region leads the provinces in the Coal Triangle Area with an electricity dispatching amount of 13 million tons of standard coal.

**Table 5** Socio-economic development parameters of the coal triangle area

| | GDP/Billion yuan | GDP growth rate | | | Population/Ten thousand people | | | |
|---|---|---|---|---|---|---|---|---|
| | 2020 year | 2020–2025 | 2025–2030 | 2030–2035 | 2020 year | 2025 year | 2030 year | 2035 year |
| Shanxi | 17,836 | 5.5% (5.2%, 5.8%) | 4.0% (3.5%, 4.5%) | 3.2% (2.7%, 3.7%) | 3490 | 3452 (3383, 3521) | 3420 (3317, 3523) | 3394 (3224, 3564) |
| Inner Mongolia | 17,360 | 5.0% (4.7%, 5.3%) | 4.0% (3.5%, 4.5%) | 3.3% (2.8%, 3.8%) | 2403 | 2432 (2383, 2481) | 2424 (2351, 2497) | 2392 (2272, 2512) |
| Shaanxi | 26,014 | 6.0% (5.7%, 6.3%) | 4.5% (4.0%, 5.0%) | 4.1% (3.6%, 4.6%) | 3955 | 4002 (3922, 4082) | 3977 (3858, 3096) | 3910 (3715, 4106) |
| Ningxia | 3956 | 6.0% (5.7%, 6.3%) | 4.5% (4.0%, 5.0%) | 4.4% (3.9%, 4.9%) | 721 | 756 (741, 771) | 781 (758, 804) | 798 (758, 838) |

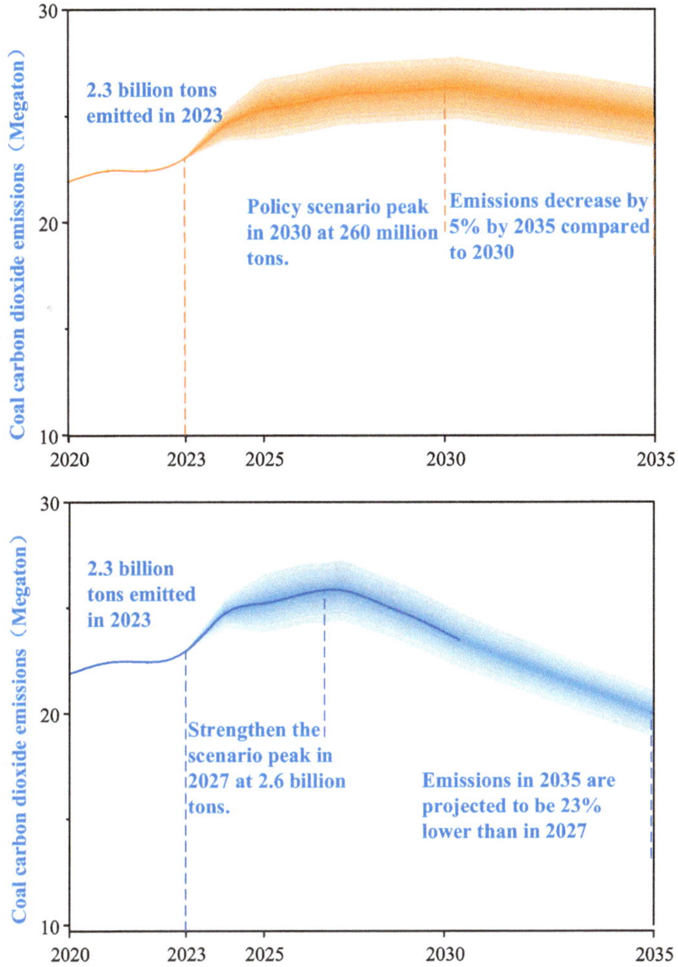


**Fig. 7** Policy scenario (left) and enhanced scenario (right) for $CO_2$ emissions in the Coal Triangle Area

### 6.2.2 Assessment of New Energy Development Potential

The Coal Triangle Area is one of the fastest-growing areas for wind and solar power generation in China. According to the research report "Assessment of Wind and Solar Power Potential in China (2024)," it is expected that by 2025, the installed capacity of solar power in the Coal Triangle Area will reach 110 million kilowatts [23]. According to the same research report, it is expected that by 2025, the installed capacity of wind power in the Coal Triangle Area will reach 150 million kilowatts [24].

### 6.2.3 New Energy Consumption and Power Grid Construction

The energy transition in the Coal Triangle Area relies on the vigorous development and absorption of new energy sources. Specifically, this includes: (1) Scientifically determining the new energy absorption targets for the Coal Triangle Area. Regularly conduct analyses of the new energy absorption capacity of the Coal Triangle Area's power grid, maintaining new energy utilization at a reasonable level, and proposing the scale of new energy that can be connected and absorbed; (2) Promoting the broader optimization of new energy allocation. Fully implement the national major strategic deployments, build ultra-high voltage DC transmission projects to ensure the outward transmission of new energy from the Coal Triangle Area, fully form a DC power transmission pattern to North China, East China, and Central China, further expand the development and outward transmission scale of new energy in the Coal Triangle Area, promote the broader optimization of new energy absorption, and accelerate the transition of energy resource advantages into economic advantages; (3) Continuously enhancing the regulation capacity of the power supply side. Improve the prediction accuracy of wind conditions and sunlight, enhancing the accuracy of power output predictions; promote applications such as electric heating, electric vehicles, and hydrogen production from electricity, expanding the space for new energy absorption in the Coal Triangle Area (Fig. 8).

## *6.3 Technological Transition Pathways*

Specific technological transition paths include: (1) High-quality intelligent coal mining technology, which deeply integrates modern information technologies such as big data, artificial intelligence, blockchain, and the Internet of Things with the coal industry, driving the coal sector towards digitalization and intelligent transformation [25]; (2) Advanced and efficient coal-fired power generation technology, promoting the development of coal-fired power generation towards high-parameter, large-capacity, and intelligent systems, advancing ultra-high parameter coal-fired power generation, new power cycle systems, highly flexible and intelligent coal-fired power generation, efficient and low-cost multi-pollutant joint control of coal-fired power, and the independent industrialization of complete sets of technologies and equipment for resource utilization, facilitating the technological upgrade and structural transition of power equipment [26]; (3) By optimizing the combination of coal, oil, and gas gas resources, breaking the traditional single processing mode of coal chemical and petrochemical industries, effectively solving the problems of "too much carbon and too little hydrogen" in coal-to-methanol and "too much hydrogen and too little carbon" in natural gas-to-methanol, achieving "carbon-hydrogen complementarity," and creating a new model for green, low-carbon, and circular development in the energy and chemical industry; (4) Green electricity to green hydrogen technology, coupling new modern coal chemical low-carbon and clean development [27]; (5) Hydrogen energy storage technology, promoting a comprehensive transition from

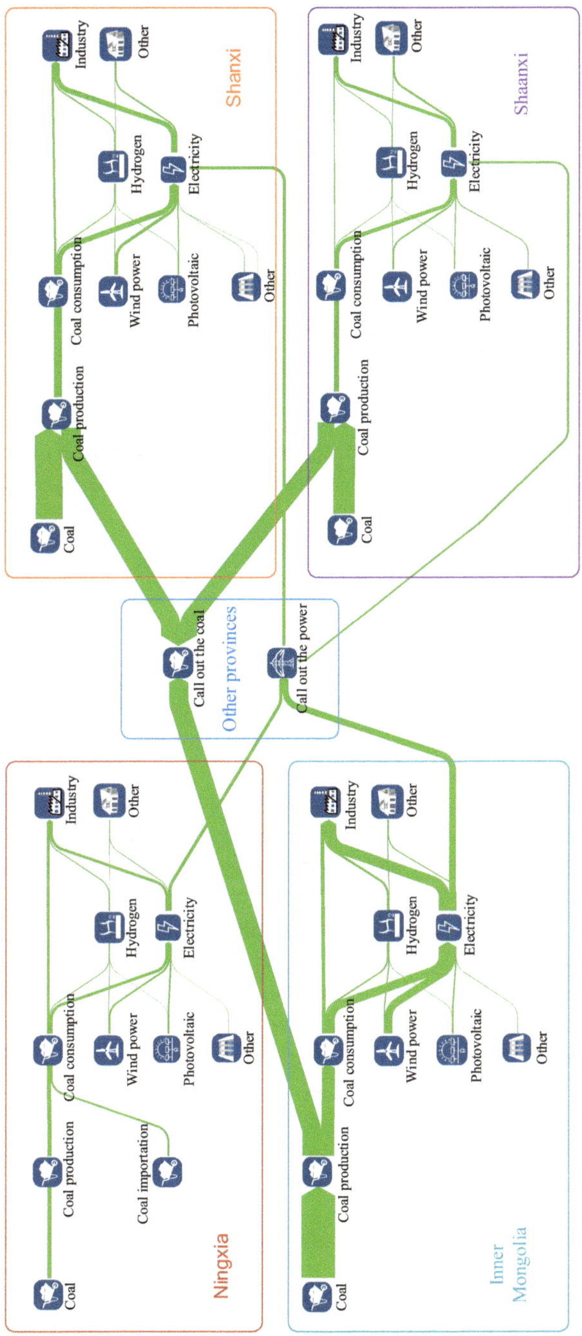


**Fig. 8** Coal and new energy production and circulation diagram for the coal triangle area in 2035

coal-fired power generation to new energy power generation [28]; (6) Utilize CCUS technology (Carbon Capture Utilization and Storage) to separate $CO_2$ from emission sources related to power generation and industrial processes and convert it into useful products or permanently store it [29, 30]; (7) The state and local governments jointly fund the establishment of the "Coal Triangle Area Energy Transition Innovation Technology Research Center".

### *6.4 Industrial Transition Path*

The traditional coal industry is facing enormous pressure to transform. The decline in coal production will lead to a decrease in the added value of traditional industries such as coal-fired power, coal coking, and steel, worsening the overall production and operational conditions of the industry and having a negative impact on economic development [31]. It will also directly affect employment in coal-related industries.

## 7 Fair Transition

The energy transition in the Coal Triangle Area has had a profound impact on the job market [32, 33]. With the decline of the traditional coal industry, a large number of workers are facing the risk of unemployment, especially those with specialized skills who find it difficult to adapt to the requirements of new industries. Moreover, although the rise of new energy industries has created new job opportunities, these positions often require higher skill levels, and the threshold for transition is high for workers from the original coal industry, further exacerbating employment equity issues. The issue of ecological equity during the energy transition process is also not to be overlooked. For residents and communities affected by coal mining, a fair compensation mechanism needs to be established to ensure that they receive reasonable compensation for any losses incurred due to ecological restoration [34]. In addition, the environmental impact assessment and regulation of new energy projects need to fairly consider all stakeholders to ensure that ecological protection measures are effectively implemented and to avoid unfair damage to the ecological environment [8, 32, 35, 36].

# 8 Policy Recommendations

## *8.1 Establish a National-Level Coordination Mechanism for the Low-Carbon Transition of the Coal Triangle Area*

Rationally determine the functional positioning of the Coal Triangle Area in the national strategy for phased peak carbon emissions, fully considering national energy security and energy demand, clarify the national demand for coal and coal-fired power transmission from the Coal Triangle Area, establish a coal production capacity reserve system, and introduce preferential policies to encourage new energy power generation and a green channel for grid absorption, to incentivize non-fossil fuel power generation [37]. Optimize the timetable, roadmap, and assessment methods for the Coal Triangle Area to reach peak carbon emissions. At the same time, fully consider the impact and effects of the low-carbon transition of traditional energy on vulnerable groups, employment, ecology, etc., and explore the establishment of a fair transition mechanism. Encourage the Coal Triangle Area to establish communication channels with international coal areas, leading the global coal areas in low-carbon transition.

## *8.2 Establish a "Zero-Carbon Power Industry and Trade Special Zone"*

It is suggested that the Coal Triangle Area, leveraging its resource endowment, comprehensively establish a "Zero-Carbon Electricity Industry Trade Zone" (referred to as "Zero-Carbon Zone") mechanism. By utilizing new microgrid business models in the region, the Zero-Carbon Zone would ensure a long-term, continuous supply of zero-carbon electricity for the entire area's power system. Additionally, it would offer preferential import and export tax policies to attract export-oriented enterprises to settle in the "Zero-Carbon Zone." Within the zone, a zero-carbon digital certification system would be built to provide full lifecycle carbon footprint certification for products, granting them traceable, internationally standard-compliant "Zero-Carbon Certification" labels.

### 8.3 Establish a Low-Carbon Transition Fund for the Coal Triangle Area

It is recommended to adopt a mixed financing model of "government-guided funds + venture capital/equity investment institutions + corporate investors," where public capital and international development capital provide capital at preferential prices to drive private capital that provides funds at market prices. A financial holding group from the provincial level of the Coal Triangle Area could jointly establish a fund management company responsible for the professional management of the transition fund. The financing model adopts a mixed financing approach, with the fund manager investing as a general partner at 1%. The transition fund is mainly used to support key technological breakthroughs during the transition of the Coal Triangle Area.

## References


1. Sanderson, K. (2023). COP28 climate summit signals the end of fossil fuels - but is it enough? *Nature, 624*(7992), 484–485.
2. Zhao, C., Dong, K., & Lee, C. C. (2024). Carbon lock-in endgame: Can energy trilemma eradication contribute to decarbonization? *Energy, 293*, 130662.
3. Zhao, C., Ma, X., Che, S., et al. (2024). Does climate aid alleviate carbon lock-in? A global perspective. *Journal of Cleaner Production, 449*, 141782.
4. Ning, S. Z., & Cao, D. Y. Coal resources in Shanxi, Shaanxi and western inner Mongolia region and its exploration and exploitation strategy-all databases[EB/OL]. 01 Jan 2010, 09 Aug 2024. https://webofscience.clarivate.cn/wos/alldb/full-record/WOS:000392561100026
5. Song, M., Wang, J., & Zhao, J. (2018). Coal endowment, resource curse, and high coal-consuming industries location: Analysis based on large-scale data. *Resources, Conservation and Recycling, 129*, 333–344.
6. Energiewende, A. (2020). The European power sector in 2019 The European power sector in 2019. agora-energiewende.org
7. Akgüç M., Arabadjieva K., & Galgóczi B. (2022) Why the EU's patchy 'just transition' framework is not up to meeting its climate ambitions, policy brief 2022.06, ETUI. https://www.etui.org/publications/why-eus-patchy-just-transition-framework-not-meeting-its-climate-ambitions
8. ILO. (2015). Guidelines for a just transition towards environmentally sustainable economies and societies for all, ILO. https://www.ilo.org/wcmsp5/groups/public/%2D%2D-ed_emp/%2D%2D-emp_ent/documents/publication/wcms_432859.pdf
9. Chen, Q., Kang, C., Ming, H., et al. (2014). Assessing the low-carbon effects of inter-regional energy delivery in China's electricity sector. *Renewable and Sustainable Energy Reviews, 32*, 671–683.
10. Chen, W., Wu, F., Geng, W., et al. (2017). Carbon emissions in China's industrial sectors. *Resources, Conservation and Recycling, 117*, 264–273.
11. Liu, B., Ding, C. J., Hu, J., et al. (2023). Carbon trading and regional carbon productivity. *Journal of Cleaner Production, 420*, 138395.
12. Sun, X., Li, J., Qiao, H., et al. (2017). Energy implications of China's regional development: New insights from multi-regional input-output analysis. *Applied Energy, 196*, 118–131.
13. Yuan, J., Na, C., Lei, Q., et al. (2018). Coal use for power generation in China. *Resources, Conservation and Recycling, 129*, 443–453.

14. Li, H., Wu, L. 2014. Analysis of Hubei province industry's carbon emissions based on the LMDI. In L. Zhang (Ed.), *Environmental technology and resource utilization II* (Vol. 675–677). Durnten-Zurich: Trans Tech Publications Ltd.
15. Ang, B. W. (2015). LMDI decomposition approach: A guide for implementation. *Energy Policy, 86*, 233–238.
16. Zhou, S., Tong, Q., Yu, S., et al. (2012). Role of non-fossil energy in meeting China's energy and climate target for 2020. *Energy Policy, 51*, 14–19.
17. Jia, Z., Liu, Y., & Lin, B. (2024). The impossible triangle of carbon mitigation policy. *Energy Policy, 189*, 114107.
18. Sun, Z., Liu, Y., & Yu, Y. (2019). China's carbon emission peak pre-2030: Exploring multi-scenario optimal low-carbon behaviors for China's regions. *Journal of Cleaner Production, 231*, 963–979.
19. Unruh G, C., & Carrillo-Hermosilla, J. (2006). Globalizing carbon lock-in. *Energy Policy, 34*(10), 1185–1197.
20. Zhang, H., Shen, L., Zhong, S., et al. (2020). Economic structure transformation and low-carbon development in energy-rich cities: The case of the contiguous area of Shanxi and Shaanxi provinces, and Inner Mongolia autonomous region of China. *Sustainability, 12*(5), 1875.
21. Gong, J. W., Li, Y. P., Lv, J., et al. (2022). Development of an integrated bi-level model for China's multi-regional energy system planning under uncertainty. *Applied Energy, 308*, 118299.
22. Zhang, J., Liu, L., Xie, Y., et al. (2022). An integrated optimization and multi-scale input-output model for interaction mechanism analysis of energy-economic-environmental policy in a typical fossil-energy-dependent region. *Energy Strategy Reviews, 44*, 100947.
23. Ministry of Ecology and Environment of China, Academy of Environmental Planning. China Wind Power and Solar Power Generation Potential Assessment Report (2024) released (EB/OL). 14 June 2024, 09 Aug 2024. http://www.caep.org.cn/sy/tdftzhyjzx/zxdt/202406/t20240614_1075803.shtml
24. Liu, F., & Lv, T. (2019). Assessment of geographical distribution of photovoltaic generation in China for a low carbon electricity transition. *Journal of Cleaner Production, 212*, 655–665.
25. Wang, J., & Huang, Z. (2017). The recent technological development of intelligent mining in China. *Engineering, 3*(4), 439–444.
26. Xie, Z., Wu, R., & Wang, S. (2021). How technological progress affects the carbon emission efficiency? Evidence from national panel quantile regression. *Journal of Cleaner Production, 307*, 127133.
27. European Commission. (2022). REPowerEU: Affordable, secure and sustainable energy. https://commission.europa.eu/strategy-and-policy/priorities-2019-2024/european-green-deal/repowereu-affordable-secure-and-sustainable-energy-europe_en
28. Zhang, S. X., Song, Q., Cheng, H. Z., et al. (2024). Key technologies and challenges of port integrated energy system planning considering energy flow-logistics coupling. *Proceedings of the Chinese Society for Electrical Engineering*, 1–20.
29. Dou, L., Sun, L., Lyu, W., et al. (2023). Trend of global carbon dioxide capture, utilization and storage industry and challenges and countermeasures in China. *Petroleum Exploration and Development, 50*(5), 1246–1260.
30. China 21st century agenda management center. China carbon dioxide capture, utilization, and storage (CCUS) Annual Report (2023) (EB/OL). 29 Dec 2023, 09 Aug 2024. http://m.tsjinquan.com/apollo/dongshow-450619.html
31. Keles, D., & Yilmaz H, U. (2020). Decarbonisation through coal phase-out in Germany and europe - impact on emissions, electricity prices and power production. *Energy Policy, 141*, 111472.
32. Feng, K., Song, K., Viteri, A., et al. (2023). National and local labor impacts of coal phase-out scenarios in Chile. *Journal of Cleaner Production, 414*, 137399.
33. Yuan, J. H., Wang, Y., Yang, X. W., et al. (2023). Employment impact and vulnerability assessment of provincial coal power phase-out under the "dual carbon" goal. *China Population, Resources and Environment, 33*(7), 67–80.

34. Tian, J. Y., & Yao, W. (2013). Research on the ecological restoration of coal mining subsidence area in Shanxi Province. *Advanced Materials Research, 827*, 384–388. https://doi.org/10.4028/www.scientific.net/amr.827.384
35. Conley, S., Konisky D, M., & Mullin, M. (2023). Delivering on environmental justice? US state implementation of the Justice40 initiative. *Publius: The Journal of Federalism, 53*(3), 349–377.
36. Galgóczi, B. (2018). Just transition towards environmentally sustainable economies and societies for all. ILO Actrav policy brief, 2018. The International Labour Office. https://www.ilo.org/wcmsp5/groups/public/%2D%2D-ed_dialogue/%2D%2D-actrav/documents/publication/wcms_647648.pdf
37. Rinscheid, A., & Wustenhagen, R. (2019). ENERGY POLICY german voters would prefer a more ambitious timeline to phase out coal. *Nature Energy, 4*(12), 1016–1017.
38. Jia, J., Tian, S. M., Jiang, E. H., et al. (2023). Research on carbon emission status in the Yellow River Basin and the path to achieve the "dual carbon" goals. *Yellow River, 45*(12), 8–13.
39. Niu, H. L. (2023). Research on measurement and temporal and spatial evolution characteristics of carbon lock-in in China's manufacturing industry. *Statistics and Decision Making, 39*(8), 153–157.

# Chapter 15
# China Council for International Cooperation on Environment and Development Issues Paper 2024

## 1 Introduction

The theme of the 2024 CCICED Annual General Meeting is *Open, Inclusive and Innovative Cooperation to Build a Clean and Beautiful World.* The 2023–2024 research program has been the most extensive of the 30+ years history of CCICED, involving 17 research groups and support working groups. This Paper highlights some of the main themes and issues of the 2024 Annual General Meeting theme.

## 2 Connected, Cascading Risks

Environmental protection has traditionally examined specific pollutants and stressors. However, a growing body of real-world events underscores the deep interconnections among climate change, biodiversity loss, and pollution.

**Climate Change**: The accelerating pace of environmental degradation increases the necessity for more urgent, ambitious action at scale to build a clean and beautiful world. Over a century of meteorological records confirm that 2023 was the hottest year in the 174-year observational record, breaking previous records for ocean heat, sea level rise, disappearing Antarctic sea ice, and glacier retreat. The world has now likely passed the 1.5 °C average global warming target so vital to the Paris Climate Agreement, with global average temperatures in 2023 reaching an average annual value of 14.98 °C, the hottest ever recorded. Some experts are preparing global overshoot scenarios, while the 2023 *Global Tipping Points* report warns that once average levels exceed 1.5 °C, the risks of cascading, non-linear, and irreversible systemic climate change increase significantly.

The content of this chapter is extracted from "China Council for International Cooperation on Environment and Development Issues Paper 2024".



*Adhering to Green and Open Cooperation to Advance Modernization*,
https://doi.org/10.1007/978-981-95-5547-5_15

**Biodiversity Loss**: Estimates of biodiversity loss vary, with recent estimates published in Science suggesting average global losses of between 2% and 11%. Among the most important drivers of biodiversity loss is land-use change, with the conversion of land for food production remaining a major driver of biodiversity loss. Underscoring the interconnection between climate change and ecosystems, recent evidence suggests that the combination of land-use change and climate change will accelerate biodiversity losses across all regions of the planet.

**Pollution**: Air, water, and other pollution sources are similarly exacerbated by warming temperatures and ecosystem losses. Air pollution has been linked to more than 6.7 million deaths in 2019. Other studies show additional effects, such as higher breast cancer risk for women living near major motorways and higher rates of lung cancer risk for populations exposed to coal-fired energy production. Water pollution remains a major source of communicable human disease, while the increased frequency and severity of extreme weather events led by flooding affected 1.6 billion people in 2022.

## 3 Risks to Actions

**Climate Mitigation**: China will likely meet its carbon peaking target before 2030, thus beginning the shift from peaking to carbon neutrality before 2060. Since this shift will likely take place during the 15th Five-Year Plan and during the updated UNFCCC nationally determined contributions (NDCs), several of CCICED's 2024 studies have made recommendations in these two critical areas.

Several CCICED working groups examined different yet complementary low-carbon measures. Examples range from speeding up the use of clean power in the electric grid—which, in addition to reducing coal usage, provides additional benefits in sectors that are hard to abate, through the use of modern industrial heating and cooling technologies—to the continued modernization of the electric grid and growth of mobility electrification, supported by the world's largest network of EV charging stations. AI is increasingly used to improve green electricity and net-zero industrial systems, while efforts need to be redoubled to reduce the growing carbon footprint of AI and data centres.

Complementary measures include the ongoing reform of China's *Energy Law*, which is expected to enhance green energy channels, bolster inter-provincial green energy trade, and support more market-oriented pricing. Such important steps can find policy synergies with the expected expansion of the national carbon market.

**Fossil Fuels and Stranded Assets**: In many ways, the 2023 Conference of the Parties, or COP28, marked a breakthrough by calling for the transition away from fossil fuels. With this decision, the longer-term implications of this decision on the fossil fuel sector have been brought sharply into focus, including the widespread use of the term "stranded assets.[1]" At the same time, there is a mismatch between the general use of stranded assets and its actual use via company financial accounting standards and rules. Currently, there are no specific standards for stranded assets, meaning a CEO who voluntarily devalues a company's asset holdings could face charges of illegal activity and be fired.

Accounting standards grapple with asset valuation, focusing, for example, on current asset valuation based on historical data or more recent fair value standards. The most common accounting rule used by all companies and households involves normal asset depreciation rates, of which none come close to the COP28 fossil fuel transition that would open stranded assets scenarios. The closest proxy are standards for impaired assets, that is, a devaluation of an asset due to three drivers: technological innovation that renders an incumbent technology obsolete, change in consumer demand to new goods and services, and mandatory government regulations that require the phase-out of the older technology.

All three are underway in the transition to carbon neutrality. However, with oil supermajors insisting that their total output will continue to increase to 2050, the role of more stringent mandatory standards covering energy efficiency, automotive fuel efficiency standards, and the planned phase-out of internal combustion engines is likely to become more important.

The international community can act together to control methane emissions. Cutting methane offers the biggest short-term climate mitigation payback, by averting as much as 0.5 °C in global warming.

**Climate Adaptation**: As global temperatures break records, more countries, communities, and families face the effects of extreme weather led by acute events such as floods, prolonged droughts, wildfires, and heat waves, as well as chronic effects, such as shifts in vector-borne diseases and growing risks of food insecurity. While climate change is a global challenge, effective climate adaptation takes place at the local level, in which adaptation actions are tailored to meet the specific characteristics of local communities and their citizens.

In 2023, the Intergovernmental Panel on Climate Change (IPCC) noted progress in adaptation planning across all regions and sectors. Examples cover many areas, from irrigation and fit-for-purpose early warning systems to agroforestry and better land-use planning, guided by accurate flood maps, often improved with the use of artificial intelligence (AI). However, maladaptation remains pervasive, hampered by a lack of information and financing and rigid regulations that require assets to be rebuilt as they were rather than providing incentives to build back better.

[1] The Background Note on **Accounting for the Enormous Potential Lost Financial Value of Fossil Fuel Physical Assets** has been commissioned to examines climate-related stranded assets, including how current financial and management accounting standards and practices might cover some aspects of stranded assets through rules like asset deprecation and asset impairment. The note also explores gaps in current standards and recent relevant trends. The note has been prepared as a background technical paper in support of CCICED's work ahead of the 2024 Annual General Meeting.

The 2023–2024 Work Plan included CCICED's first climate adaptation work. A key feature is its identification of key adaptation capacities or capabilities that together comprise effective actions. Another feature involves working with natural systems under the umbrella term nature-based solutions (NbS) that include sponge cities and other green infrastructure. Among the key challenges include learning from past mistakes, strengthening links between central agencies and local government, and boosting climate adaptation finance.

**Biodiversity and Oceans**: As noted, the Kunming-Montreal Global Biodiversity Framework (KMGBF) sets out the blueprint to halt and reverse biodiversity loss. Work continues towards the "30x30" conservation target for terrestrial and marine areas by 2030. The use of such tools as remote sensing and the ecological redline continue to support meeting this key goal within China, in which its current network of roughly 10,000 natural protected areas covers 18% of its land area. More work will be needed in most countries to ensure high levels of ecological integrity within protected areas, as well as match progress on land with marine protected areas. Challenges remain around developing measurement systems that provide comparable quantitative data to help determine progress toward the 2030 goal.

In addition to the "30x30" goals, CCICED's work on the GBF implementation has initiated research related to wider sustainable use challenges, including in the agricultural and natural resource sectors. A growing body of successful regenerative food system approaches demonstrates practical solutions to reducing pollution and soil loss, embedding climate resilience, and working with ecosystems in ways that ensure farmers, their families, and communities benefit from such transformation. However, these and other GBF targets will be stymied by the persistence of environmentally harmful subsidies, estimated in 2023 to be USD 635 billion a year, leading to the overuse of fertilizers, pesticides, and other inputs that continue to increase soil contamination and land degradation.

In many ways, oceans are at the forefront of the triple crises: from the overexploitation of fish stocks and loss of key marine habitat to land-based sources of pollution and plastics and the carbon footprint of the marine sector. With sea level rise accelerating, coastal communities are increasingly at risk of extreme weather events like typhoons and flooding. CCICED's work continued to examine opportunities to shift to a more sustainable blue economy, in which carbon-neutral marine operations, as well as marine carbon sequestration systems, will play a vital role in China's dual control and biodiversity goals.

**International Cooperation and Rule-Based Competition**: As geopolitical tensions and uncertainty increase, the importance of international cooperation to meet the Sustainable Development Goals, Paris Climate Agreement, GBF, and other goals has never been more important or challenging. An important focus of CCICED's 2024 research involves promoting south-south cooperation channels, through which China works in partnership with developing countries to help meet climate and other goals. As China is likely to exceed its carbon peaking timeline, it can share its strategy with other countries—especially developing countries—as a means to bolster their confidence in their ability to meet NDC targets ahead of schedule.

To date, the bulk of trade in such green goods and services as renewable energy or net-zero industrial technologies has involved Organisation for Economic Co-operation and Development member countries and China. By contrast, African countries collectively received 77.5% of all technologies, and the Association of Southeast Asian Nations countries received 65%. CCICED's work focuses on ways to increase cooperation, including examining significant bilateral opportunities involving China-Brazil and China-Indonesia, in areas such as the sustainable sourcing of soft commodities like soy and palm oil, increased financing for clean technologies, and cooperation in global supply chains of the minerals and metals vital to the green transition.

CCICED also looked at the growing role of trade in green technologies, led by solar and electric vehicles. Estimates by the World Trade Organization (WTO) show that the pace of trade in environmental goods and services has marginally outpaced other traded goods and services. The total proportion of trade in green goods could reach 15% of all merchandise trade by 2030. China has become a leading exporter of several green technologies.

The rule-based trading system is increasingly strained. While national security was used in less than a handful of cases during the decades of the General Agreement on Tariffs and Trade (GATT) and the early years of the WTO, there are now hundreds of export restrictions under national security provisions, many of which involve the materials at the centre of the green transition. Other measures, such as the Carbon Border Adjustment Mechanism (CBAM) and the EU Deforestation Regulation (EUDR), have raised the profile of the trade-environment nexus. A key finding of CCICED's work is the preference for multilateral solutions, such as the WTO or regional agreements that support rule-based systems.

## 4 Comprehensive Solutions

**Synergistic Solutions**: CCICED's work has been guided by the triple crises of climate change, biodiversity loss and pollution. Among the findings of nearly all working groups is the need for policy solutions that magnify synergies linking low-carbon green pathways, the protection of ecosystems, and the means to further reduce pollution. Examples range from the use of green infrastructure and nature-based solutions to co-control systems that simultaneously reduce criteria air pollutants, such as NOx (nitrogen oxides), SOx (sulfur oxides), and PM2.5 (particulate matter), as well as greenhouse gases.

**Green Industrial Policy**: There has been a resurgence of industrial policies focused on the low-carbon transition in a growing number of countries, with the International Monetary Fund's Global Alert Trade tracking more than 2500 industrial policy measures in 2023, of which a large number are tied to climate mitigation and supply chain resilience. Some common industrial policy characteristics to enhance competitiveness and innovation comprise a mix of an investment-led push and demand-side pull to scale up green technologies and systems, coupled with tighter regulations and standards, market-based pricing measures, green public procurement, and other measures. A shared challenge is to address the trade-distorting effects shared by many industrial policies.

Recently, China has advocated "new quality productive forces" as the new foundation of its economic development. High-quality green development is at the centre of this concept, led by the new trio of green technologies—solar panels, electric vehicles, and long-term battery storage technologies. Targets to scale up green innovation as the major engine of economic growth and productivity continue to transform China's electricity and automotive sectors, with China now producing roughly 80% of all solar panels and a major global leader in electric vehicles.

A challenge is maintaining the pace of innovation. Among the findings of a recent report on EU competitiveness from the European Commission, authored by the previous Italian Prime Minister and President of the European Central Bank, Mario Draghi, are steps to narrow the innovation gap through multiple measures, including increased public financing, the streamlining of regulatory and standard systems, and realizing synergies through scale effects. A key opportunity rests with those companies able to pioneer the next generation of green technologies at scale, from green hydrogen and net-zero steel to carbon storage. For China, increasing public financing for green technology research and development should be linked with the greater role of capital market activity through venture capital to accelerate the commercialization of new green technologies.

**Private-Sector Innovation**: CCICED's research in 2024 has emphasized the vital role of the private sector in green outcomes. Examples range from the role of commodity producers in implementing zero-deforestation global soft commodity supply chains as a means of protecting biodiversity and reducing the major climate repercussions of tropical forest loss, to steps by innovative fashion and textile companies in cutting their carbon, pollution, and water footprints while anticipating a growing demand among consumers—especially younger consumers—for green lifestyles. A recurring emphasis of CCICED's work is matching the promise of AI and digitization with green innovation.

**Closing the Implementation Gap**: A unique feature of the current multilateral agenda is the clarity of action needed toward 2030 and mid-century targets. The Paris Climate Agreement target of 1.5 °C, together with complementary targets such as tripling renewable energy, doubling energy efficiency, and increasing climate financing, set the direction both for domestic action and international cooperation. The 23 targets of the Global Biodiversity Framework are intended to work in tandem to halt and reverse biodiversity loss. Pledges, phase-out schedules, and funding to developing countries to tackle hydrofluorocarbons (HFCs) under the Montreal Protocol are already producing co-benefits to protect stratospheric ozone and cut a potent short-lived climate pollutant. Similar benefits can be accrued by addressing methane, a highly potent greenhouse gas. International air quality standards, drinking water standards, and other standards play similar roles in benchmarking action.

The Sustainable Development Goals encompass these and other actions, underscore the important synergies, co-benefits and interconnections across policies, and the need for international cooperation to deliver results.

A major focus of CCICED's recent work remains implementation. As the green transition and related industrial policies cover most economic sectors, economywide measures, and demand-side drivers, the need for coherent, coordinated, efficient, and cost-effective implementation outcomes is paramount.

Yet, as policies and measures become more comprehensive, evidence suggests they are falling short in many jurisdictions. In the case of climate change, greenhouse gas emissions have never been higher than in 2023 (International Energy Agency), average global temperatures were the highest on record (World Meteorological Organization), and disasters driven by extreme weather events increased.

A welcome focus of recent work is unpacking why climate mitigation and other measures are not delivering the expected results. In 2024, the International Organization of Supreme Audit Institutions launched its Climate Scanner to share the results of implementation audits covering public policies, governance systems, and finance. A common challenge for most jurisdictions is policy coherence between and across ministries and across federal, sub-federal, and municipal jurisdictions. Recent analysis led by the Potsdam Institute examined some 1500 climate mitigation measures across 41 countries, concluding that only 63 delivered GHG mitigation outcomes in an effective way, while analysis of implementation results classified through four clusters of climate mitigation policies has begun unpacking what works, what doesn't, and why.

Obstacles to effective implementation include the persistence of billions in annual subsidies to fossil fuel producers as well as other environmentally harmful subsidies in the agricultural, natural resource, and other sectors. Policy ambition, measured through performance standard requirements or carbon price stringency, is an important determinant of outcomes. The analysis also shows that more policies do not always mean stronger outcomes: a key determinant of effective climate mitigation measures is the interaction of market-based pricing measures with a small set of well-designed green subsidies, performance standards, and others.

**Building Inclusion and Equity**: One of the most important lessons in actions aimed at cutting pollution, tackling climate change, and conserving nature is that no policies will work unless they are designed at the outset to be inclusive, build equity, seek input to the co-design of policies, and provide opportunities by way of good jobs, clean communities, and wider equity returns, including concrete steps to enhance gender equity. These values resonate especially strongly with young people, who look for more ambitious environmental action by their employers.

Among the pledges adopted at the *UN Summit of the Future* meeting in September 2024 is the commitment to significantly increase investments to close the gender gap, including by strengthening institutional support for gender equality and the empowerment of women, including advancing reforms giving women equal rights to economic resources, access to ownership, financial services, inheritance, natural resources, and appropriate new technologies. CCICED also maintained its focus on gender equity throughout its working groups, as reflected in the 2024 Gender Report.

The Summit of the Future also pointed to the role of Beyond GDP measurement systems in order to advance a comprehensive system that places as much weight on social capital, human capital, and natural capital as it does on produced capital and financial capital. By going beyond GDP indicators, such a framework can help support the 2035 *Beautiful China* transition target.

# Chapter 16
# China Council for International Cooperation on Environment and Development (CCICED) Policy Recommendations to the Chinese Government at the 2024 Annual General Meeting

## Further Deepen Comprehensive Reforms, Seize New Opportunities Brought by the Technological Revolution and Industrial Transformation, and Accelerate the Comprehensive Green Transition of the Economy and Society

The World Meteorological Organization (WMO) reports that the decade from 2014 to 2023 has been the warmest on record globally. In 2023, the temperature increase reached approximately 1.45 °C, and the time between now and reaching the global tipping points is shrinking. The implementation of the Paris Agreement and other multilateral agreements is facing urgent crises. However, the global economy is experiencing multiple challenges, including insufficient growth momentum, fragmented international trade, and geopolitical tensions. The global gap in achieving the United Nations Sustainable Development Goals (SDGs) is widening, further emphasizing the importance of international cooperation. The "UAE Consensus," adopted at the 28th Conference of the Parties to the United Nations Framework Convention on Climate Change (COP 28), explicitly calls for tripling global renewable energy capacity and doubling energy efficiency by 2030, as well as the transition away from all fossil fuels in energy systems. This reflects the high degree of consensus among all parties on the urgency of addressing climate change.

---

The content of this chapter is extracted from "China Council for International Cooperation on Environment and Development (CCICED) Policy Recommendations to the Chinese Government at the 2024 Annual General Meeting".



*Adhering to Green and Open Cooperation to Advance Modernization*,
https://doi.org/10.1007/978-981-95-5547-5_16

At the same time, a new wave of technological revolution and green industrial transformation characterized by digitalization, greening, and low carbonization is advancing rapidly. New frontiers for future growth and development are beginning to emerge, exploring new pathways to address major challenges facing human society, such as climate change and energy crises. China's renewable energy and energy storage technologies are developing at an unprecedented pace, achieving the country's renewable energy installation capacity target 6 years ahead of schedule and reaching a 50% penetration rate for new energy vehicles 11 years ahead of schedule. The high-quality and affordable green technologies provided by China have also instilled confidence in the global green transition process and achievement of the Paris Agreement goals.

CCICED Members highly commend China's remarkable achievements in building an ecological civilization and its key role in global environmental governance on issues such as climate change and biodiversity. They recognize that China is providing the world with leading green technologies and products, making significant contributions to the global green transition.

Based on the joint policy research conducted by Chinese and international experts and discussions during the 2024 Annual General Meeting, CCICED Members put forward the following recommendations in this document, titled ***Further Deepen Comprehensive Reforms, Seize New Opportunities Brought by the Technological Revolution and Industrial Transformation, and Accelerate the Comprehensive Green Transition of the Economy and Society***, for reference by governments and the international community:

**Adopt a systematic and integrated approach to address environmental, climate, and development challenges and promote global sustainable development**. In response to the urgent need to achieve the 2030 SDGs, it is essential to rely on the coordinated efforts of proactive governments and effective markets. Integrate global environmental issues into the planning of various sectors, including the economy, society, energy, transportation, food, health, industry, agriculture, trade, and finance to enhance policy synergy and efficiency, enabling high-quality sustainable development. Align the new round of economic stimulus policies more closely with pollution reduction, carbon mitigation, environmental quality improvement, energy transition, and climate adaptation to ensure precise and targeted actions.

**Address the significant externalities of climate change with long-term, phased, and quantitative emission reduction targets to provide certainty, which is critical for economic development**. Both China and the rest of the world have invested vast resources in achieving carbon neutrality, making it essential to maintain the strategic focus on the "dual carbon" goals. In this context, new nationally determined contributions (NDCs) could include quantitative targets for 2030 and 2035 to send strong demand signals to the market. Policy tools such as total emission quota settings can demonstrate a broad and systemic commitment to the green transition, establishing clear medium- and long-term expectations for investment, research and development, and innovation in green and low-carbon industries. These measures will also help expand green consumption and foster sustainable lifestyles.

**In a timely manner, raise renewable energy development goals to achieve a win-win outcome for the economy and climate.** Set renewable energy targets that align with economic development patterns, and accelerate their deployment. Translate the "dual-control" carbon emission targets into specific quantitative indicators, breaking them down by key sectors, enterprises, and various levels of government to stimulate demand and establish effective climate target monitoring and accountability mechanisms. Accelerate the iteration of green technologies and products both domestically and internationally. Create economies of scale in green industries to enhance green competitiveness.

**Make climate adaptation a political and governance priority, and develop short-, medium-, and long-term plans.** Central and local governments should take a leading role by formulating systematic action plans, providing climate vulnerability assessment frameworks, and encouraging active participation from all stakeholders. Establish rapid data-sharing platforms, such as early warning systems, to enhance climate adaptation efforts. Proactively identify resilience gaps in urban and rural areas, prioritizing densely populated cities and vulnerable coastal regions as key areas for climate adaptation.

**Deepen international cooperation on climate and environmental issues, and improve the global governance systems.** The world is fast approaching a series of critical tipping points at which damage to the global ecological environment may become irreversible. However, trade protectionism targeting green products and technologies in some countries risks delaying the global green transition, making the Paris Agreement goals increasingly difficult to reach. Such trade barriers must be avoided. Strengthened international cooperation is essential to align actions on climate, biodiversity, oceans, and environmental protection. Leverage initiatives such as the Green Belt and Road, South–South, and Triangular cooperation to uphold international trade rules and collectively build a clean and beautiful world.

Specific recommendations are as follows:

# 1 Promote Comprehensive Socio-Economic Green Transformation during the 15th Five-Year Plan Period

1. **Leverage the update of NDCs, setting ambitious, quantitative, and time-bound targets to accelerate the green and low-carbon transition.** The new round of NDCs will send a strong signal that all governments, including China, are steadfast in their commitment to achieving the Paris Agreement goals, promoting high-quality development, and advancing the comprehensive green transformation of the economy and society. Prioritize setting emission reduction targets for carbon dioxide and non-carbon dioxide gases such as methane for 2030 and 2035. Strengthen the synergy between renewable energy and new power grids to facilitate the integration and consumption of wind and solar energy. Accelerate the development of key technologies, such as solar, wind, green hydrogen, large-scale energy storage, electric

vehicles, and smart grids. Explore developing a strategic roadmap for the application of new carbon dioxide removal technologies, such as direct air capture (DAC). Further strengthen the role of nature-based solutions.

2. **Enhance synergy and efficiency between the forthcoming "Ecological and Environmental Code" and related regulations to comprehensively deepen ecological civilization reform**. Advance the formulation of laws related to climate change and the "dual carbon" goals. Ensure coordination between the "Ecological and Environmental Code" under development and other related regulations, and explore incorporating climate change response into the environmental public interest litigation system. Further reform the carbon market, studying the establishment of a total emissions cap and a 10-year declining emissions trajectory for the power sector. Gradually expand the coverage of the carbon market in a phased and orderly manner, following principles such as "starting with easier tasks, prioritizing sectors with high emission reduction potential and sectors with significant international trade impacts, and offering multiple low-carbon technology options." Improve the carbon pricing mechanism, and promote the gradual alignment of domestic carbon market rules with international standards.

3. **Emphasize the government's ability to integrate and disseminate green technology-related information, and formulate a new round of green industrial policies driven by innovation**. Establish a public information monitoring platform for green and low-carbon industries and technologies, and release data and information on industrial capacity utilization, market supply and demand, as well as technology investments and applications. Develop a green and low-carbon technology directory that balances emission reduction benefits, costs, and industrial innovation, aligning it with subsidy, financing, and other policies. Conduct regular assessments, and establish mechanisms for adjusting and phasing out technologies in the directory. Revise subsidies and tax policies with harmful environmental impacts, and strengthen government procurement support for green and low-carbon products. Consider pilot projects for green and sustainable consumption in the fashion sector, engaging youth to lead the trend toward green consumption.

4. **Promote the coordinated development of the domestic and international green supply chains**. It is recommended that the country establish cross-departmental working groups to study and promote the synergies between climate and trade policies. Restructure the industrial organization upstream in the industrial chain, improve industrial efficiency, and strengthen entry thresholds to maintain moderate competition and reasonable pricing. Encourage technological innovation to ensure the healthy and orderly development of the industry. Accelerate global cooperation in renewable energy and other green industrial supply chains, support low-carbon energy transition in developing countries, and contribute to achieving the goal of tripling global renewable energy installed capacity by 2030.

5. **Elevate the National Climate Change Adaptation Strategy to one of the highest-level national development strategies to reflect the urgency of climate adaptation**. Incorporate climate adaptation goals and requirements into the 15th

Five-Year Plan, particularly in comprehensive plans such as spatial planning. Promote the development of foundational institutional frameworks to support the climate adaptation strategy, including legal and regulatory policies, fiscal and financial frameworks, and knowledge and data systems. Fully promote the establishment of a multi-level climate adaptation assessment mechanism from national to local levels. Develop short-term action plans for climate adaptation in response to climate risks, and outline medium- and long-term plans and visions for 2035, 2050, and 2100.

## 2 Vigorously Promote the Development of Green and Low-Carbon Industries

6. **Accelerate the development of green and low-carbon industries, and expand the economies of scale of the green economy.** Promote the organic connection between green electricity markets, green certificate markets, renewable energy power quotas, green procurement, and the carbon markets. Fully leverage economies of scale of the green economy to support the development of green, low-carbon industries. Enhance the role of artificial intelligence in the decarbonization process. Consider setting clear development targets for renewable energy to stabilize market expectations and expand market size. For example, by 2030, achieve the installed capacity of wind and solar power of over 2400 GW, accounting for over 50% of total power generation; and by 2040, increase the installed capacity of wind and solar power to over 6000 GW, accounting for over 80% of total power generation.

7. **Promote the application of innovative technologies such as industrial heat pumps, thermal batteries, and zero-carbon steel, utilizing fiscal and tax policies, extended producer responsibility, and circular economy tools to accelerate the decarbonization of key industries while enhancing resource utilization efficiency.** Provide research funding and fiscal and tax policy support for electrification technologies like heat pumps and thermal batteries. Gradually phase out industrial self-owned fossil-fuel power plants, and implement increasingly stringent energy efficiency and emissions standards to expedite industrial decarbonization. Accelerate grid decarbonization through innovative measures such as promoting interprovincial electricity trading, adopting smart grids, implementing market-driven green power pricing, and integrating vehicle-to-grid technologies. Promote best practices in the circular economy, and implement standards for product eco-design. Explore extended producer responsibility systems for key industries such as batteries, plastic products, textiles, and electronics. Reduce the carbon footprint of data centres, artificial intelligence, and the digital economy. Develop regulatory frameworks for algorithm and data governance to ensure transparency and compliance during usage.

8. **Establish a just transition mechanism to strengthen the green and low-carbon transformation of industries in traditional energy regions.** For regions

heavily dependent on coal and other traditional energy sources, establish a national-level coordination mechanism to clarify coal development expectations under the "dual carbon" goals. Accelerate the green transition in areas with renewable energy advantages. Establish funding mechanisms such as a "Coal Transition Fund" to support the development of low-carbon technologies in traditional energy regions, including green hydrogen, energy storage, coal-to-chemical technology, carbon capture, utilization, and storage, and ecological restoration. Emphasize resource redistribution during the green transition and climate change responses by establishing a just transition mechanism to mitigate the socio-economic impacts of the low-carbon transition on vulnerable groups, industry workers, women, and affected communities. Strengthen international exchanges with regions facing similar challenges to encourage countries to explore just transition solutions suited to their own circumstances, ensuring that no one is left behind in the green transition.

**Continue to leverage China's role as the COP 15 Presidency to the UN Convention on Biological Diversity, actively implementing the Kunming** 9. **Montreal Global Biodiversity Framework (KMGBF) and its associated initiatives.** Expand national parks, build and optimize the protected area system, and improve the management and governance mechanisms of protected areas. Expand the coverage of marine protected areas, and develop pilot areas that integrate coastal climate adaptation. Strengthen the research and sharing of policy tools such as protected area systems, ecological conservation red lines, and other effective area-based conservation measures (OECMs). Promote the establishment of an ecological conservation red-line model that conforms to the international perspective to address biodiversity loss and other critical challenges in a coordinated manner.

10. **Achieve a balance between environmental and social development through integrated ocean management, using the "dual carbon" goals and technological innovation to drive the green transformation of marine industries.** Define the concept, principles, and industry classifications of the sustainable blue economy, incorporate its development into the 15th Five-Year Plan, and develop relevant incentive policies. Integrate environmental sustainability into approval processes and monitoring plans for rapidly growing sectors such as coastal infrastructure, marine energy, shipping, and marine food. Establish specific indicators and tracking mechanisms. Strengthen strategic planning and spatial management for offshore aquaculture by comprehensively considering ecosystems and incorporating the views of stakeholders across the industry value chain. Pay particular attention to interactions with ecological hotspots, such as the peripheries of marine protected areas, critical biological habitats, and migration corridors. Place greater focus on women and vulnerable groups involved in the fisheries and aquaculture industry chain.

11. **Strengthen top-level planning for the marine renewable energy industry, and build the foundation for its industrial design and transformation.** Expand the application of mature marine renewable energy technologies, including offshore wind power, and promote demonstrations for emerging and strategic marine renewable energy technologies, such as floating wind, wave, and tidal energy, and offshore green

hydrogen production. Make the marine energy industry's contribution to the sustainable blue economy clearer and more visible. Advance the sustainable development of the marine renewable energy industry in the context of carbon neutrality.

12. **Accelerate the development of new green and sustainable agricultural productive forces, enhance the ecological functions and biodiversity of agriculture, and ensure ecological safety in food production.** Adopt crop diversification to increase farmland biodiversity and soil productivity. Incorporate soil biodiversity indicators into farmland soil health and agricultural ecological assessment systems. Actively explore transition pathways for high-yield and stable-yield regenerative agriculture and modern green organic agriculture, aiming to accelerate the development of new productive agricultural forces. Fully integrate agricultural production, consumption, and dietary structure to conduct assessments and management of farmland soil health. Encourage the use of organic fertilizers. Develop ecological farming practices, and establish integrated spatial layouts of farmland and ecological islands to promote the transformation of agricultural production methods.

## 3 Leverage Green Finance to Promote Sustainable Development across Sectors

13. **Fully mobilize public and private sector funds to establish a comprehensive and long-term financing policy system for green and low-carbon technological innovations**. Mobilize diverse funding sources, including bank loans, equity investments, and green bonds, to form a multi-tiered, multi-channel green finance system. Develop a financial service system centred on venture capital and supported by the stock market to drive original and transformative green and low-carbon technological innovations. Enhance the inclusiveness of the stock market for initial public offerings to build an innovation ecosystem that encourages mutual advancement across various fields and technological pathways. Develop more flexible evaluation standards for technological attributes to better support innovative small and medium-sized enterprises. Actively explore low-carbon procurement pilot programs targeting green technologies and product services.

14. **Encourage long-term or patient capital funds to become the main source of green investment**. Encourage bank funds to support the green industry venture capital market, and guide high-net-worth group capital, pension funds, and other long-term funds into low-carbon technologies. Optimize the evaluation and guidance mechanisms for green industry funds, enhancing their investment and management capabilities. Address the potential market segmentation that may arise from industrial guidance funds, increase the proportion of national-level funds, and reduce the proportion of municipal and county-level funds when setting up such funds.

15. **Strengthen the consistency and alignment of climate risk assessment and information disclosure standards for domestic and international enterprises**. Encourage enterprises to implement and adhere to the International Sustainability Standards Board (ISSB) standards for sustainability information disclosure, actively considering disclosing other indirect greenhouse gas emissions generated across the full value chain as well as climate information disclosures for overseas projects. Accelerate integration into international green finance markets, and expand climate and green financing channels. Increase financial openness to attract high-quality foreign general partners to enter China's venture capital market, and value the positive roles of high-quality foreign venture capital and corporate venture capital (CVC). Optimize the antitrust concept, and value the positive role of CVC in investment and management.

16. **Actively participate in building global sustainable trade and green value chain systems**. Adhere to the principles of shared benefits by transferring low-carbon production capacity and technologies to developing countries, emphasizing low-carbon sustainable development across the entire industrial chain during implementation. Support green growth in developing countries by establishing dedicated funds to provide capital for overseas investments in low-carbon technologies, complemented by risk mitigation tools such as insurance and guarantees. Create a favourable policy environment for outward direct investment, and expand trade in low-carbon technologies and related minerals with key trade partners. Diversify export destinations, and expand trade with emerging economies in low-carbon and environmental goods and services. Support technology transfer and enhance market access for green technologies and services.

17. **Utilize multiple financing channels to support biodiversity conservation**. Expand private sector financial support for biodiversity conservation through platforms such as the Kunming Biodiversity Fund, the Global Environment Facility, and the UNDP Biodiversity Finance Initiative (BIOFIN). Enhance the disclosure of biodiversity-related information by listed companies in line with the KMGBF and international standards and practices. Integrate biodiversity financing into China's carbon market to promote the synergistic development of biodiversity conservation and climate change response efforts, making biodiversity conservation an integral part of the socio-economic development and transformation.

## 4 Deepen International Cooperation on Environment and Climate, and Promote Global Environmental Governance

18. **Actively lead and promote global environmental and climate governance processes.** Facilitate multilateral cooperation under Article 6 of the Paris Agreement. Through multi-level financing models, integrate global climate finance with

poverty alleviation, infrastructure development, and SDGs. Strengthen international cooperation and research on climate adaptation and exchanges on experience, practices, and lessons learned. Work toward achieving a pragmatic, balanced, and legally binding international treaty on plastic pollution by the set deadlines.

19. **Actively advocate for and promote the greening of global trade and investment.** Assess the impact of the Carbon Border Adjustment Mechanism (CBAM) on key industries. Utilize existing multilateral mechanisms to establish exchange platforms for low-carbon industry investment and trade, strengthen dialogues, and seek solutions acceptable to all parties. Actively participate in or initiate negotiations for multilateral and bilateral investment and trade protection agreements. Promote the resumption of the World Trade Organization's "Environmental Goods Agreement" negotiations and the expansion of environmental products under the Asia-Pacific Economic Cooperation. Launch a low-carbon product trade liberalization and zero-tariff alliance, and enhance international trade cooperation and technology exchange to accelerate the global dissemination and application of low-carbon technologies.

20. **Enhance the evaluation, management, and financing support for green Belt and Road projects.** Leverage the Green Investment and Finance Partnership and tools such as blended finance to increase private sector investment in green financing projects in emerging economies, and promote increased leverage of public financing in low-income countries. Further integrate climate elements into export credit policies and overseas investment and financing decisions, encouraging the use of climate stress tests for the balance sheets of overseas projects. Promote diversified financing methods such as low-cost capital, patient capital, joint ventures, venture capital, and equity investment.

# Chapter 17
# Progress on Environment and Development Policies in China and Impact of CCICED's Policy Recommendations (2023–2024)

## 1 Foreword

In the past year, faced with an unusually complex international environment and the arduous tasks of reform, development, and maintaining stability, the Chinese government withstood external pressure, overcame internal difficulties, made arduous efforts, and took solid steps in building a modern socialist country in an all-round way. The next 5 years will be an important period for the development of a Beautiful China. The Chinese government will continue to strengthen the development of ecological civilization and promote green and low-carbon development as one of the important contents of its work tasks and deeply implement the ecological civilization thought of socialism with Chinese characteristics in the new era.

In the past year, there has been a milestone reached in the field of ecological and environmental protection. The national ecological and environmental system firmly implemented Xi Jinping's thought on ecological civilization, coordinated the promotion of high-quality economic development and high-level protection of the ecological environment, vigorously promoted the construction of a Beautiful China, and achieved new results in ecological environment governance. The ecological environment quality improvement indicators have all successfully completed the annual targets, exceeding the timeline requirements of the 14th Five-Year Plan.

In the past year, China's ecological environment legislation and supervision and law enforcement have been continuously strengthened, and the efficiency of ecological environment governance has continued to improve; it has deepened energy conservation and carbon reduction in key areas, accelerated the innovation and application of green and low-carbon technologies, promoted the greening and low-carbonization of economic and social development, and steadily promoted carbon peaking and carbon neutrality. Carbon monitoring and assessment pilot projects have

The content of this chapter is extracted from "Progress on Environment and Development Policies in China and Impact of CCICED's Policy Recommendations (2023–2024)".



*Adhering to Green and Open Cooperation to Advance Modernization*,
https://doi.org/10.1007/978-981-95-5547-5_17

been expanded, the national voluntary greenhouse gas emission reduction trading market has been launched, and the second compliance cycle of the national carbon emissions trading market has been successfully concluded. These actions are part of the deeper implementation of China's national climate change strategy.

In the past year, China has insisted on precise, scientific, and legal pollution control, steadily and deeply advancing the battles to protect blue skies, clear waters, and clean land, strengthening the management of solid waste and new pollutants, and achieving new and significant results in the battle against pollution; it has accelerated the promotion of major projects for the protection and restoration of important ecosystems, implemented major projects for biodiversity protection, and made institutional progress in cross-departmental ecological supervision, such as desertification prevention and control and wetland protection.

In the past year, China has deeply participated in global environmental and climate governance. It has actively and steadily carried out bilateral and multilateral environmental and climate exchanges, deeply participated in global climate negotiations, put forward the implementation initiative of the "Kunming-Montreal Global Biodiversity Framework," and striven to promote the implementation of COP15 outcomes. It has also continued to promote the implementation of international environmental conventions and international negotiations.

As a link, bridge, and window for promoting exchanges and cooperation in the fields of environment and development between China and foreign countries, CCICED continues to leverage its direct, international, and comprehensive advantages, bringing together the collective wisdom of top experts at home and abroad to implement the "dual-carbon" strategic goals, explored multi-objective collaborative innovation paths, and accelerated the promotion of green, low-carbon, and high-quality development. It has put forward many policy recommendations that have been referenced or adopted by the Chinese government, making contributions to China's sustainable development.

## 2 Environmental and Development Planning

### *2.1 Fully Promoting the Building of Beautiful China 2035 to Achieve Harmony Between Humans and Nature*

#### 2.1.1 National Level

In December 2023, the Central Committee of the CPC and the State Council issued the *Opinions on Fully Promoting the Building of a Beautiful China*. Building a Beautiful China is an important goal in the comprehensive construction of a modern socialist country and a significant part of realizing the Chinese Dream of the great rejuvenation of the Chinese nation. On the new journey, it is imperative to place the building of a Beautiful China in a prominent position within the efforts to build a strong nation and achieve national rejuvenation. We must maintain the strategic determination to strengthen ecological civilization and unswervingly follow the path of civilized development characterized by developed production, affluent living, and sound ecology, creating a beautiful homeland with blue skies, green lands, and clear waters.

According to the *Opinions*, by 2027, significant progress will be made in the building of a Beautiful China. Green and low-carbon development will be deeply advanced, with a continuous reduction in the total emissions of major pollutants and ongoing improvements in ecological and environmental quality. The layout for the development and protection of national land space will be optimized, the ecological system's service functions will be continuously enhanced, urban and rural living environments will be significantly improved, national ecological security will be effectively guaranteed, and the ecological and environmental governance system will be more complete, forming a number of practical models. By 2035, the goal of a Beautiful China will be basically achieved. Green production and living styles will be widely adopted, carbon emissions will peak and then steadily decline, and fundamental improvements in the ecological environment will be realized. A new pattern of national land space development and protection will be fully formed, with significant improvements in the diversity, stability, and sustainability of ecosystems, more secure national ecological security, and the basic realization of the modernization of the ecological and environmental governance system and governance capabilities. Looking ahead to the middle of this century, a Beautiful China will be fully built. Ecological civilization will be fully enhanced, green development methods and lifestyles will be fully established, key areas will achieve deep decarbonization, the ecological environment will be healthy and beautiful, and the modernization of the ecological and environmental governance system and governance capabilities will be fully realized.

The *Opinions* point out that the comprehensive promotion of a Beautiful China aims to achieve harmony between humans and nature and to promote green and low-carbon development. This will be achieved through strengthened ecological protection, optimized national land space layout, promotion of green transformation in the economy and society, and acceleration of the process of ecological civilization construction. With green and low-carbon development as the mainline, the goal is to promote comprehensive improvements in the ecological environment, laying a solid foundation for a Beautiful China.

### 2.1.2 Local and Industry Level

All localities have fully implemented the core idea of building a beautiful China in practice and continued to promote the implementation of ecological civilization.

In January 2024, the National Development and Reform Commission and the People's Government of Hebei Province issued the *Opinions on Promoting Xiong'an New Area as a Model for Green Urban Development,* proposing to promote the green development of Xiong'an New Area and build a modern city that is ecological, livable, low-carbon, and intelligent, with the goals of achieving green development and modernization by 2030 and 2035 respectively.

In order to play the leading and driving role of ecological civilization, in February 2024, the General Office of the Ministry of Ecology and Environment revised and issued the *Construction Indicators of Ecological Civilization Demonstration Zones (Cities)*, *Construction Indicators of Ecological Civilization Demonstration Zones (Counties)*, *Management Regulations for Ecological Civilization Demonstration Zones*, and *Management Regulations for the Construction of 'Lucid Waters And Lush Mountains Are Invaluable Assets' Practice and Innovation Bases*. By building an indicator system, standardizing the construction and management of ecological civilization demonstration zones, summarizing and promoting typical experience models, and actively exploring effective transformation paths for "Lucid Waters And Lush Mountains Are Invaluable Assets."

In February 2024, the National Development and Reform Commission and the People's Government of Beijing issued the *Notice on the Implementation Plan for the Construction of the National Green Development Demonstration Zone in Beijing's Sub-Center*, aiming to achieve carbon peaking and carbon neutrality goals. The plan focuses on three key areas: buildings, transportation, and industry, and strengthens three major supports: energy, ecology, and culture, to promote the construction of the national green development demonstration zone in Beijing's sub-centre.

In February 2024, the Ministry of Ecology and Environment issued the *Notice on Strengthening Environmental Impact Assessment Management to Promote the Green Development of Civil Transport Airports*, which requires strengthening the management of environmental impact assessments for civil transport airports to ensure ecological protection, reduce environmental impacts, and promote green and low-carbon development.

In March 2024, the National Development and Reform Commission and other departments issued the *Notice on Supporting the Green, Low-Carbon, and High-Quality Development of Inner Mongolia*, which supports the green, low-carbon, and high-quality development of Inner Mongolia. The content covers various aspects such as energy transition, industrial upgrading, technological innovation, and ecological and environmental protection.

## 2.2 *Clarifying the New Era Biodiversity Conservation Strategy and Deploy an Action Roadmap*

In January 2024, with the approval of the State Council, the Ministry of Ecology and Environment issued the *China Biodiversity Conservation Strategy and Action Plan (2023–2030)*. The *Action Plan* implements a series of major decisions and deployments, clearly outlining the biodiversity conservation work and priority areas for China in the coming period. It emphasizes systematic planning, clear objectives, key priorities, and phased advancement. The plan sets forth detailed biodiversity conservation targets for 2030, such as the continued promotion of baseline biodiversity surveys and assessments in priority conservation areas and national strategic regions, aligned with China's "30 × 30" target in the Kunming-Montreal Global Biodiversity Framework. It also proposes medium- and long-term goals for 2035, including the comprehensive enhancement of conservation capacity, a unified and orderly conservation framework, and a stable improvement in ecosystem carbon sequestration capabilities. Looking ahead to 2050, the plan envisions the realization of a Beautiful China where humans and nature coexist in harmony, achieving the vision of harmonious coexistence between people and nature.

## 2.3 Launching the First Batch of Carbon Peaking Pilot Projects and Promoting Comprehensive Green Transformation in Various Regions

In October 2023, the National Development and Reform Commission issued the *National Carbon Peaking Pilot Construction Plan.* To implement the carbon peaking action plan, 100 representative cities and industrial parks nationwide were selected to carry out a carbon peaking pilot construction. This initiative aims to explore carbon peaking pathways for cities and parks with different resource endowments and development foundations, providing practical, replicable, and scalable experiences and practices for the entire country.

In November 2023, the General Office of the National Development and Reform Commission issued the *List of the First Batch of Carbon Peaking Pilot Projects*, identifying 25 cities, including Zhangjiakou, and 10 industrial parks, including Changzhi High-Tech Industrial Development Zone, as the first batch of carbon peaking pilot cities and parks. Each pilot city and park must fulfill its primary responsibility, treat the carbon peaking pilot construction as a crucial lever to promote comprehensive green transformation of local economic and social development, plan key tasks in a coordinated manner, study and introduce reform measures, and steadily advance major projects. They should also scientifically draft pilot implementation plans based on their actual conditions.

## 2.4 CCICED Policy Recommendations

For the building of a Beautiful China, in 2023, CCICED recommended that China should continue to maintain strategic focus, using green and low-carbon development as an endogenous driving force. It should promote high-quality development through multi-goal coordination, and while advancing carbon reduction and pollution reduction simultaneously, it should also balance carbon reduction, pollution reduction, energy, the supply chain, and food security. Accelerating the digital and green transformation of traditional industries, building a green financial system and a new power system, improving legal guarantees, and promoting harmony between humans and nature are essential, along with fostering international cooperation. In 2021, CCICED suggested that urban renewal should be viewed as a significant opportunity for the green urbanization transformation. Setting green urban renewal goals and design and construction standards is important, and best practices that bring multiple benefits, including ecological protection, public health, human capital, and improved quality of life, should be encouraged.

For biodiversity protection, in 2023, CCICED recommended that from agreements to synergies, implement the Kunming–Montreal Global Biodiversity Framework (KMGBF) and swiftly update the National Biodiversity Strategy and Action Plan (NBSAP) alongside corresponding policy measures and roadmaps in line with the KMGBF. As the Presidency of COP 15, China should continue to communicate with parties to the convention, observer states, and other stakeholders to promote cutting-edge biodiversity conservation initiatives. It should also take early action toward the "30x30" and all the other GBF targets to achieve rapid early results while boosting confidence in the implementation of the GBF. In 2022, CCICED recommended mainstreaming nature-based solutions (NbS) and continuously promoting bilateral and multilateral climate and biodiversity dialogues. In 2021, CCICED recommended promoting the full implementation of the global biodiversity framework post-2020, sharing innovative mechanisms for achieving targets such as ecological redlines and ecological compensation, and establishing a reliable, operational progress evaluation indicator system. The value of natural capital and ecosystem services should be fully considered to contribute positively to the goals of the United Nations Decade on Ecosystem Restoration.

For environmental infrastructure construction, in 2023, CCICED recommended continuing to improve the policy mix for promoting new-energy heavy-duty trucks. This includes developing industry standards, implementing fiscal incentives such as vehicle purchase tax exemptions, non-fiscal incentives like road rights priority, and specifying procurement requirements for commercial fleets' new-energy vehicles. The construction of infrastructure for new-energy heavy-duty trucks, such as charging stations and battery swapping stations, should be accelerated. Long-term sales proportion targets for new-energy heavy-duty trucks should be set at 45% by 2030, 75% by 2035, and 100% by 2040. An off-grid energy storage system for road charging should be established, and a "dual-credit" policy for new-energy heavy-duty trucks should be introduced. In 2021, CCICED recommended intensifying efforts to renovate urban green and low-carbon infrastructure, accelerating the construction and management of urban rainwater systems, including adopting nature-based solutions, such as sponge cities. In 2020, CCICED recommended promoting green travel by encouraging walking, cycling, and public transportation as low-carbon travel modes and increasing the promotion and use of new-energy vehicles. A comprehensive green policy system for the entire automotive industry chain should be established, the construction of electric vehicle charging infrastructure should be intensified, and a recycling system for electric vehicle batteries should be established. Tax incentives should be strengthened to encourage energy conservation and emission reduction in the automotive industry, and economic incentives for purchasing and using green vehicles should be expanded. Support measures for the automotive and aviation industries must consider ecological and environmental requirements. Strengthening rail freight and sustainable urban logistics, such as the digitalization and automation of freight railways, is also necessary.

For the carbon peaking pilot projects, in 2023, CCICED recommended selecting certain provinces, cities, and key industries to conduct carbon emission dual control pilots in the mid-to-late stage of the 14th Five-Year Plan. In the early stage of the 15th Five-Year Plan, a carbon emission dual control system should be trialed nationwide, using carbon intensity as a binding indicator and total carbon emissions as an indicative indicator. After 2030, a carbon reduction system primarily based on total carbon emissions control should be improved.

## 3 Governance and Rule of Law

### *3.1 High-Quality Legislative Guarantees to Promote Green Development*[1]

The steady progress in promoting the high-quality development of legislative work is a solid guarantee of the rule of law for the modernization of China to comprehensively promote the construction of a strong nation and the great rejuvenation of the Chinese nation. Strengthen organizational leadership, closely collaborate and cooperate, broadly absorb public opinion and gather public wisdom, systematically integrate and compile existing ecological and environmental laws and regulations, and form a high-quality draft *Ecological and Environmental Code*, as well as closely follow the legislative needs of the Chinese-style modernization, accurately grasp the "the top priorities of the country," and adhere to the principles of focusing on key issues, prioritizing urgent needs, and striking a balance between overall considerations, and scientifically and reasonably arrange legislative projects.

On May 9, 2024, the General Office of the State Council issued the *Legislative Work Plan for 2024*, which specifies a number of legislative objectives centred on promoting green development. The plan proposes to consider the draft *National Parks Law*, formulate the *Regulations on the Administration of Carbon Emissions Trading* and the *Regulations on Compensation for Ecological Protection*, and prepare to formulate the *Regulations on Ecological Environment Monitoring*. The enactment and consideration of these laws and regulations are aimed at strengthening ecological environmental protection and promoting green and low-carbon development. Currently, China has formulated and revised laws such as the Environmental Protection Law, Environmental Protection Tax Law, Yangtze River Protection Law, and the laws on the prevention and control of air, water, and soil pollution, creating a "1

[1] http://en.moj.gov.cn/2023-11/14/c_938670.htm

+ N + 4" legal framework for ecological and environmental protection. The scope of legislation, the strictness of enforcement, and the remarkable effectiveness are unprecedented.

## 3.2 Regulating Ecological and Environmental Planning

As one of the special plans in the national planning system, ecological environmental protection planning, systematically analyzing its development history, summarizing the development experience, and further adhering to and improving the ecological environmental planning system, is very important for promoting the development of the national planning system, supporting Beautiful China, and synergistically promoting the high-quality development of the economy and high-level protection of the ecological environment. The preparation of ecological and environmental planning supports the overall planning for promoting high-quality economic and social development, focuses on the scientific application and practical deepening of the system concept in ecological environmental protection, follows the wholeness and systemic nature of ecosystems and their intrinsic developmental laws, continuously improves the systemic, holistic, and synergistic nature of ecological and environmental governance, and continuously promotes the management of the source, the management of the system, and the management of the entirety of the ecological and environmental environment.

On May 11, 2024, MEE issued the announcement of national ecological and environmental standard *Technical Guidelines for Ecological and Environmental Planning General Outline*, in order to implement a number of laws on environmental protection, to guide and standardize the preparation of ecological and environmental planning, The General Outline issued as a national ecological and environmental standard, and to stipulate the general principles, procedures, contents, methods and requirements for the preparation of ecological and environmental plans. It also stipulates the general principles, procedures, contents, methods and requirements for the preparation of ecological environment plans.

For example, Hefei high-tech zone in Anhui province adheres to the direction of "high-quality, high-tech, high-efficiency, low-pollution and ecological" for industry, compiles green development planning for industrial parks, and innovatively establishes the evaluation index system for green and low-carbon development. It carries out comprehensive design for the integration of industry and city, promotes the revitalization and upgrading of urban space, and promotes the intensive and compact space utilization, functional composite, and low-carbon efficiency. It has taken pollution reduction and carbon synergy as the focus point for optimizing industrial structure and achieving high-quality development. Taking Panyu district as a case study, the Guangzhou Municipal Bureau of Planning and Natural Resources explores the practice of sustainable green infrastructure planning and construction, combining the natural endowment and ecological characteristics of the Lingnan water town area, and taking the implementation of the integrated protection of mountains, water, forests, fields, lakes, and grasses and the systematic governance as the lead, constructing a sustainable green infrastructure planning and implementation pathway, which will positively support a green and beautiful Guangzhou.

## 3.3 Strengthening Ecological and Environmental Zoning Management and Control

The implementation of ecological zoning control is a major exploration in the field of spatial control of the ecological environment and a fundamental, forward-looking and long-term endeavour in the field of ecological environmental protection. Since 2012, President Xi Jinping has repeatedly emphasized the need to accelerate the delineation and strict enforcement of ecological protection redlines, environmental quality baselines, and resource utilization ceilings. The general requirements for the construction of the ecological environment zoning control system, the overall requirements, mechanism construction, implementation and application, supervision and assessment, and organizational protection of the whole chain of system construction system. Realize the harmonious coexistence of humans and nature, promote the construction of ecological civilization to a new level, help the building of a Beautiful China.

On March 18, 2024, the General Office of CPC and the General Office of the State Council issued the *Opinions on Strengthening Ecological and Environmental Zoning Control*, proposing to strengthen ecological and environmental zoning control to safeguard ecological functions and improve environmental quality, and to promote high-quality development. Through the formulation of control programs, the identification of control units, the preparation of access lists, information sharing, dynamic updating and other measures, it serves major national strategies, promotes green and low-carbon development, realizes a high level of ecological environmental protection, and strives to achieve the beautiful China initiative in which human beings coexist harmoniously with nature.

For example, in Fengjie County, Chongqing, the Rapeseed Dam pumped storage power station is under construction. When selecting the site for the power station, there were several options. The local use of ecological environment zoning control results for project access and site selection provided comprehensive guidance for project approval and site selection, avoiding environmentally sensitive areas, such as nature reserves and drinking water source protection zones. The final site for the power station was determined based on these considerations. For key controlled units, the focus is on addressing prominent ecological and environmental issues, promoting optimization of industrial layout and structural upgrading, and effectively curbing the blind development of "high energy consumption, high emissions, and low efficiency projects."[2] For the priority protection unit, it focuses on ecological environmental protection to ensure the overall stability of the ecological security pattern.

[2] https://www.mee.gov.cn/xxgk2018/xxgk/xxgk03/202105/t20210531_835511.html

Shunde District, Foshan City, Guangdong Province, the "capital of China's household electrical appliances," combined the requirements of industry development and the ecological environment endowment of controlled units with the development of a targeted ecological access list, such as in key water environment control areas; projects that directly discharge wastewater into rivers or streams are prohibited if the receiving water bodies or monitored sections do not meet the required standards for new construction or expansion. In Xiamen, Fujian Province, the application of ecological environment zoning control has deepened departmental collaboration, guiding nearly 10,000 projects to optimize their layout or adjust their processes, avoiding investment losses of nearly 800 million yuan.

## 3.4 Regulating the National Carbon Market and Voluntary Emissions Trading

The development of a unified national carbon market is a major institutional innovation to promote the greening and low-carbon development of China's economy and society. The national carbon market can not only give play to the decisive role of the market in the allocation of carbon emission resources, but also realize the organic combination of an effective market and a competent government and is a policy tool for realizing low-cost carbon reduction for the whole society. It will promote the achievement of national carbon peaking and carbon-neutral goals, promote environmental technology innovation and green industry development, push China's ecological civilization to a new level, and help achieve the goals of global climate governance.

On February 4, 2024, the State Council issued the *Interim Regulations on the Administration of Carbon Emission Trading*, as the first specialized regulation in the field of addressing climate change in China. The Regulation focuses on clarifying the institutional mechanisms, standardizing trading activities, guaranteeing data quality, and punishing illegal acts, etc. It is the first time that the carbon emissions trading system is clarified in the form of an administrative regulation, which is of great significance in realizing China's dual-carbon goal and promoting the green and low-carbon transformation of the whole society. Since the launch of trading, the national carbon emission trading market has operated smoothly overall, covering an annual average of about 5.1 billion tons of carbon dioxide emissions, accounting for more than 40% of the country's total emissions. By the end of 2023, a total of 2257 power generation enterprises had been included in the national carbon emissions trading market, with a cumulative trading volume of about 440 million tons and a transaction value of about 24.9 billion yuan. The policy effects of carbon emissions trading have begun to show initial results.

On October 19, 2023, MEE issued the *Greenhouse Gas Voluntary Emission Reduction Trading Management Measures (Trial)*. The document outlines regulations for voluntary greenhouse gas emission reduction trading activities across the country. It aims to promote the development of market-based emission reduction mechanisms by encouraging companies and other entities to participate in voluntary emission reduction trading, thus driving low-carbon development and the transition to a green economy. It regulates voluntary emission reduction trading, ensures fairness and openness, sets the responsibilities of regulatory bodies and trading rules, and promotes carbon-peaking carbon-neutral targets. Launching the voluntary emission reduction trading market is conducive to supporting the development of forestry carbon sinks, renewable energy, methane emission reduction, energy-saving and efficiency projects, and is also conducive to incentivizing a wider range of industries, enterprises and all sectors of society to participate in greenhouse gas emission reduction actions, which is of positive significance in promoting green and low-carbon economic and social transformation and realizing high-quality development. At the same time, the *Notice on the Arrangement of Matters Related to the National Greenhouse Gas Voluntary Emission Reduction Trading Market* was issued, with specific details and arrangements for the management of matters.

Ministries and enterprises have actively responded to the policy of reducing greenhouse gas emissions and promoting the establishment of a national carbon trading market. For example, on October 8, 2023, MEE issued a notice regarding reporting and verifying the greenhouse gas emissions of enterprises in some key industries in 2023–2025, requiring local ecological and environmental departments to organize enterprises in key industries to report their greenhouse gas emissions data. Key industries, such as petrochemicals, chemical industry, building materials, steel, nonferrous metals, papermaking, and civil aviation, with annual greenhouse gas emissions of 26,000 tons of carbon dioxide equivalent (comprehensive energy consumption of about 10,000 tons of standard coal) and above, are included in the scope of annual greenhouse gas emissions reporting and verification as outlined in this notice. The accuracy of the reports will be verified to support the development of the national carbon market. It is necessary to ensure that the task is completed on time, to arrange work funding in advance, and to strengthen supervision and capacity building. On May 29, 2024, MEE issued the *Technical Guidelines for Environmental Impact Assessment of Greenhouse Gas Emissions from Construction Projects in the Thermal Power Industry (Trial)*, which incorporates greenhouse gas control into the management of EIA and standardizes the technical methods for EIA of greenhouse gas emissions in key industries. It further strengthens greenhouse gas emission control and requirements, promotes the process of carbon peaking and carbon neutrality, and contributes to the Beautiful China initiative.

## 3.5 Legislation to Preserve and Improve Marine Ecosystems

On October 24, 2023, the sixth meeting of the Standing Committee of the 14th National People's Congress voted to adopt the newly revised *Marine Environmental Protection Law*. The revised *Marine Environmental Protection Law* covers marine environmental management, pollution prevention and control, ecological protection and legal responsibility, adheres to land and sea integration and regional coordination, comprehensively strengthens the prevention and control of marine environmental pollution, improves marine ecological protection, strengthens the supervision and management of the marine environment, and promotes the extraterritorial application of marine environmental protection laws. This law came into force on January 1, 2024.

## 3.6 *Establishment of Basic System Rules for Compensation for Ecological Protection*

The ecological protection compensation system is an important part of the ecological civilization system. Since 2012, the Chinese government has taken the establishment and improvement of the ecological protection compensation mechanism as an important initiative for ecological civilization and has coordinated the planning and comprehensively pushed forward the reform of the ecological protection compensation system and related areas, with the work on ecological protection compensation accelerating. At present, China has basically built the ecological protection compensation mechanism with the widest coverage, the largest beneficiary population and the largest investment in the world, which has played an important role in consolidating the achievements of ecological civilization, consolidating and expanding the results of the fight against poverty and promoting coordinated and sustainable development in the region.

On April 6, 2024, the State Council promulgated the Regulations on Ecological Protection Compensation. These regulations legally regulate ecological protection compensation with the aim of galvanizing the enthusiasm of all parties and fostering a favorable scenario where the entire society engages in ecological protection. By elucidating the legal foundation and specific measures, the Regulations establish a multi-level and multi-channel ecological protection compensation system, highlighting a comprehensive mechanism propelled by government leadership, market regulation, and social participation. Through the concerted efforts of the financial, market, and social sectors, the Regulations seek to elevate China's ecological protection to a new echelon, achieve the integration of ecological, economic, and social benefits, and offer robust assurance for the Beautiful China initiative.

China's ecological protection compensation system can be traced back to 1998, when the launching of key ecological projects such as the return of farmland to forests and grasslands marked the beginning of China's ecological protection compensation. After more than two decades of unremitting efforts, China has built the ecological protection compensation mechanism with the widest coverage, the largest beneficiary population, and the largest investment in the world, realizing the "price" of green mountains and the "gold" of green water, and increasing the sense of achievement of ecological protection subjects, effectively implemented the rights and responsibilities of ecological protection, effectively mobilized all parties to participate in ecological protection, and contributed a particularly important institutional strength to the construction of ecological civilization and the drawing of a new picture of beautiful China initiative. For example, since 2003, China has implemented the project of returning pasture to grassland, and then implemented the grassland protection subsidy and reward mechanism, which has now been implemented in the third round, significantly improving the quality of grassland. This has enabled China to move from an area of grassland degradation accounting for 90% of the total area to an area of vegetation restoration accounting for 48% of the total area, with a 5.26% increase in the comprehensive index cover of grasslands and a 9.2% increase in the net primary productivity of northern grassland vegetation.

At present, the scale of annual national financial compensation funds has reached the level of 200 billion yuan, the level of protection of important ecosystems and ecologically functionally important areas across the country has continued to improve, about half of the forests and grasslands have been included in the scope of compensation, one-third of the counties have received transfers to key ecological functional areas, and 21 provinces have signed agreements on cross-provincial watersheds to compensate for horizontal ecological protection, involving 20 cross-provincial watersheds (river segments).

## 3.7 Improving Ecological and Environmental Risk Management Systems

Improving ecological and environmental risk management is an important foundation for realizing the Beautiful China initiative, respecting nature, adapting to nature, and protecting nature, insisting that ecological protection comes first, natural restoration is the mainstay, and that natural ecological security boundaries are kept, and insisting on overall coordination, classified policy-making, scientific prevention and control, and systematic governance. The establishment of these principles will play an important guiding role in optimizing the ecological security pattern, protecting important ecosystems as a whole, systematically repairing degraded ecosystems, preventing and controlling ecological risks, and reducing the ecological impact of human activities, etc., and will provide legal support for safeguarding ecological security, building an ecological civilization highland, promoting sustainable economic and social development, and realizing the harmonious coexistence of human beings and nature.

On March 21, 2024, MEE issued a public notice on the national ecological and environmental standard *Nomenclature Specification for Environmental Management of Chemical Substances*, which further regulates the naming of chemical substances and supports environmental management registration and risk assessment.

China has carried out two national surveys on the production and use of chemicals in key industries in 2013 and 2016, and a large number of chemical substances filled in by enterprises had non-standardized names, such as detergent 209, purifying agent

6501, additives, plasticizers, pearlescent powder, etc., which resulted in the inability to judge their active ingredients, and the inability to further identify and characterize their hazards and environmental risks. Therefore, the standardized names of chemical substances are the basic prerequisite for carrying out the collection of information related to chemical substances, as well as the important basis for the environmental risk assessment and environmental risk control of chemical substances.

### 3.8 Promoting Innovation in Ecological Environment-Oriented Development Patterns in a Standardized and Orderly Manner

Ecological environment-oriented development model innovation is the key to achieving green development and ecological civilization. Ecological environment-oriented development model innovation is the key to achieving green development and ecological civilization. Led by Xi Jinping's ecological civilization thought, the project promotes the effective integration of ecological environmental governance with strong public welfare and related industries with better returns, value-added feedback, coordinated promotion, market-oriented operation, integrated implementation, and sustainable operation through industry chain extension, combined development, and joint operation, so as to enhance the operating returns of related industries by ecological environmental governance, and feedback the inputs of ecological environmental governance by value-added returns of the industries, and realize the innovative project organization and implementation mode of internalizing the external economic nature of ecological environmental governance. The innovative project organization and implementation method of internalizing the external economy of ecological environmental governance is a project practice of practicing the concept that lucid waters and lush mountains are invaluable assets, which is conducive to actively and steadily promoting the operation and development of ecological products and the effective realization of the value of ecological products.

On December 27, 2023, MEE and four other ministries issued a notice on the issuance of *Guidelines for the Implementation of Ecological and Environmentally Oriented Development (EOD) Projects (Trial)*, which recommends in a standardized and orderly manner to promote the innovation of ecological and environmentally oriented development mode, the quality of projects, and to promote the coordination of ecological environmental protection and economic development. The notice emphasizes the importance of project planning and quality control, and requires localities to strengthen resource security, accelerate project implementation, and promote synergistic development of projects with neighbouring projects. Local governments need to strengthen project tracking and summarization, promote typical cases, and ensure the demonstration effect of project implementation. Risk prevention and control is the key, and all parties need to strictly abide by laws and regulations, strengthen financial supervision, ensure the reasonable use of project funds, and guarantee the sustainable development of the project. This guideline will help realize the goal of green development, improve the quality of the ecological environment, and promote the high-quality development of the industry.

Various regions have responded positively and developed policy innovations in line with the actual situation. Anhui, Fujian, Jiangsu, Shandong, Zhejiang, Hubei, Guangxi and other provinces have launched provincial EOD pilots and introduced innovative policies. Jiangsu Province gives priority to EOD projects to provide "environmental loans" and "environmental protection guarantee" support, and the provincial financial incentives. Guangxi developed "Gui Hui loan-ecological loan", the interest rate of 3%, applicable for one year, and a maximum subsidy of up to five million yuan. Anhui, Fujian and other provinces have given green financial support to EOD projects and prioritized resources. Anhui Province is carrying out a project on the integrated development of ecological and environmental management and rural revitalization of Tianhe Lake in Yuhui District, Bengbu City, with a total investment of about 3.2 billion yuan. The project addresses the management needs of the Tianhe Lake area, the current situation of the surrounding area and sources of pollution, and the demand for water ecological and environmental protection; it is carrying out the project for the expansion of the capacity of the Tianhe Lake by clearing and expanding the capacity of the lake and for the comprehensive treatment of water ecology and the environment, and is relying on the 10,000 acres of wetland farmland around the Tianhe Lake to explore in depth the culture of Dayu and the Huaihe River. It also explores initiatives such as "water resources + cultural tourism + eco-agriculture + eco-industry" to accelerate the integration of water tourism, water agriculture, and water industry, and to form a virtuous cycle of ecological environment and industrial linkage.

## 3.9 CCICED Policy Recommendations

In 2023, CCICED made a number of valuable policy recommendations on governance and the rule of law; these are detailed below:

Promote the formation of synergies among legislative, judicial and administrative organs to practice ecological civilization, establish a sound and modernized environmental governance system, and improve the coordination and efficiency of green governance: explore more scientific, rational and practical methods of accounting for the value of natural capital and mechanisms for realizing it. Expand vision and deepen understanding and integrate environmental factors into broader economic and social planning and policies. Promote the construction of green market systems, such as carbon emissions trading. Improve the green standard system, green fiscal and taxation system, and green financial system, form policy incentives coordinated with green development, and promote policy implementation through policy compliance and regulatory enforcement.

Improve the governance system for green and low-carbon transformation and strengthen institutional capacity for innovation and flexibility: prioritize the enactment of climate change-related laws, explore preventive public interest litigation, and study the formulation of a framework law on combating climate change or a carbon-neutral promotion law. Incorporate carbon emission reduction measures into laws and regulations in the fields of transportation, construction and urban-related areas. Formulate provincial and municipal action plans for the implementation of the "1 + N" policy system and improve the management mechanism in terms of target setting, data analysis technology, public participation, continuous monitoring and evaluation, and dynamic adjustment. Include climate change litigation in the scope of environmental public interest litigation, establish a preventive environmental public interest litigation system, and formulate judicial guidelines for adjudicating climate change cases, with special attention to short-lived greenhouse gases. Further promote the improvement of energy utilization efficiency and the strengthening of demand-side management. Include high-emission industries in the carbon market as soon as possible. Provide long-term, clear, and stable market expectations and an effective price transmission mechanism and improve the construction of a national carbon emissions trading system.

Identify environmentally harmful incentives, regulations, spatial planning, subsidies, etc. and promote reforms: e.g., eliminating environmentally harmful agricultural subsidies and supporting low-carbon, regenerative agricultural practices. Introduce convenient and quantifiable criteria for identifying subsidies, tailored to local conditions. Increase investment in green technology innovation and diffusion in agriculture. Focus on supporting women farmers with targeted knowledge and skills training. Strictly adhere to the red line of ecological protection and spatial planning, protect ecosystems and ecological corridors, set up targets for building ecological corridors based on the red line of ecological protection and protected areas, build an efficient and stable ecological safety network, and protect the integrity of ecosystems.

Make ecological conservation, restoration, and regeneration an important area of green finance: further identify the necessary initiatives, conduct ecological conservation finance pilots, risk disclosure, reporting, stress testing and public accountability for private sector investment, etc., to ensure that public and private sector financial flows are consistent with biodiversity targets. Promote the use of fintech in biodiversity conservation, including through the establishment of pilot demonstration zones for "fintech + biodiversity conservation." Ensure that local communities benefit from nature conservation. Pay attention to green debt arrangements and other emerging international practices in eco-conservation finance. Actively build a diversified financial investment and financing mechanism. Based on climate, biodiversity, pollution risk disclosure and transition timetable, use transition finance to assist enterprises in green transition, and establish a blue finance framework to strengthen financial support for sustainable blue economy. Strengthen the financing and coordination of international scientific research cooperation related to the sustainable blue economy and marine carbon emission reduction.

## 4 Energy, Environment, and Climate

### *4.1 Systematic Deployment of Methane Control and Enhancement of Carbon Reduction and Pollution Reduction Synergies*

Methane is the second largest greenhouse gas globally, after carbon dioxide. According to the *Intergovernmental Panel on Climate Change (IPCC) Sixth Assessment Report*, methane's global warming potential (GWP) over a 20-year period is about 81.2 times that of carbon dioxide, and over a 100-year timescale, it is about 28 times that of carbon dioxide. Due to methane's relatively short atmospheric lifespan (around 12 years), many experts believe that reducing methane emissions can deliver quicker results compared to carbon dioxide reductions, making it critically important in mitigating the urgent climate crisis.

In November 2023, the United States and China issued the *Sunnyside Statement on Strengthening Cooperation to Address the Climate Crisis*, which specifically mentions methane and other non-carbon dioxide greenhouse gas emissions, suggesting that the two countries will implement their respective National Methane Action Plans and plan to refine further measures as appropriate.

At the end of the same month, the Ministry of Ecology and Environment and other 11 departments jointly issued the *Action Program for Methane Emission Control* (Methane Program). The Program is the first policy document of the Chinese government focusing on methane emission control in the whole field, which puts forward a general plan and specific action deployment for the overall methane emission control in China in the future.

For the first time, the Methane Program specifies the methane emission control targets for China's key areas by 2030 and puts forward the main targets for methane emission control during the 14th Five-Year Plan and the 15th Five-Year Plan. At the same time, the Methane Program puts forward eight key tasks, such as strengthening the methane monitoring, accounting, reporting and verification system, accelerating emission controls in energy, agriculture, garbage and sewage treatment, strengthening the synergistic control of pollutants and methane, strengthening the technological innovation and supervision of methane emission control, and accelerating the construction of regulations, standards and policy system. From strengthening monitoring, accounting, reporting, regulatory system construction and other top-level involvements, to specific actions in key areas, incentive policies and social mobilization, to the implementation of safeguard policies, such as synergistic mechanisms among various government departments, the Methane Program presents a complete blueprint for China's methane emission reduction for the first time, which will surely have an important impact on China's comprehensive green and low-carbon transformation, and the construction of modernization of the harmonious coexistence of human and nature.

## 4.2 *Increasing Efforts to Save Energy and Reduce Carbon Emissions and Maximizing Efforts to Achieve the 14th Five-Year Plan Targets*

Energy conservation and carbon reduction are important initiatives to actively and steadily promote carbon peaking and carbon neutrality, comprehensively advance the construction of a Beautiful China, and promote a comprehensive green transformation of economic and social development.

In March 2024, the State Council issued the *Action Plan for Promoting Large-scale Equipment Renewal and Consumer Goods Replacement*, proposing that by 2027, the scale of investment in equipment in the fields of industry, agriculture, construction, transportation, education, culture and tourism, and medical care will increase by more than 25% compared with 2023; the energy efficiency of major energy-using equipment in key industries will basically reach the energy-saving level, and the proportion of production capacity whose environmental performance reaches the level of A will increase significantly (and other targets). These goals signal a clear commitment to enhancing energy-saving and carbon reduction efforts across industries, driving a new wave of society-wide energy conservation and carbon reduction initiatives.

The Plan emphasizes the promotion of new industrialization with a focus on energy saving and carbon reduction, ultra-low emissions, safe production, digital transformation, and intelligent upgrading. It targets key industries such as iron and steel, non-ferrous metals, petrochemicals, chemicals, building materials, electric power, machinery, aviation, shipping, textiles, and electronics, vigorously promoting the updating and technological transformation of production equipment, energy-using equipment, power generation, transmission and distribution equipment. The Plan aims to accelerate the promotion of energy-using equipment with advanced energy efficiency and energy-saving levels and implement energy-saving and carbon-reducing reforms by industry and field.

In May 2024, the State Council issued the *Action Plan on Energy Conservation and Carbon Reduction for 2024–2025*. It is proposed that by 2024, energy consumption and carbon dioxide emissions per unit of GDP will be reduced by about 2.5% and 3.9% respectively, the energy consumption per unit of value added in industries above designated size will be reduced by about 3.5%, the proportion of non-fossil energy consumption will reach about 18.9%, and energy-saving and carbon-reducing reforms in key areas and industries will result in energy savings of about 50 million tons of standard coal and carbon dioxide emission reductions of about 130 million tons.

In the 14th Five-Year Plan, China has set a binding target of reducing energy consumption per unit of GDP by 13.5%. In the first 3 years of the 14th Five-Year Plan, China's national energy consumption intensity has been reduced by a cumulative total of about 7.3%, but due to the impact of the new COVID-19 epidemic and other factors, the reduction of the national energy consumption intensity is still lagging behind the progress of the time schedule, and the situation of energy conservation and carbon reduction in some areas is relatively serious.

Against this background, the Chinese government has formulated the Action Plan, the main purpose of which is to anchor the target tasks of the 14th Five-Year Plan, implement energy-saving and carbon-reducing special actions in different areas and industries, and intensify energy-saving and carbon-reducing efforts, so as to complete the targets of the 14th Five-Year Plan to the best of its ability. The Action Plan focuses on key areas and industries such as energy, industry, construction, transportation, public institutions, energy equipment, etc., and deploys 10 major actions for energy conservation and carbon reduction, and puts forward six work initiatives, such as improving the system and standards, perfecting the price policy, strengthening financial support, strengthening scientific and technological leadership, improving the market-oriented mechanism, and implementing universal action, etc., so as to provide support for the implementation of energy conservation and carbon reduction goals and tasks.

## *4.3 Accelerating the Establishment of a Carbon Footprint Management System and Promoting the Application of Green and Low-Carbon Technologies*

### 4.3.1 Accelerate the Establishment of a Carbon Footprint Management System

As the global green and low-carbon transition continues to advance, the concept of a product's carbon footprint is being strengthened in the production practices of various countries. The EU took the lead in declaring the launch of the Carbon Border Adjustment Mechanism (CBAM) in 2022 and published the *EU Batteries and Waste Batteries Regulation* in 2023, which stipulates the requirements related to the carbon footprint of battery production. Some experts believe that this move opens the era of global carbon tariffs. Under the new international situation, managing the carbon footprint of production products is a new issue and challenge that China must deal with in order to carry out international trade cooperation. The Carbon Peaking Action Program by 2030 issued by the Chinese government proposes to "explore the establishment of carbon footprint standards for the whole life cycle of key products," and the government work report of 2024 clearly puts forward the task of "establishing a carbon footprint management system."

In November 2023, the National Development and Reform Commission and other departments issued the Opinions on Accelerating the Establishment of Product Carbon Footprint Management System, proposing that "By 2025, the national level will have introduced about 50 key product carbon footprint accounting rules and standards, a number of key industries will have completed the initial construction of carbon footprint background databases, the national product carbon labeling system will be basically established, and international mutual recognition will be achieved for a number of key products.... By 2030, about 200 key product carbon footprint accounting rules and standards will be introduced at the national level…the main product carbon footprint accounting rules, standards and carbon labels will be widely recognized internationally, and the product carbon footprint management system will provide a strong guarantee for the comprehensive green transformation of economic and social development."

In April 2024, the Certification and Accreditation Administration of China (CNCA) issued the Notice of CNCA on Clarifying the Filing Requirements for Certification Rules Directly Involving Carbon, which further clarifies the filing requirements for certification rules directly related to carbon categories, making the certification about carbon footprint and other GHG emission quantification standards more standardized, and laying the foundation for the next step to strengthen carbon footprint management.

In June 2024, the Ministry of Ecology and Environment, the Ministry of Finance, the People's Bank of China and other 15 departments jointly issued the *Implementation Program on Establishing Carbon Footprint Management System*. The Program states that by 2027, the carbon footprint management system will be initially established; by 2030, the carbon footprint management system will be more complete, and the application scenarios will be more abundant. The Implementation Program clarifies the carbon footprint work objectives and task division in four aspects, namely, establishing a management system, building a work pattern, promoting international mutual trust in rules and strengthening capacity building, with more detailed key tasks and more focused measures, which is the "task book" and "construction plan" for the construction of a carbon footprint management system for Chinese products in the coming period.

### 4.3.2 Promoting the Application of Green and Low-Carbon Technologies

In the 2023 Central Economic Work Conference, it was noted that economic work in 2024 will focus more on science and technology leadership, industrial innovation, and expanding domestic demand, and emphasized the dual-wheel-drive role of green low-carbon and science and technology innovation. According to the International Energy Agency's assessment, 50% of the key technologies to achieve net-zero emissions by 2050 globally are not yet mature. Accelerating the innovation and application of green and low-carbon technologies is an inevitable requirement for China to realize the dual-carbon goal, and an important means to promote the development of new quality productivity.

In September 2023, the Ministry of Science and Technology (MOST) issued the *National Green and Low-Carbon Advanced Technology Achievement Catalogue 2024*. In February 2024, the Ministry of Ecology and Environment (MOE) and six other departments issued the *Implementation Plan for the Collection and Promotion of National Key Low-Carbon Technologies*, which aims to provide strong support for the publicity and promotion of the application of advanced and applicable low-carbon technologies, and to give full play to the important role that low-carbon technologies play in promoting the realization of the goal of peak carbon attainment and carbon neutrality.

In March 2024, the National Development and Reform Commission issued the *List of Demonstration Projects of Green and Low-carbon Advanced Technologies (First Batch),* aiming to implement the demonstration projects of green and low-carbon advanced technologies, lay out a number of demonstration projects with leading technological level, outstanding emission reduction effect, synergistic pollution reduction and carbon reduction, and obvious demonstration effect, and accelerate the development of the world's leading green and low-carbon technologies and industrial highland, so as to provide strong support for the realization of the goal of peak carbon and carbon neutrality, and provide strong support for the economic and social development of high quality. It will provide strong support for realizing the goal of carbon peaking and carbon neutrality and provide green kinetic energy for high-quality economic and social development.

## 4.4 Deepening Response and Adaptation to Climate Change and Enhancing Climate Resilience

As the largest developing country vulnerable to the adverse effects of climate change, climate change has become an important risk in the process of building a beautiful China and advancing Chinese-style modernization. In recent years, China's weather and climate have become more complex, with a high incidence of extreme precipitation, typhoons, high temperatures, cold waves, and other extreme weather and climate events. Since the start of the flood season, regional and periodic flooding and drought disasters have been particularly evident. Proactive adaptation to climate change is a realistic and urgent task that China and the world are currently facing.

In December 2023, the *Opinions of the CPC Central Committee and the State Council on Comprehensively Promoting the Construction of a Beautiful China* included "effectively responding to the adverse impacts and risks of climate change" as one of the important aspects of "guarding the safety bottom line of the construction of a beautiful China," and explicitly proposed to "Adhere to the equal importance of mitigation and adaptation, and enhance the ability to adapt to climate change."

On June 7, 2024, *China's Climate Change Adaptation Progress Report (2023)* was officially issued, systematically combing the progress and effectiveness of China's climate change adaptation work since the issuance of the National Climate Change Adaptation Strategy 2035 from the dimensions of policy construction, monitoring and early warning, natural ecology, economic and social development, regional planning, and safeguard mechanisms.

## 4.5 *Improving the Climate Investment and Financing System and Strengthening Financial Support for Green and Low-Carbon Development*

On March 27, 2024, the People's Bank of China (PBOC), the National Development and Reform Commission (NDRC), the Ministry of Industry and Information Technology (MIIT), the Ministry of Finance (MOF), the Ministry of Ecology and Environment (MOE), the General Administration of Financial Supervision (GAFSP), and the China Securities Regulatory Commission (CSRC) issued the "Guidance Opinions on Further Strengthening Financial Support for Green and Low-Carbon Development," which emphasizes the importance of strengthening financial support for green and low-carbon development, and, by doing so, aims to promote China's economic transformation in a green and low-carbon direction, respond to the challenges of climate change, and achieve sustainable economic development.

## 4.6 *Promoting Investment and Consumption and Enhancing Recovery and Recycling*

On February 6, 2024, the State Council issued the *Opinions on Accelerating the Construction of a Waste Recycling System*, which is guided by Xi Jinping's thought of socialism with Chinese characteristics in the new era, comprehensively implements the idea of ecological civilization, and pushes the overall requirements of high-quality development. Efforts are made to promote fine waste management and effective recycling, cultivate the resource recycling industry, improve the policy mechanism, and strengthen the organization and implementation.

On March 7, 2024, the State Council issued a notice on the *Action Plan for Promoting Large-scale Equipment Renewal and Consumer Goods Trade-in*, which aims to promote investment and consumption, accelerate industrial restructuring, and implement four major actions: equipment renewal, consumer goods trade-in, recovery and recycling, and standards upgrading. By 2027, the goal is to achieve growth in the scale of investment in equipment, an increase in the proportion of production capacity, and a significant increase in the amount of used products recycled.

On April 17, 2024, the General Office of the Ministry of Industry and Information Technology (MIIT), the General Office of the Ministry of Housing and Urban-Rural Development (MOHURD), the General Office of the Ministry of Agriculture and Rural Development (MARD), the General Office of the Ministry of Commerce (MOFCOM), the General Office of the State Administration for Market Supervision and Administration (SAMSA), and the General Office of the State Administration of Radio and Television (SARA) jointly issued a notice on the comprehensive implementation of the green building materials activities in the countryside, which stated that the departments will jointly carry out activities of the green building materials activities in the countryside to advance the high-quality green building materials industry. The notice indicates that all departments will work together to carry out green building materials activities in the countryside, promote the high-quality development of the green building materials industry, accelerate the certification and promotion of products, explore new forms and modes of business, promote green construction in urban and rural areas, and help revitalize rural areas.

## 4.7 Standardizing Ecological Monitoring and Improving the Scientific Nature of Environmental Quality Assessment

On October 13, 2023, the State Council issued the Notice on *Issuing the Working Rules for the Preliminary Study on National Ecological and Environmental Monitoring Standards (Trial).* These rules clarify the responsibilities of the management department, technical support unit, and undertaking unit. They also emphasize the significance of each step in the process, including soliciting opinions, researching, finalizing argumentation, and project establishment. The rules are designed to standardize the preliminary research workflow, enhance the quality and efficiency of standard formulation and revision, and ensure the scientific soundness and practicality of the standards, thus better supporting ecological environment monitoring.

On June 7, 2024, the Ministry of Ecology and Environment issued a letter on *China's Progress Report on Adaptation to Climate Change (2023)*, which aims to strengthen climate change monitoring, early warning and risk assessment, and promote response strategies and technical support. It establishes provincial action programs, strengthens regional disaster prevention and mitigation, improves the accuracy of meteorological warnings, safeguards public safety, focuses on scientific and technological innovation, promotes green and low-carbon development, accelerates energy transformation, and promotes sustainable economic and social development.

## 4.8 CCICED Policy Recommendations

CCICED 2023 makes a number of valuable policy recommendations on energy, environment, and climate, as outlined below:

Special attention to short-lived GHGs. CCICED 2023 policy recommendations specifically call for a special focus on short-lived greenhouse gases. CCICED's Preliminary Research Report on Innovative Technologies for Monitoring Greenhouse Gas Emissions and Carbon Sequestration suggests that accurate measurement of methane is the key to China's realization of its dual-carbon goal and proposes a new-generation top-down monitoring system guided by remote sensing and satellite monitoring platforms, which will provide a powerful complement to the bottom-up monitoring system.

Establishing a green and low-carbon production indicator system. Establish a green and low-carbon production indicator system centered on core indicators such as carbon productivity, energy efficiency, water and material consumption. Use digitalization to strengthen continuous carbon monitoring and identify carbon reduction priorities; establish carbon asset management systems for key manufacturing industries and gradually promote corporate climate-related information disclosure; and use the supply chain as a system to incentivize upstream and downstream companies to track carbon emissions data and product carbon footprints. The CCICED 2023 report "Digitization and Green Technology for Sustainable Development" proposes to manage the full life-cycle carbon emissions data of electronic equipment.

Relying on digital technological innovation, accelerate the digitalization and green upgrading of traditional industries. Promote the application of green innovation across the industrial chain and accelerate the construction of a new power system. Adopt policy measures to incentivize the development of digital solutions for climate change adaptation.

Utilizing digital technology to enhance climate adaptation capacity. Increase climate modeling, simulation and climate risk assessment, and enhance meteorological forecasting and disaster monitoring capabilities. Accelerate the formulation of the Yangtze River Basin Development Plan and Territorial Spatial Plan, establish a comprehensive assessment mechanism for the basin, and systematically assess the impact of short-term impacts and long-term pressures of climate change. Incorporate climate adaptive capacity assessment into the decision-making processes of policy formulation and construction projects.

Shape a diversified green climate investment and financing mechanism by utilizing comprehensive incentives including tax, price, compensation, and procurement. Accelerate the promotion of categorization catalogs and the formulation of rules and standards for transformational finance, emphasize the disclosure of information on the risk of climate, environmental and biodiversity loss, and regulate the environmental, social and governance (ESG) investment market. Maintain the consistency of green categorization standards with international standards, expand the scope of disclosure, refer to the International Financial Reporting Standards of the International Council on Sustainability Standards on ESG, and prepare for the upcoming disclosure standards on biodiversity risks.

Increase the promotion of circular economy development. Mandate implementation of the extended producer responsibility system in industries such as electronics, home appliances and express logistics, which generate large amounts of waste at the consumption level. At the same time, extend producers' resource and environmental responsibilities for their products from the production link to the entire life cycle of product design, production process control, intelligent logistics, recycling, waste treatment and disposal, and promote the intensification of resource consumption, minimization of pollutant and greenhouse gas emissions, and affordability of residents' consumption of green consumer goods and services through measures such as eco-design, cleaner production, and green supply chain management, and thereby form the habit of green consumption.

Promote green home appliances. Encourage consumers to choose green products, such as energy-saving home appliances, high-efficiency lighting products, water-saving appliances and green building materials, encourage enterprises to provide and allow consumers to choose reusable, durable and repairable products, support the development of the sharing economy, encourage the effective reuse of personal idle resources, improve the social recycling system of renewable resources, promote the greening, minimization and recycling of express delivery packaging, and strictly implement the government's preferential procurement and mandatory procurement system for energy-saving and environmentally friendly products. Strictly implement the government's preferential and mandatory procurement system for energy-saving and environmentally friendly products and expand the scope and scale of government green procurement.

Emphasize the green and low-carbon investment potential of sovereign assets, mainly sovereign wealth funds and social security funds. Encourage sovereign asset owners to carry out pilot demonstrations of sustainable investment and financing, incorporate the value of climate and ecological environment into the performance assessment system, and provide flexible support for relevant investment and financing activities in terms of investment return assessment and the use of risk-sharing tools. Establish sustainable investment principles for sovereign asset owners, including clear strategic objectives and organizational guarantees. Encourage sovereign asset owners to carry out sustainable investment and financing exchanges and cooperation with international partners.

Study the formulation of a framework law on combating climate change or a carbon-neutral promotion law. Incorporate carbon reduction measures into laws and regulations in transportation, construction, and urban-related areas. Formulate action plans for the implementation of the "1 + N" policy system at the provincial and municipal levels and improve the management mechanism in terms of target setting, data analysis techniques, public participation, continuous monitoring and evaluation, and dynamic adjustment.

## 5 Pollution Prevent and Control

### *5.1 Further Advancing Improvement of Air Quality*

The quality of the atmospheric environment is closely related to the health of the people. The Chinese government attaches great importance to the prevention and control of air pollution, and in recent years has continued to improve China's ambient air quality through the formulation and implementation of the *Action Plan for the Prevention and Control of Air Pollution* and the *Three-Year Action Plan for Winning the Battle for the Blue Skies*.

In November 2023, the State Council issued the *Action Plan for Continuous Improvement of Air Quality*, as the third "*Ten Atmospheric Rules,*" the Plan takes continuous improvement of air quality as the core, vigorously promotes the synergistic emission reduction of multi-pollutants, such as PM2.5, NOx, VOCs, etc., highlights the precise, scientific and law-based pollution control, and pushes forward the green and low-carbon transformation of industry, energy and transportation green and low-carbon transformation, and continue to fight a good battle for the blue sky in depth. To implement the plan, in December 2023, MEE, together with 10 units including the National Development and Reform Commission (NDRC) and 7 local governments, including the Beijing Municipal People's Government (PMG), issued the *Program for the Comprehensive Control of Air Pollution in the Autumn and Winter of Beijing-Tianjin-Hebei and its Surrounding Areas and the Fenwei Plain for the Period of 2023–2024*. The Program focuses on key areas, time periods, and fields, with the main objectives of reducing heavily polluted weather and lowering fine particulate matter concentrations, strengthening multi-pollutant synergistic control, and continuing to carry out autumn and winter comprehensive air pollution control campaigns.

In January 2024, MEE issued the *Guidance on Further Optimizing the Response Mechanism for Heavily Polluted Weather*, which proposes to optimize the mechanism for responding to heavily polluted weather, including optimizing emergency response rules, standardizing response work, strengthening emergency response linkage, revising emergency response plans and strengthening effect assessment, so as to respond to heavily polluted weather in a scientific, lawful, and precise manner, and to ensure that people's health and the environment continue to improve.

According to data released by MEE, the national air quality will maintain a long-term positive trend in 2023, with the annual average PM2.5 concentration in cities above the prefecture level nationwide reaching 30 micrograms per cubic meter, the proportion of good days being 85.5%, and the proportion of days with heavy and above pollution being 0.8%.

To adapt to the current environmental management needs of sewage discharge permits, in April 2024, MEE issued the *Measures for the Administration of Sewage Discharge Permits*. The Measures take the inclusion of sewage registration units into the scope of management, standardize the management process, and refine the content of supervision in accordance with the permit as the focus of the revision, strengthen the main responsibility of sewage disposal units, and provide comprehensive guidance to achieve the full coverage of sewage permits. In September 2023, the General Office of MEE issued the *Implementation of the management of industrial noise sewage permits notice* focused on the promotion of the implementation of the inclusion of industrial noise in the management of sewage permits, the scope of implementation timeframe, implementation mode and audit management.

## 5.2 *Further Strengthening the Precise Prevention and Control of Water Pollution*

China continues to keep waters clear, marked by the implementation of the *Action Plan for the Prevention and Control of Water Pollution*, to promote synergistic control of water environment management, and constantly improve the quality of the water environment while integrating water resources, water environment, and water ecological resilience.

In August 2023, the General Offices of the Ministry of Ecology and Environment (MEE), the Ministry of Water Resources, and the Ministry of Natural Resources jointly issued t*he Technical Guidelines for Delineating Key Zones for Groundwater Pollution Prevention and Control.* These guidelines clarify the classification and meaning of key groundwater pollution prevention and control areas, optimize the technical methods and workflow for zoning and delineation, and effectively protect groundwater quality and its sustainable use. Local governments have actively advanced the implementation of delineating these key zones. As of October 16, 2023, 83 prefectures-level cities across the country have completed the delineation of key groundwater pollution prevention and control areas at the prefectural- and -municipal level. Provinces like Heilongjiang, Shanghai, Jiangsu, and Zhejiang have finished the provincial-level delineation of such key areas.

In December 2023, the General Office of MEE led the General Office of the Ministry of Water Resources and two other departments to jointly issue the "*Guidelines for the Treatment of Rural Black and Stinking Water Bodies,*" which clarified the requirements for the identification of black and stinking water bodies in rural areas, and carried out list management; and established a long-term management and care mechanism for the treatment of black and stinking water in rural areas to put an end to the return of black and stinking situations.

In December 2023, the National Development and Reform Commission (NDRC) led the Ministry of Housing and Urban-Rural Development (MOHURD) and two other departments to jointly issue the *Implementing Opinions on Promoting Pollutant Reduction, Carbon Reduction, and Synergistic Efficiency of Sewage Treatment* to synergistically promote the reduction of pollutants and the reduction of GHG emissions in the whole process of sewage treatment and to carry out the efficiency of water conservation at the source, the energy saving and carbon reducing of the treatment process, the resourceful utilization of sewage sludge, and to comprehensively increase the comprehensive efficiency of sewage treatment.

## 5.3 Accelerating the Promotion of Soil Pollution Control and Remediation

China has been continuously upgrading the technology and management level of soil pollution control, comprehensively improving the level of investigation and assessment of soil pollution status through the whole process of quality control and supervision, promoting the integration of multidisciplinary and multi-method investigation techniques, and accurately portraying the scope of contamination and the degree of contamination.

In December 2023, the General Office of MEE issued the *Guiding Opinions on Promoting Soil Pollution Risk Control and Green and Low-Carbon Remediation* to promote the risk control and green and low-carbon remediation of soil pollution in China's construction land use, to enhance the efficient utilization of resources and energy and the prevention and control of secondary pollution, and to guide the risk control and remediation activities toward the goals of greenness, low-carbon and sustainability.

In April 2024, the Ministry of Natural Resources (MNR) led six departments to jointly issue the *Circular on Further Strengthening the Construction of Green Mines*, emphasizing the concept of green development throughout the entire process of mineral resources exploration and development, accelerating the green, low-carbon transformation and development of the mining industry, and comprehensively advancing the construction of green mines through government guidance, departmental collaboration, enterprise master building, and social supervision.

## 5.4 Regularizing Solid Waste Assessment and Management

China has continued to deepen the management of chemicals and solid waste, including strengthening the management of new pollutants and continuing to promote solid waste control and piloting in the process of negotiating an international convention on plastic pollution control.

In November 2023, the General Office of the Ministry of Ecology and Environment issued the *Notice on Further Strengthening the Standardized Environmental Management of Hazardous Waste*. The notice focuses on improving the standardized management of hazardous waste by enhancing three key areas: deepening standardized assessments, utilizing information technology, and strengthening the application of assessment results. These efforts aim to push local governments and relevant departments to fulfill their regulatory responsibilities and ensure that hazardous waste-related entities comply with legal regulations and systems.

In January 2024, the Ministry of Ecology and Environment (MEE) issued the *Catalogue of Solid Waste Classification and Codes*. This catalogue sets requirements for solid waste transfer management, information disclosure, industrial solid waste management accounts, sewage permits, environmental statistics, and more. It enhances the standardization, refinement, and informatization levels of solid waste management.

## 5.5 CICED Policy Recommendations

In the policy recommendations submitted by CCICED to the State Council in 2023, it is proposed to continuously optimize the industrial structure and promote the synergy of carbon reduction and pollution reduction:

Continuously adjusting and optimizing the structure of energy, industry, transportation, and land use, accelerating the shift from end-to-end to source management, and promoting synergies in carbon reduction and pollution reduction in multiple fields. Accelerating the reduction of emissions at the source, process control, end-of-pipe treatment and comprehensive utilization in the industrial sector, and promoting green development of the entire process.

Strengthening synergies in pollution prevention and control in the fields of air, water, soil and solid waste, and continuing the nationwide battle to prevent and control pollution. Promote in-depth control of air pollution in key industries and energy-saving and carbon reduction actions as a whole. Establish a synergistic target and evaluation system for environmental quality, pollution control and greenhouse gas emission reduction, and strengthen the synergistic control of fine particulate matter and ozone. Utilize digital technology to improve the enforcement efficiency of pollution control. Promote synergistic control of water environment governance and integrate water resources, water environment and water ecology resilience. Strengthen synergistic control of soil pollution management and encourage green and low-carbon soil remediation. Promote synergistic control of solid waste pollution prevention and control, and strengthen the construction of "waste-free cities." Promote the organic combination of natural and artificial means of restoration.

Establish a new type of cooperative relationship between importing and exporting jurisdictions and companies to optimize the layout of the global industrial and supply chains and collectively ensure the supply of critical minerals, materials, and components in green and low-carbon industries. Establish BRI green innovation partnerships. Create cross-departmental coordination mechanisms for a resilient and sustainable development of the industrial and supply chains.

Comprehensive and integrated policy recommendations on pollution control and management in China have been put forward, which have an important role to play in promoting the continuous upgrading of environmental quality in China, reducing pollution in the atmospheric and water environments, as well as in solid waste and upgrading the management of chemicals.

# 6 Ecosystem and Biodiversity Conservation

China has launched the initiative to implement the KMGBF, updated and released *China's Biodiversity Conservation Strategy and Action Plan (2023–2030)*, launched the Kunming Biodiversity Fund. CCICED carried out a special policy study on biodiversity conservation and the implementation of the KMGBF, which has called on all parties to join hands to promote the comprehensive and effective implementation of the KMGBF. It called on all parties to make concerted efforts to promote the full and effective implementation of the KMGBF and to work together to promote effective, inclusive and sustainable global biodiversity conservation actions. The main newly released biodiversity conservation policies are listed below.

On December 27, 2023, the Central Committee of the Communist Party of China and the State Council issued the *Opinions on Comprehensively Promoting the Construction of a Beautiful China*, making the enhancement of ecosystem diversity, stability and continuity one of the key tasks in solidly promoting the construction of a beautiful China. It is required to comprehensively promote the construction of a nature reserve system with national parks as the main body, accelerate the implementation of major projects for the protection and restoration of important ecosystems, and major projects for the protection of biodiversity. It has strengthened the unified supervision of ecological protection and restoration, carried out monitoring and assessment of ecological conditions and assessment of the effectiveness of ecological protection and restoration, improved the mechanism for realizing the value of ecological products, and promoted compensation for the protection of key ecological function zones, the red line of ecological protection, and important ecosystems.

## *6.1 Systematic Construction of a Strategic System for Ecological Management and Biodiversity Conservation*

On January 18, 2024, the MEE released *China's Biodiversity Conservation Strategy and Action Plan (2023–2030)*, which, in accordance with the latest requirements of the KMGBF, clearly defines four priority areas, including mainstreaming of biodiversity, responding to the threat of biodiversity loss, sustainable use and benefit sharing of biodiversity, and modernization of biodiversity governance capacity; each priority area is divided into six to eight priority actions, covering laws and regulations, policy planning, law enforcement and supervision, publicity and education, social participation, investigation, monitoring and assessment, protection and restoration, biosafety management, sustainable management of biological resources, realization of the value of eco-products, urban biodiversity, benefit sharing, climate and environmental governance, investment and financing, international implementation and cooperation, etc., which provide guidance for the work of biodiversity conservation in all sectors and regions. It provides guidelines for all departments and regions to promote biodiversity conservation.

The Plan of Action makes it clear that by 2030, the trend of biodiversity loss will be effectively mitigated, the level of biodiversity protection and management will have been significantly raised, and a favourable situation will have been created in which all people will participate in biodiversity protection. At least 30% of land, inland waters, coastal, and marine degraded ecosystems will be effectively restored, and at least 30% of land, inland waters, coastal and marine areas will be effectively protected and managed. By 2035, a national spatial pattern of biodiversity protection that is unified, orderly, structurally connected and dynamically adjusted will have been formed; by 2050, a green mode of development and lifestyle will have been fully formed, and a beautiful China in which human beings live in harmony with nature will have been built, realizing the beautiful vision of the harmonious coexistence of human beings and nature.

On April 6, 2024, the 26th executive meeting of the State Council adopted the *Regulations on Compensation for Ecological Protection.* As the first law dedicated to ecological protection compensation, the Regulations establish the basic rules of the ecological protection compensation system in the form of legislation, clarify the system and mechanism of ecological protection compensation, and compensate the units and individuals who carry out ecological protection by means of financial vertical compensation, inter-regional horizontal compensation, and compensation through the market mechanism, so as to fully mobilize all parties to participate in ecological protection. At present, China has basically established the ecological protection compensation mechanism with the widest coverage, the largest beneficiary population, and the largest investment in the world.

## 6.2 Effective Integrated Management of Mountains, Forests, Fields, Lakes, Grasses, and Sands

The period of 2021–2030 is the construction period of the six phases of the "Three Norths" project,[3] the key period of consolidating and expanding the achievements of sand control and preventing sand, and the period of promoting the high-quality development of the "Three Norths" project. In 2023, the "Three Norths" project battle was launched in full swing, in accordance with the integrated management of mountains, water, forests, lakes, grasses and sands, the strengthening of regional joint prevention and joint treatment, and the role of science and technology and mechanism innovation to support the fundamental guidelines, over the past year, various regions within the "Three-North" program areas have balanced efforts to combat desertification with economic development and have simultaneously increased both

[3] Three-North Shelterbelt Program:a series of human-planted windbreaking forest strips (shelterbelts) in China

greenery and income, achieving remarkable results in sand control and desertification management, resulting in a win-win situation for ecology and economy. The battle of the Yellow River "several bends," the battle of the two major sands of Horqin and Hunshandak, and the battle of the three major landmarks of the Hexi Corridor-Taklamakan Desert Edge Blocking Battle have achieved a good start.

Important results have been achieved in the integrated protection and systematic management of mountains, water, forests, fields, lakes, grasses, and sands. By the end of 2023, the State had deployed and implemented more than 50 projects for the integrated protection and restoration of mountains, water, forests, fields, lakes, grasses, and sands around important ecological security barriers, with a cumulative total of 80 million mu of treatment area completed.

In March 2024, the General Office of the Central Committee of the Communist Party of China (CPC) and the General Office of the State Council issued the *Opinions on Strengthening Ecological Environment Zoning Control.* Based on the complexity and diversity of ecosystems, the characteristics of spatial heterogeneity of ecological environment, and to meet the needs of refined ecological environmental protection and governance, the Opinions refine the application subject, application direction, and application path from three aspects: policy formulation, environmental access, and environmental management.

### *6.3 Promoting Aquatic Biodiversity Restoration and Watershed Ecological Restoration*

In March 2024, the General Office of the State Council issued the *Opinions on Unswervingly Promoting the Ten-Year Fishing Ban on the Yangtze River*, which consolidated the results of the fishing ban and retreat work in terms of optimizing the coordinated and integrated work-promoting mechanism, sustaining the resettlement guarantee, continuously strengthening the law enforcement and supervision, and speeding up ecological restoration, so as to accelerate the promotion of the Yangtze River's aquatic biodiversity and the ecological restoration of its waters. At present, under the joint efforts of multiple departments, the 10-year ban on fishing in the Yangtze River has achieved important milestones: the number of illegal fishing reported by the public dropped by 28% year-on-year in 2023, and the high and frequent incidence of illegal fishing has been effectively curbed; and the population of the Yangtze River dolphin, an iconic species, reached 1249 in 2022, an increase of 237 over the 2017 census, realizing the historic stop of the decline and rise; the Yangtze River cutlassfish has migrated anadromously to Dongting Lake and has reached the farthest distribution waters in history.

## 6.4 Regulating the Identification and Assessment of Damage to Agro-Ecosystems

In March 2024, MEE issued the national ecological environment standard *"Technical Guidelines for Ecological Environment Damage Appraisal and Assessment of Ecosystems Part 1: Farmland Ecosystems."* As China's first national standard supporting ecosystem damage identification and assessment, the Standard stipulates the content, working procedures, methods and technical requirements for farmland ecosystem damage identification and assessment, and standardizes the work of identification and assessment of farmland ecosystem damage caused by ecological destruction or pollution of the environment. It provides technical and institutional support for the protection of farmland ecosystems and agricultural biodiversity.

## 6.5 CCICED Policy Recommendations

In 2023, CCICED made a number of recommendations to the Chinese government in the area of ecosystem and biodiversity conservation.

Ensure the applicability of the green finance classification criteria to the biodiversity conservation financing targets of the KMGBF. Gradually implement mandatory biodiversity disclosure standards at the market level in line with international standards. Build a systematic and comprehensive methodology to assess the ecological impacts of subsidy policies, using agriculture, forestry, and fisheries as a pilot.

Link with economic sectors to advocate for systematic sustainable land use approaches and promote mainstreaming of biodiversity conservation. Reassess and optimize land use practices based on ecosystem services and incorporate science-based climate and nature objectives into decision-making and operations. Use agriculture as an entry point to study pathways and methods for achieving the sustainable use action goals under the KMGBF. Implement pilot projects on regenerative agriculture and conservation farming and summarize experience in a timely manner.

Establish an expert group on biodiversity to support participation, coordination, and implementation at the national and international levels. For challenging targets, encourage the development, promotion and application of methods and tools that will help achieve the targets, and assist in their realization with appropriate incentive mechanisms.

# 7 Regional and International Engagement

## *7.1 Strengthening Cooperation to Advance Global Climate Governance*

Climate change is a common challenge for all humanity. The Chinese government attaches great importance to addressing climate change, deeply engaging in multilateral climate processes, and actively contributing to the promotion of a fair, reasonable, and cooperative global climate governance system.

On October 18, 2023, at the opening ceremony of the third Belt and Road Forum for International Cooperation, President Xi Jinping stated that China will continue to deepen cooperation in green infrastructure, green energy, green transportation, and other fields. China will increase support for the Belt and Road International Green Development Coalition, continue to host the Belt and Road Green Innovation Conference, establish a dialogue mechanism for the photovoltaic industry, and create a network of green and low-carbon experts. Additionally, China will implement the Belt and Road Green Investment Principles and train 100,000 people from partner countries by 2030.

On October 25, 2023, President Xi Jinping, during his meeting with California Governor Gavin Newsom, mentioned that there is great potential for cooperation between China and the United States in green development and climate change. He emphasized that both sides should strengthen this cooperation, making it a new highlight in the development of China-U.S. relations.

On November 15, 2023, China and the United States issued the "Sunny lands Statement on Strengthening Cooperation to Address the Climate Crisis," deciding to launch the "Enhanced Climate Action in the 2020s Working Group." This group will accelerate concrete actions in key areas such as energy transition, methane and other non-$CO_2$ greenhouse gas emissions, circular economy and resource efficiency, local cooperation, forests, synergies in reducing greenhouse gases and air pollutants, 2035 nationally determined contributions, and COP28. This includes developing feasible and tangible cooperation plans and projects.

On December 1, 2023, Vice Premier Ding Xuexiang, as a special representative of President Xi Jinping, attended the World Climate Action Summit in Dubai and delivered a speech. He emphasized China's vigorous promotion of green development, energy revolution, and international cooperation on climate change. He called on all countries to strengthen their determination and capability to jointly address climate change, practice multilateralism, accelerate green transformation, and enhance implementation actions.

On December 2, 2023, Vice Premier Ding Xuexiang attended the "G77 and China" Climate Change Leaders' Summit in Dubai and delivered a speech. He stated that China is steadily advancing its carbon peaking and carbon neutrality goals, firmly promoting South-South cooperation on climate change, and building a green Belt and Road. He called for the G77 and China to jointly promote sustainable development, strengthen strategic alignment for green transformation, enhance mutual assistance, and explore new paths for coordinated development and protection.

On February 28, 2024, Minister of Ecology and Environment Huang Runqiu, during the sixth session of the United Nations Environment Assembly, stated that global environmental challenges such as climate change, biodiversity loss, and pollution are interrelated and mutually impactful, requiring comprehensive, coordinated, and holistic responses. Multilateral environmental agreements and the synergies between these agreements are crucial methods for advancing global environmental governance and addressing global environmental challenges.

## 7.2 Actively Promoting the Implementation of Multilateral Environmental Agreements

Climate change, biodiversity loss, desertification, and the increasing frequency of extreme weather events pose significant challenges to human survival and development. In this context, China has actively engaged in global governance, advocating for maintaining multilateral consensus, focusing on pragmatic actions, and accelerating green transformation to jointly build a clean and beautiful world.

On September 20, 2023, China signed the *Agreement on the Conservation and Sustainable Use of Marine Biological Diversity of Areas Beyond National Jurisdiction* at the United Nations Headquarters in New York. This Agreement covers marine genetic resources access and sharing, the establishment of marine protected areas, environmental impact assessments, capacity building, and marine technology transfer. China's signing of this important legal document on the first day it was open for signature reflects the nation's commitment to actively and deeply participating in global environmental governance, marine protection, and sustainable use.

On December 9, 2023, China's Minister of Ecology and Environment, Huang Runqiu, virtually attended a high-level event to promote the "30x30" target under the KMGBF. As the president of the fifteenth meeting of the Conference of the Parties (COP15) to the Convention on Biological Diversity (CBD), Huang announced that China has officially launched the initiative to implement the KMGBF. He expressed China's willingness to work with the international community to address global challenges such as biodiversity loss and climate change, collectively promote the success of COP28, and build a harmonious coexistence between humans and nature on Earth.

From February 26 to March 1, 2024, the sixth session of the United Nations Environment Assembly (UNEA-6) took place at the United Nations Environment Programme (UNEP) headquarters in Nairobi, Kenya. The Chinese delegation, led by Minister of Ecology and Environment Huang Runqiu, included representatives from the Ministry of Ecology and Environment, the Ministry of Foreign Affairs, the Ministry of Water Resources, the China Meteorological Administration, the Chinese

Embassy in Kenya, and the Permanent Mission to UNEP. In his speech at the conference, Huang stated that China is willing to work with all parties to uphold the concept of a community with a shared future for mankind, maintain and practice true multilateralism, and promote the establishment of a fair, reasonable, and cooperative global environmental governance system. The theme of this year's assembly was "Effective, Inclusive, and Sustainable Multilateral Actions to Address Climate Change, Biodiversity Loss, and Pollution."

From April 23 to 29, 2024, the fourth session of the Intergovernmental Negotiating Committee (INC-4) on the international legally binding instrument on plastic pollution, including marine plastic pollution, was held in Ottawa, Canada. The Chinese delegation, composed of representatives from the Ministry of Ecology and Environment, the Ministry of Industry and Information Technology, and other relevant departments, attended the meeting. The INC-4 session primarily reviewed the revised draft text of the legally binding international instrument on plastic pollution issued after INC-3. The focus was on the first and second parts of the revised zero draft text, covering 13 implementation-related issues, including national reporting, compliance and enforcement mechanisms, financial arrangements, and other negotiation compliance and legal matters.

On May 3, 2024, the think tank report on *China-EU Environment and Climate Cooperation: Progress and Prospects* was released. China-EU environmental and climate cooperation has become a highlight of the bilateral strategic partnership, characterized by broad consensus and deep cooperation. Both sides have a strong foundation for cooperation in green and low-carbon development, which is significant for global environmental governance. The report calls for strengthening cooperation mechanisms and innovating cooperation models to contribute to global environmental protection.

On May 28, 2024, Ding Xuexiang, Vice Premier of the State Council, attended the signing ceremony of the Kunming Biodiversity Fund cooperation agreement in Beijing and delivered a speech. Ding emphasized that under the guidance of Xi Jinping's thought on ecological civilization, China is firmly pursuing a path of high-quality development that prioritizes ecological conservation and green, low-carbon growth, leading to significant achievements in biodiversity conservation. As the chair country of COP15, China led the adoption of the KMGBF and took the lead in establishing the Kunming Biodiversity Fund, making significant contributions to global biodiversity conservation. The Ministry of Ecology and Environment of China has signed cooperation agreements with relevant United Nations agencies.

### *7.3 Deepening South-South Cooperation*

On December 2, 2023, at the G77 and China Leaders Summit on Climate Change, Chairman Xi Jinping's special representative, Vice Premier of the State Council, Ding Xuexiang, emphasized in his speech the ongoing challenges in improving global climate governance. He stressed that the G77 and China should further unite to safeguard common rights and interests, promote sustainable development together, strengthen alignment on green transition strategies, enhance mutual assistance, and explore new paths that integrate development and protection.

On November 21, 2023, the tenth meeting of the China-Kazakhstan Environmental Protection Cooperation Committee was held via video conference. The meeting was co-chaired by Zhao Yingmin, Vice Minister of Ecology and Environment of China and co-chair of the Chinese side of the China-Kazakhstan Environmental Protection Committee, and Erlan Aubakirov, Vice Minister of Ecology and Natural Resources of the Republic of Kazakhstan and co-chair of the Kazakh side. The meeting reviewed the progress of cooperation since the ninth meeting of the Committee, acknowledged the completion of annual plan tasks, and approved the 2024 work plan. It emphasized enhancing ecological environmental cooperation and technological exchanges, and elevating cooperation levels under multilateral frameworks such as the Belt and Road Initiative, Shanghai Cooperation Organization, and Asia Cooperation Dialogue to jointly address global environmental challenges.

On December 8, 2023, the Chinese delegation successfully hosted the 11th High-Level Forum on South-South Cooperation on Climate Change Response at the 28th Conference of the Parties to the United Nations Framework Convention on Climate Change (COP28) in China. The forum was attended by ministers or ministerial-level representatives from countries including Botswana, Chad, Kiribati, the Laos, Malawi, Nigeria, Cambodia, and Ghana. Simon Stiell, Executive Secretary of the UNFCCC Secretariat and Director of the UN Office for South-South Cooperation, emphasized the importance of South-South cooperation in addressing climate change. He highlighted it as a crucial pathway to help developing countries achieve their nationally determined contributions. He stressed the need to further focus on the climate change needs of developing countries and increase support for them. Dima Al-Khatib, Director of the UN Office for South-South Cooperation, emphasized the significance of South-South cooperation in enabling developing countries to achieve green and low-carbon transitions. She noted that China's Belt and Road Initiative provides a robust platform and opportunities for enhancing climate change cooperation among developing countries, serving as a model and demonstration. Foreign ministers attending the forum highly praised China's efforts in South-South cooperation on climate change and appreciated the assistance provided by China. They expressed anticipation for continued cooperation with China in areas such as new energy, climate mitigation and adaptation, climate finance, and other fields related to addressing climate change.

## 7.4 CCICED Policy Recommendations

In recent years, CCICED has put forward numerous policy recommendations to strengthen regional cooperation in the fields of environment and climate, and to promote international participation, as follows:

Upholding green openness to facilitate global low-carbon transformation. In 2022, CCICED proposed utilizing mechanisms such as the EU-China Environmental and Climate High-Level Dialogue and Climate Action Ministerial Meetings to actively engage in dialogues on climate "Track 2" and "Track 1.5," exchanging measures for reducing carbon dioxide and non-carbon dioxide greenhouse gas emissions. In 2023, CCICED proposed establishing a "Belt and Road" climate investment and green credit system to lower financing costs for low-carbon projects. Through multilateral cooperation platforms like the Belt and Road Initiative International Green Development Alliance, it aims to enhance dialogue and communication. Seizing the opportunity of the third Belt and Road Forum for International Cooperation, it puts forward initiatives for international cooperation in green and low-carbon development. Coordinating resources from all parties, it promotes demonstration cooperation in innovative application scenarios such as "photovoltaic +" to explore profitable models for green cooperation projects tailored to the characteristics of developing countries.

Enhancing environmental governance systems and improving synergies among multilateral environmental conventions. CCICED's 2019 recommendation supports global innovative marine governance. It suggests that China actively engage in the formulation and revision of international norms related to deep-sea resource development, with a focus on sustainable ocean economic development along the Belt and Road Initiative.

In 2023, CCICED recommended identifying key steps to further tackle plastic pollution. Develop an action plan to implement the global plastics treaty after its adoption. Consider launching a series of pilot projects designed to reuse, reduce, and recycle plastics. In 2021, CCICED emphasized the need to strengthen control over marine plastic pollution and microplastics at their sources, aiming to reduce plastic pollution and improve waste management and disposal capabilities.

CCICED's 2022 recommendation emphasized strengthening policy coordination to establish a multi-target synergistic mechanism aimed at maximizing synergies across carbon reduction, pollution reduction, green expansion, and economic growth efforts. It advocates for the ongoing promotion of bilateral and multilateral dialogues on climate and biodiversity and integrating global climate and biodiversity governance into broader global development initiatives.

In preparation for COP15 of the Convention on Biological Diversity, the recommendation encourages readiness for implementing the "Post-2020 Global Biodiversity Framework," including updating National Biodiversity Strategies and Action Plans (NBSAPs). It calls for enhanced discussions on coordinated climate change governance during the second phase of COP15 and anticipates greater progress in synergizing biodiversity and climate change efforts at COP27 of the UNFCCC. Prior to the formal adoption of a global plastic pollution governance treaty, CCICED encourages plastic waste reduction, reuse, recycling, and substitution initiatives, and emphasizes the need for strengthened international cooperation to launch pilot projects in this regard.

In 2023, CCICED recommended that China maintain a strategic commitment to using green and low-carbon approaches as internal driving forces. It advocates for coordinated efforts to achieve high-quality development across multiple targets. This includes updating policy measures and roadmaps for National Biodiversity Strategies and Action Plans in line with the goals of the "Kunming-Montreal Framework." As chair country, China is encouraged to continue engaging with parties, observer countries, and other stakeholders to promote biodiversity protection initiatives and early actions toward meeting targets like "30x30." CCICED also supports the establishment of the Kunming Biodiversity Fund, calling for contributions from parties to integrate and coordinate international financing to support biodiversity conservation in developing countries.

Establishing a platform for developing green projects under the framework of South-South cooperation. The CCICED's recommendations in 2021 and 2022 propose further implementation of capacity-building projects under initiatives like the Belt and Road Initiative's South-South Cooperation on Climate Change and the Green Silk Road Envoy Program. These recommendations emphasize actively promoting the advantages of green solutions to partner countries, aiming to build closer partnerships in green and sustainable development. The goal is to assist partner countries in enhancing their capabilities to address climate change and achieve inclusive, resilient recovery.

# 8 Conclusions

At present, China's ecological civilization construction is in a critical period of superimposed pressure and heavy burden. CCICED continues to play the role of a high-end think tank to promote China's environment and development, and has put forward many forward-looking and constructive suggestions, which fully reflect the superb predictive ability of Chinese and foreign members on domestic and international situations and their insight into the laws of environment and development.

In the past year, the forward-looking suggestions put forward by CCICED in the fields of pollution control, clean production, biodiversity protection, energy and environment, low-carbon economy, ecological compensation mechanism, green finance, etc. have been highly valued by the Chinese government and have important reference significance for future ecological civilization construction. China has actively and steadily carried out bilateral and multilateral environmental and climate exchanges, achieved a major transformation from a global environmental governance participant to a leader, and made progress in green development, responding to global climate change, South-South cooperation, compliance with international environmental conventions and international negotiations. In the future, China will continue to actively carry out international cooperation with a more open attitude and contribute to promoting the green prosperity and development of the world.

The achievements of ecological civilization construction in the new era have attracted worldwide attention. Over the past year, China's environmental quality has continued to improve, and it has taken the lead in the world to achieve "double reduction" in the area of desertified land and sandy land; green transformation has been steadily advancing, and the world's largest carbon market and clean power generation system have been built; the institutional system of the "four beams and eight pillars"[4] of ecological civilization has basically been formed. At the same time, China stands at the height of responsibility for human civilization and actively builds a clean and beautiful world.

2024 marks the 75th anniversary of the founding of the People's Republic of China, and it is also an important year for comprehensively promoting the building of a Beautiful China. Looking forward, CCICED will more accurately analyze the overall trends and directions of China's domestic and international environment and development in its research, focus on the innovation and foresight of research results, and put forward innovative and leading policy recommendations on issues such as environmental governance innovation, green governance system, sustainable production and consumption, and low-carbon inclusive transformation that are of great concern at home and abroad.

[4] In September 2015, the *Overall Programme for the Reform of the Ecological Civilization System* was announced, proposing eight systems.

# Chapter 18
# Report on Gender Mainstreaming in SPS Research for the Period 2023–2024

## 1 Introduction

Gender mainstreaming has been a key commitment of the CCICED Executive Committee since 2018. To support the integration of gender considerations as part of the Special Policy Studies (SPS), each SPS research team receives gender mainstreaming guidance at the outset of the research process, as well as at various touchpoints during the drafting process. The guidance is meant to provide possible research avenues related to gender equality and highlight potential entry points for considering equity and diversity perspectives within each theme.

In addition to targeted guidance, CCICED updated its *Toolkit on Gender Equality and Environmental Sustainability* in 2022, which proposes practical tools for SPS teams to ensure gender equity, diversity, and inclusion are given due consideration in the course of their work to improve the effectiveness and efficiency of policies and programs, enhance and promote sustainability, and safeguard the human rights of all in the context of sustainable development.

This report provides a detailed overview of the integration of gender perspectives as part of CCICED's 2023–2024 SPS final research papers. The report also identifies best practices for the mainstream gender perspectives.

---

The content of this chapter is extracted from the main findings of the 2023-2024 CCICED Special Policy Study (SPS) on "Gender Mainstreaming".



*Adhering to Green and Open Cooperation to Advance Modernization*,
https://doi.org/10.1007/978-981-95-5547-5_18

## 2 Gender Equality and Environmental Sustainability: Recent Developments

In recent years, the attainment of the 2030 Agenda for Sustainable Development has been overshadowed by what is referred to as an unprecedented triple planetary crisis characterized by climate change, pollution, and biodiversity loss. As each issue must be addressed, there is growing recognition of the interdependence between the three and the exacerbating effect each has on the other and on the rest of the Sustainable Development Goals (SDGs). Given its cross-cutting nature, gender equality must remain a priority of any transformative environmental and climate action. In March 2023, the United Nations Special Rapporteur on human rights and the environment published an annex on *Women, Girls and the Right to a Healthy Environment,* in which he stated that "Women and girls continue to bear the brunt of the triple planetary crisis while being excluded from most environmental and climate decision-making processes."[1] He added that "immediate gender-transformative action at the intersection of gender equality, environmental action and climate action is a prerequisite to remedying the trifold nature crisis."[2] In this vcin, important advancements were made at COP 26 in November 2023 with the adoption of the Glasgow Climate Pact, which called on Parties to respect obligations on human rights and promote and respect obligations related to gender equality and the empowerment of women, including women's meaningful and equal participation in climate action, as well as those related to youth, Indigenous Peoples, and local communities.

## 3 CCICED's 2023–24 Special Policy Studies Gender Equality-Related Recommendations

In total, CCICED conducted nine SPSs focusing on four key topics: "Innovation in Global Environmental Governance," "National Green Governance System," "Sustainable Production and Consumption," and "Low-Carbon Inclusive Transition." The SPSs included several key recommendations specific to gender equality.[3] These recommendations provide important entry points for further research, particularly

[1] Office of the High Commissioner for Human Rights, *Women, girls and the right to a healthy environment: Good practices*, Supplementary information to the report of the Special Rapporteur, David R. Boyd, on the issue of human rights obligations relating to the enjoyment of a safe, clean, healthy and sustainable environment, March 2023, p. 3, https://www.ohchr.org/sites/default/files/documents/issues/environment/srenvironment/reports-annex/Annex1-to-A-HRC52–33.pdf

[2] Ibid.

[3] It should be noted that while this report focuses only on the SPSs, several special scoping studies have also highlighted gender. For example, the CCICED special scoping study on carbon pricing underscored the challenges of many fiscal policies in gender-related issues linked to distributional effects at the household and income levels.

looking at how the recommendations can be translated into practice that can improve outcomes for all.

## 3.1 *Task Force on Innovation in Global Environmental Governance*

- **SPS on China's Pathways for Achieving Carbon Neutrality and Global Climate Governance**: Develop policies that offer retraining, skills upgrading, and social security protections for women. Promote targeted efforts to encourage women's participation and skills development in emerging green technologies and new energy sectors, including through regional and international technical cooperation and training programs. Consider the social and gender impacts of climate projects in cross-border investments.
- **SPS on Sustainable Oceans Management under the Vision of Carbon Neutrality**: Offer technical training and create inclusive spaces for learning to enable women's increased participation in various aquaculture sectors. Promote women's participation in the industrial transformation of offshore aquaculture to create more equitable and sustainable outcomes.
- **SPS on Biodiversity Conservation and Implementation of the Kunming-Montreal Global Biodiversity Framework**: Improve the evaluation system for the effectiveness of protected areas, including issues relating to social equity, gender equity, and human rights.

## 3.2 *Task Force on National Green Governance System*

- **SPS on Collaborative Mechanism for Carbon Reduction, Pollution Reduction, Green Expansion and Growth**: Thoroughly consider gender mainstreaming when formulating emission mitigation policies and designing mechanisms. Doing so includes encouraging women's participation in emission mitigation policy formulation and implementation and addressing gender discrimination within employment by emphasizing women's roles and specifically targeting them for skills development and training.
- **SPS on Green Development and Climate Adaptation for Urban and Rural Areas**: Pay attention to social equity and gender issues in climate adaptation (including assessments, policies, and programs) and accelerate the enhancement of the climate adaptation capacity of less-developed regions and vulnerable groups.

### *3.3 Task Force on Sustainable Production and Consumption*

- **SPS on Digitalization and Green Technologies for Sustainable Development**: Leverage ICT applications to increase the participation of diverse stakeholders in informing the gender-responsiveness and inclusivity of service delivery; include a broader range of institutions representing social and environmental interests within smart city development efforts; include rural areas within smart city catchment areas to ensure that urban-rural inequalities are not exacerbated.
- **SPS on Innovative Mechanisms of Sustainable Investment in Environment and Climate**: Promote opportunities for women's education, training, employment, and entrepreneurship within green technology, removing barriers that hinder their participation and actively incorporating their perspectives in the design and implementation of green technologies.

### *3.4 Task Force on Low-Carbon Inclusive Transition*

- **SPS on Sustainable Trade and Supply Chains**: Recognize that higher levels of gender equality within the workforce can enable countries to capitalize on the economic and environmental opportunities offered by sustainable trade and supply chains.
- **SPS on Green Opening-up and South-South Cooperation**: Recognize that South–South collaboration can have a tremendous impact when it comes to advancing gender equality.

## 4 CCICED's 2023–2024 Gender-Related Work through Special Policy Studies: Key Observations

This section highlights the research conducted by each SPS in the areas of gender equality and social equity, as well as best practices in the integration of gender-related concepts. The section also identifies opportunities to further advance gender equality within environmental and climate change policy development and governance.

### *4.1 SPS on China's Pathway for Carbon Neutrality and Global Climate Governance (SPS Climate)*

As recognized by the global community within the framework of the SDGs, the United Nations Framework Convention on Climate Change (UNFCCC), and other international conventions, gender equality and climate change are interconnected in significant ways. The societal and economic roles traditionally assigned to women and men lead to gender-specific impacts of climate change, with women's perspectives, participation, and leadership in addressing these issues often excluded. Persistent gender stereotypes and discrimination often leave women with less access to and control over resources needed to mitigate and adapt to climate change, limit their input in climate action decision-making, and restrict their access to rights. In recognition of the need to promote gender equality to fulfill human rights and to accelerate effective climate action, international frameworks aim to enhance women's participation and leadership in the transition to a green economy, within clean energy investments and across all areas of climate action.

SPS Climate reviews recent advances in climate governance, focusing on China's efforts and challenges in achieving a green, low-carbon transition while also assessing China's role in promoting the global green energy shift and the development of sustainable supply chains. As part of this research topic, the SPS explores the importance of gender equality in climate governance and provides gender mainstreaming recommendations.

The SPS is highly effective in integrating relevant gender equality dimensions into the research topic. It does so through the inclusion of a substantive gender-focused chapter entitled *Pathways to Gender Mainstreaming in Climate Change Analysis*, which explores various aspects of the intersection between gender equality and climate change and draws on examples of frameworks, policies, and practices from both the international realm and within China. The chapter emphasizes the need to integrate gender considerations into global climate policies due to women's greater vulnerability to climate change, resulting from socio-economic disparities and gender norms that limit their access to resources, rights, and decision-making. It also recognizes that women possess valuable and unique knowledge and skills that are essential for climate adaptation and mitigation, necessitating their participation in climate governance.

The SPS highlights that in recent years, the incorporation of gender issues has gained increasing importance within global climate governance frameworks, such as the UNFCCC, and within national climate policies, including National Adaptation Plans (NAPs) and Nationally Determined Contributions (NDCs). It recognizes that promoting gender equality within multilateral climate frameworks is vital, including through women's representation in international climate conferences, the provision of financial and capacity-building support to integrate gender equality into climate change initiatives, and funding to support women's participation in climate action. The SPS asserts that enhancing women's adaptability and influence in green and low-carbon sectors is important to advancing climate goals and ensuring gender equality. It lauds China's progress in promoting gender equality in its green transition through policies that emphasize women's participation in economic and social development.

To address economic costs borne disproportionately by women in certain sectors impacted by the green transition, the SPS recommends the development of policies that offer retraining, skills upgrading, and social security protections for women. It also recommends targeted efforts to encourage women's participation and skills development in emerging green technologies and new energy sectors, including through regional and international technical cooperation and training programs. In terms of international cooperation, the importance of considering the social and gender impact of climate projects in cross-border investments is recognized.

## *4.2 SPS on Sustainable Oceans Management under the Vision of Carbon Neutrality (SPS Oceans)*

Promoting gender equality, as emphasized in the SDGs, is essential for the effective management and protection of ocean and marine resources. The distinct roles of men and women, who use and are impacted by marine ecosystems differently, often put women at a disadvantage, limiting their engagement in consultations and access to their rights and participation in decision-making processes. Environmental challenges and climate change affect women and men differently, meaning their needs and contributions to solutions also vary. Though they are present throughout the ocean economy, women tend to be concentrated in lower-paid industries and excluded from more technical fields, as well as from leadership and governance positions. To ensure a sustainable blue economy that benefits women and men from diverse backgrounds, including Indigenous Peoples, barriers to women's participation and leadership throughout related initiatives must be identified and addressed.

SPS Oceans studies and recommends systems for ocean governance, blue finance, and green technologies for carbon neutrality to promote sustainable blue economic development in China, with a focus this year on offshore aquaculture, deep seabed mining, ocean energy industry transition, and ocean economy and blue finance. The SPS is effective in presenting the relevance of gender and equity considerations in two focused sections of the paper.

The SPS's first gender-focused section, explores the importance of equity considerations, with a focus on gender, to a sustainable blue economy and carbon neutrality goals. The second focus area relates to gender equality in offshore aquaculture. The SPS recognizes that gender equality is a key issue when evaluating the social risks of the aquaculture industry, though there is limited research on gender dynamics in offshore aquaculture. Across the entire aquatic food production sector, women make up about half of the workforce, yet they remain underrepresented in decision-making and leadership roles—often occupying lower-paid and low-skilled positions due to gender-based barriers. Engaging in offshore aquaculture, which requires higher investment and offers greater returns, poses further challenges for women, who, due to gender-discriminatory social norms and other barriers, face limited opportunities in this male-dominated industry.

The SPS asserts that offshore aquaculture's reliance on policy-driven transformation offers an opportunity for reform, through which governments can support women by offering technical training and creating inclusive spaces for learning, enabling their increased participation in various aquaculture sectors. Additionally, because women often make more sustainable and environmentally conscious decisions, the SPS puts forth the idea that promoting their participation in the industrial transformation of offshore aquaculture can lead to more equitable and sustainable outcomes.

## *4.3 SPS on Biodiversity Conservation and the Implementation of the Kunming-Montreal Global Biodiversity Framework (SPS GBF)*

The Kunming-Montreal Global Biodiversity Framework recognized the critical role that gender equity plays via its Target 22, which seeks to ensure the full, equitable, inclusive, effective, and gender-responsive representation and participation in decision-making by Indigenous Peoples and local communities, and Target 23, which calls on the implementation of the Framework through a gender-responsive approach. The focus is on recognizing that women and men generally have different roles in the use and management of natural resources due to societal gender norms and that understanding these roles is key to combating biodiversity loss and ensuring equitable access to limited and life-giving resources. While gender-differentiated responsibilities vary by region, in many communities around the world, women act as primary caretakers and natural resource managers—for example, procuring water and firewood, managing waste, and providing health care, often through plant-based medicines. As a result, biodiversity loss directly affects women's activities, and the related impacts on women's lives in terms of health, safety, and education are reinforcing unequal rights and access to decision-making.

SPS GBF focuses on strengthening biodiversity conservation and China's commitment to and strategy for implementing the Kunming-Montreal Global Biodiversity Framework. The study recognizes that within the Framework, Target 23 looks specifically at gender issues and, in doing so, calls on the importance of greater stakeholder participation and influence in biodiversity governance, including women, youth, Indigenous Peoples, and local communities, to ensure the long-term effectiveness of any initiative. For example, the SPS identifies the involvement of Indigenous Peoples and local communities as crucial when considering the development of biodiversity credit markets. Similarly, protected areas and OECMs also require involvement and support from surrounding communities to be effective and sustainable.

The SPS also focuses on the importance of considering equity and gender equality, particularly when it comes to the rights and social equity of Indigenous Peoples and local communities, which, as stated by the study, "requires refinement and wide application of social impact and governance assessments." The SPS includes a specific section on *Challenges and best and practices in selecting areas and building the 'social license to protect', leveraging support from and generating socio-economic benefits to local communities*, which looks at the importance of engaging Indigenous Peoples and local communities and addresses the inherent tension between what they refer to as "the need for speed," referring to the 2030 deadline, juxtaposed with the need to bring people along with any conservation initiatives. The report also points to the fact that there are still important lessons to be learned when it comes to the benefits of a bottom-up approach to conservation—an opportunity for future focus when it comes to this SPS. In addition to the above, the SPS report identifies a gender-specific recommendation calling for the improvement of the evaluation system for the effectiveness of protected areas by including issues related to gender and social equity, as well as human rights. This recommendation provides an additional opportunity for future research when it comes to biodiversity conservation.

Finally, a few areas were identified that present an opportunity for a gender-responsive approach. The first is that when discussing support provided by international organizations, it is important to ensure that any support provided by international organizations is done with a gender perspective in mind and in consultation with the developing country (including those involved in biodiversity conservation). A second opportunity is, when carrying out assessments of the benefits of ecological redlining, it would be important to ensure that any assessment includes a gender analysis to understand the gendered impacts of ecological redlines and identify any inequities when it comes to their benefits, as well as possible drawbacks or gendered risks.

## *4.4 SPS on Collaborative Mechanism for Carbon Reduction, Pollution Reduction, Green Expansion and Growth (SPS co-Management)*

Integrating gender considerations into low-carbon development policies is necessary to ensure equitable access to affordable, reliable, sustainable and clean energy by all. These are preconditions for gender equality, health equity and well-being. We know that climate change and related policies impact women, men, and social groups differently, based on their different needs, perspectives, and gendered realities. In addition, a lack of access to modern energy resources disproportionately affects low-income households and disadvantaged groups, including women, as well as those in rural areas, thereby exacerbating poverty and women's marginalization. As such, the development of policies and laws that build equality, prohibit discrimination, and empower women to both participate in and benefit from emission reduction and economic growth is key.

SPS Co-Management seeks to challenge the traditional concept that emission reduction is an impediment to economic growth and focus our attention on the mutually reinforcing relationship between emission reduction and economic growth. In doing so, the SPS calls on the recognition of the role of gender mainstreaming in reducing emissions and promoting economic growth. In other words, to further enhance the intrinsic power of emission reduction for economic growth, it is imperative to "fully consider the various asymmetric impacts that emission reduction may bring about, and take into account the practical interests of the main parties, encourage women's participation in emission reduction policy formulation and implementation, and give full play to the roles of women's participation and gender mainstreaming in emission reduction to promote economic growth."

The SPS includes a specific section on the role of gender mainstreaming in reducing emissions and in the promotion of economic growth, which identifies several entry points for enhancing gender equality and social equity. These are:

a. **The importance of encouraging women's participation in emission mitigation policy formulation and implementation**; not only does women's participation bring more diverse opinions and experiences, but it also improves policy formulation and implementation.
b. **The importance of addressing gender discrimination within employment and leveraging social capital for emission reduction**; emphasizing women's roles in emission mitigation policies and specifically targeting them for skills training and employment support can help alleviate gender discrimination and increase women's involvement in promoting social capital for emission reduction.

In terms of recommendations, the SPS calls on the need to strengthen societal support and participation, including women's participation, in emission mitigation policies to form a consensus on green development involving the entire society. The SPS also calls on the importance of thoroughly considering gender mainstreaming when formulating policies and designing mechanisms to ensure the pursuit of a truly green just transition.

## *4.5 SPS on Green Development and Climate Adaptation for Urban and Rural Areas (SPS Adaptations)*

Equity must be a central part of building a climate-adapted future if adaptation to climate change is to be transformative and sustainable. To do so, it is important that adaptation measures consider the disproportionate impacts of climate change due to factors such as disparities in income and wealth, access to land and resources, representation in government, health status, age, gender, indigeneity, etc. Measures that do not take these factors into consideration risk exacerbating social injustices and inequalities, which further undermine long-term adaptation goals. As more initiatives are shared in which equity perspectives are integrated at the community and regional levels, opportunities for learning and knowledge sharing on adaptation best practices will increase.

SPS Adaptations considers the escalating risks posed by climate change and the urgency of implementing more systematic and urgent climate adaptation measures. The SPS also considers the importance of social and gender equity as part of climate adaptation. This is done by introducing a separate section that looks at social inequities in climate change with a specific look at urban-rural and regional disparities. The section discusses how existing climate adaptation actions may, in fact, be exacerbating existing inequities due in part to government bias, neglect of disadvantaged areas and groups, the unbalanced distribution of resources, and limited public participation in the decision-making and implementation process.

Indeed, one of the risks identified by the SPS relates to the body of evidence that demonstrates how socially and economically disadvantaged regions and populations are disproportionately affected by the adverse impacts of climate change, leading to further inequity. As part of the research, the study considers four case studies highlighting climate adaptation efforts and identifying gaps in policies and practices. While the case studies do not specifically address gendered considerations, the examples identify low levels of awareness of and attention to climate change among the population and local communities, as well as limited public participation and grassroots-level engagement in climate adaptation assessments, disaster and emergency management, and the implementation of adaptation efforts as an important gap. The case studies also point to the importance of public participation in the co-creation of a shared vision for the future to increase support and effectiveness of policies. The SPS introduces two additional examples of social equity done well in climate adaptation: the Canadian Healthy Plan City tool and the Chinese "photovoltaic poverty alleviation" and green finance policy, which allow for greater prioritization of climate adaptation efforts to communities who would benefit the most.

The study calls on a more inclusive planning process to ensure equitable policy outcomes. This means assessing how adaptation policy decisions affect vulnerable groups by identifying those who benefit from these decisions and those who are disadvantaged by them. As such, an important recommendation put forward by the SPS is to "pay attention to social equity and gender issues in climate adaptation (including assessments, policies and programs), and accelerate the enhancement of the climate adaptation capacity of less developed regions and vulnerable groups."

## *4.6 SPS on Digitalization and Green Technologies for Sustainable Development (SPS Digitization)*

The digital-green dual transformation and sustainable development of cities creates new opportunities for the economic empowerment and increased well-being of women and groups often marginalized based on intersectional identity factors such as gender, age, and ethnicity, but also presents risks of their exclusion from emerging employment and economic opportunities given gender-based barriers. To harness the potential of digitization to achieve gender equality and enable women's empowerment, the way digitalization affects women and men differently must be recognized, and gender-based barriers that prevent them from equitably benefiting from digital advancements must be addressed. This includes considering the perspectives of diverse women and men, ensuring that women have access to digital tools and employment in STEM fields, and considering the structural and transformational changes necessary within digital governance and the broader digital ecosystem to make them more gender-responsive and reduce the gender digital divide.

SPS Digitization explores how China can best achieve its digital and green transformation goals while integrating economic, social, and ecological considerations into the sustainable development of cities. The SPS effectively integrates relevant considerations in a gender-focused section and at various points within the research.

The SPS explores gender equality and intersectionality in the context of China's rapid development of smart cities and emphasizes that understanding intersectionality is essential for identifying vulnerable groups and effectively implementing measures to address inequality. The paper demonstrates that in China, the development of smart cities risks worsening existing inequalities, especially for women, the elderly, and migrant workers. Over the past decades, gender inequality in China has increased despite improvements in women's access to education, healthcare, and wealth. Women's ability to benefit from smart city developments is limited due to their underrepresentation in high-paying jobs, such as those found in STEM fields. Rural households, particularly those of migrant workers, also face significant challenges, particularly the risk of job displacement due to technological advances. Female migrant workers, who often hold lower-skilled jobs, are especially vulnerable, while the elderly, particularly women, often struggle with low pensions, chronic health issues, and difficulties in using digital technologies.

The SPS also shows that a large proportion of China's workforce lacks higher education. This education gap particularly impacts women, especially female migrant workers, who lack the higher education required to effectively adapt to the demands of smart cities. The gig economy integral to smart cities provides new opportunities through flexible work options but also job insecurity and isolation, particularly for women, necessitating gender-sensitive designs and support mechanisms for gig workers to address these challenges. The paper also highlights the lack of sex-disaggregated data in China, the scarcity of which is exacerbated by strict data privacy laws. The scarcity of such data hinders efforts to identify and address gender bias and inequality—a gap that could be addressed by improving data transparency and sharing information with research institutions.

Ultimately, the paper concludes that smart city developments that fail to consider and address existing inequalities are at risk of creating a future with even greater social disparities. For smart cities to be sustainable, they need to identify and address gender and other inequalities. The SPS recommends that ICT applications be leveraged to increase the participation of diverse stakeholders in informing the gender-responsiveness and inclusivity of service delivery; that a broader range of institutions representing social and environmental interests be included within smart city development efforts; and that rural areas be included within the catchment areas for smart cities to ensure that urban-rural inequalities are not exacerbated.

## *4.7 SPS on Innovative Mechanisms of Sustainable Investment in Environment and Climate (SPS Finance)*

Gender considerations are important in green technology innovation, as women are often marginalized within development and deployment processes as employees, entrepreneurs, and leaders within green technology fields. The gender dimensions of green technology innovation relate to the need to address existing inequalities in women's and men's access to resources, including financing, to participate in green technology innovation and acquisition; their inclusion within livelihood opportunities related to green technology development; their ability to benefit from innovation processes; and consideration of their diverse perspectives as green technology users in designing and implementing innovation processes and outcomes that support their inclusion and ability to benefit from innovation. Considering gender equality within such processes also has a multiplier effect on promoting green technology innovation among a wider population of women and men.

SPS Finance examines China's capital market's role in supporting technological innovation, focusing on the investment process to enhance its effectiveness. As part of this analysis, the SPS includes a gender-focused section on the gender dimensions of talent within the green technology innovation ecosystem.

The SPS asserts that the construction of a green technology innovation ecosystem heavily relies on talent and that gender inclusion is vital in this process, despite women's current low representation within the scientific research and development global workforce. Various factors, including discriminatory social norms and limited access to STEM education, hinder women's participation in male-dominated sectors like green technology. The environment is often unsupportive or hostile, creating additional challenges for women entering and advancing in this field. The SPS continues by explaining that the exclusion of women from green technology innovation can limit the incorporation of women's perspectives, which may prevent technological innovations from fully addressing the needs of society, thereby reducing their overall societal impact.

To create a more inclusive and effective green technology innovation ecosystem, the paper asserts that it is essential to promote opportunities for women's education, training, employment, and entrepreneurship within green technology. This would remove barriers that hinder their participation and actively incorporate their perspectives in the design and implementation of green technologies. Doing so would ensure that green technology innovation serves the interests of all members of society, maximizing the potential benefits of green technology.

## 4.8 SPS on Sustainable Trade and Supply Chains (SPS Trade)

Trade in low-carbon technologies can bring new economic development opportunities for women, such as more access to low-carbon technologies, reduced time burdens, and new job opportunities. However, the potentially negative impacts associated with increased trade in low-carbon technologies may also be most keenly felt among women, who start at an economic disadvantage due to discriminatory gender norms. These gender norms result in women's lower access to resources, such as training and skills development, particularly in STEM fields; concentration in lower-paid and lower-skilled jobs; and lesser access to information and resources to enable them to benefit from low-carbon technologies—all of which give them less flexibility than men in adapting to economic shifts and climate change and risk their exclusion from the economic and environmental benefits of trade in low-carbon technologies.

In recognition of the importance of a global shift to sustainable trade to address climate change, SPS Trade analyzes China's bilateral trade in low-carbon technologies (LCTs) and their associated minerals, exploring the environmental impacts and economic factors influencing them. This includes a gender-focused section analyzing the role of gender equality in the labour force and its impacts on trade in LCTs and their associated minerals.

The quantitative analysis the SPS carried out for this section examined factors such as the ratio of female employers, the female employment rate, and female employment in services against international trade in LCTS and minerals. In the countries examined outside of China, these gender-based factors were found to be positively correlated with LCT exports, suggesting that female representation and leadership in the workforce were strongly linked to LCT exports to China, highlighting the importance of female leadership in driving trade. However, the SPS finds that LCT imports from China were less correlated with gender-specific performance, except for a negative association with female employment in industry. For mineral trade, the SPS finds that both imports from and exports to China were significantly related to gender variables. Female leadership and participation in the workforce, particularly in industry and services, were found to positively influence mineral exports to China, while the ratio of female employers and the female-to-male employment rate were found to positively impact mineral imports from China.

Based on the results of the overall analysis, the SPS concludes that gender equality in the workforce enhances countries' competitiveness in international trade in LCTs and minerals. This suggests that countries with higher levels of gender equality may be better positioned to capitalize on the economic and environmental opportunities offered by sustainable trade and supply chains.

## *4.9 SPS on Green Opening-up and South–South Cooperation (SPS South-South)*

South–South cooperation can play a critical role in promoting gender equality and advancing the broader SDGs. South–South cooperation can facilitate the exchange of knowledge, expertise, and resources, enhancing the capacity of countries to implement gender-responsive policies and programs. South—South cooperation can focus on empowering women and mainstreaming gender into activities across all SDGs. To ensure the adoption of inclusive and impactful South–South cooperation models; however, it is imperative to recognize the unique challenges faced by women and girls and ensure women's leadership and active participation, including that of women's rights and feminist organizations, at all levels and in all stages of the South–South cooperation practices.

The SPS South-South report sought to better understand the importance of green opening-up and South–South cooperation for China by identifying the needs and challenges of developing countries in responding to climate change, renewable energy development, and other fields, as well as the requirements for ensuring win–win cooperation. Through the report, it is clear that collaborative efforts among countries are key to green development. The leading role that China plays in the Global South and its promotion of South–South cooperation means that it can offer valuable lessons and provide a model for sustainable development through green policies. It also means that it can have a tremendous impact when it comes to advancing gender equality.

The SPS report includes several specific paragraphs on the role of women and gender equality, which are important contributions to the theme at hand. The SPS also references the importance of engaging a wide range of partners, including government departments, enterprises, international organizations, non-governmental organizations (NGOs), and diverse social groups.

Finally, at the time of review, a few areas were identified that present an opportunity for a gender-responsive approach as part of a long-term vision for China's involvement in Green Opening-Up and South–South cooperation. The first area relates to the opportunity to deepen the research on the importance of gender mainstreaming as part of South–South cooperation, looking specifically at how South–South cooperation can benefit women and support gender equality. The second area relates to the SPS's recommendation to include comprehensive environmental, social, and governance (ESG) indicators in the evaluation system of South–South cooperation projects. In addition to doing so, there are important linkages between ESG and gender equality that should be explored to better understand how ESG can be leveraged to support gender equality. A final area relates to the importance of carrying out gender analyses to ensure that any initiative or effort undertaken considers the differentiated impacts on women and diverse social groups and how they can contribute to the design and implementation of sustainable solutions. For example, when looking at collaboration between Chinese and African companies, a gender analysis would help identify specific barriers facing women-led companies, which would allow the establishment of specific support or measures to facilitate their participation. Similarly, when looking at support for engineers and technicians, a gender analysis would help identify specific barriers faced by women engineers and technicians and would help to create targeted solutions to ensure their equitable access to training, employment, and career development.

# 5 Good Practices in the Integration of Gender Equality Perspectives

The following section highlights several best practices from the SPS final research papers that should continue to be incorporated into SPS research and development.

- **Carrying out a preliminary gender analysis early in the planning process**. Conducting an initial gender analysis early in the research planning process can enable the identification of the gender equality issue(s) most pertinent to the research topic and with adequate existing research to draw upon. This can then provide the basis for resourcing the researching and drafting of this section appropriately by either an existing member of the SPS team or an external expert. A more in-depth gender analysis could then be carried out by the researcher responsible for that section.
- **Including a gender-focused section in the SPS**. Including a gender-focused section within the SPS improves the visibility of gender equality-related considerations and enables the provision of adequate context to raise awareness of the relevance of gender equality to the research topic. Previous SPS have done this in the form of text boxes of gender-focused chapters or sections of the SPS.
- **Integrating gender considerations throughout the SPS**. In addition to a gender-focused section, it is good practice to also integrate gender considerations at relevant points throughout the SPS. The integration of gender within the text can follow from the context provided within the gender-focused section, or it can be independent of the context provided in the gender-focused section. There are gender dimensions in most research topics, so the integration of gender considerations throughout is generally relevant.
- **Including diverse demographic groups in stakeholder engagement**. When discussing the engagement of stakeholders, it is good practice to promote the meaningful participation of women and diverse demographic groups, particularly those marginalized based on their intersectional identities. Such participation is important as part of policy development and decision-making, as well as within consultations.

CCICED believes the work of SPS research teams around gender equality can have an important impact on policy direction and future environmental efforts related to climate adaptation, carbon reduction, biodiversity conversation, south-south cooperation, and more. As such, CCICED will continue to support SPS research teams in integrating gender equality considerations and in reinforcing their gender mainstreaming approaches.

# Annex CCICED Phase VII Composition (as of February 2025)

## CCICED Phase VII Members (as of February 2025)

### Chinese Members

| | | |
|---|---|---|
| 1. | Mr. Xuexiang Ding | Chairperson of CCICED |
| 2. | Mr. Runqiu Huang | **Executive Vice Chairperson of CCICED**<br>Minister, Ministry of Ecology and Environment |
| 3. | Mr. Zhenhua Xie | **Vice Chairperson of CCICED**<br>Former Special Envoy on Climate Change of China |
| 4. | Mr. Shengxian Zhou | **Vice Chairperson of CCICED**<br>Former Minister of Environmental Protection |
| 5. | Mr. Yingmin Zhao | **Secretary General of CCICED**<br>Vice Minister, Ministry of Ecology and Environment |
| 6. | Mr. Shijin Liu | **Chinese Chief Advisor of CCICED**<br>Former Vice President of the Development Research Center of The State Council |
| 7. | Mr. Zhaoxu Ma | Executive Vice Minister of Foreign Affairs (full minister rank) |
| 8. | Mr. Shouwen Wang | China International Trade Representative (full minister rank), and Vice Minister of Commerce, Ministry of Commerce |
| 9. | Mr. Pengcheng Yan | Deputy Director (Vice Minister), Office of the Central Financial and Economic Affairs Commission |
| 10. | Mr. Chenxin Zhao | Vice Chairman, National Development and Reform Commission |
| 11. | Mr. Min Liao | Vice Minister, Ministry of Finance |
| 12. | Mr. Shuxian Sun | Vice Minister, Ministry of Natural Resources; Director of State Oceanic Administration |
| 13. | Mr. Yang Li | Vice Minister, Ministry of Transport |
| 14. | Mr. Xingwang Zhang | Vice Minister, Ministry of Agriculture and Rural Affairs |

(continued)



*Adhering to Green and Open Cooperation to Advance Modernization*,
https://doi.org/10.1007/978-981-95-5547-5

(continued)

| | | |
|---|---|---|
| 15. | Mr. Yanshun Xiao | Member of the Leading Party Members Group and Deputy Director, State Council Research Office |
| 16. | Mr. Yuansheng Xie | Member of the Leading Party Members Group, Ministry of Industry and Information Technology |
| 17. | Mr. Yong Ren | Chief Engineer and Director General, Department of Institutional Administration and Human Resources Management, Ministry of Ecology and Environment |
| 18. | Mr. Shuhua Dou | Vice-chairperson, The Environmental Protection and Resources Conservation Committee of the 13th National People's Congress |
| 19. | Mr. Yaping Zhang | Former Vice President, Academician, the Chinese Academy of Sciences |
| 20. | Mr. Fang Cai | Member of the Monetary Policy Committee of the People's Bank of China, chief expert of the National Think Tank at CASS, academician of CASS |
| 21. | Mr. Yuyan Zhang | Dean, School of International Politics and EconomicsUniversity of Chinese Academy of Social Sciences |
| 22. | Mr. Zuqiang Zhang | Deputy Administrator, China Meteorological Administration |
| 23. | Mr. Xiuxin Deng | Vice President and Academician, the Chinese Academy of Engineering; Vice president, China Association for Science and Technology |
| 24. | Mr. Baoxing Qiu | Academician of the International Eurasian Academy of Sciences;Former Vice Minister of Ministry of Housing and Urban-Rural Development, P. R. China |
| 25. | Ms. Hui Luo | Director General of Department of International Affairs (Hong Kong, Macao and Taiwan Exchange Office) |
| 26. | Mr. Lan Xue | Dean of Schwarzman College at Tsinghua University; Co-Chair of the Leadership Council of the UN Sustainable Development Solution Network (UNSDSN); Distinguished Professor of Arts, Humanities and Social Sciences at Tsinghua University |
| 27. | Mr. Kebin He | Member of Chinese Academy of Engineering; Professor of the School of Environment and Dean of the Institute for Carbon Neutrality at Tsinghua University |
| 28. | Mr. Yuanhang Zhang | Professor, College of Environment Sciences and Engineering, Pecking University; Academician, the Chinese Academy of Engineering |
| 29. | Mr. Minhan Dai | Chair Professor, Xiamen University; Academician of Chinese Academy of Sciences |
| 30. | Mr. Jingyun Fang | Professor, College of Urban and Environmental Sciences, Pecking University; Academician of Chinese Academy of Sciences |
| 31. | Mr. Jinnan Wang | Academician of Chinese Academy of Engineering; Deputy Chairmen of Population, Resources and Environment Committee of the Chinese |
| 32. | Mr. Xiaoye Zhang | Dean of Yulin College of Carbon Neutrality, Northwest University; Academician of Chinese Academy of Engineering |

(continued)

(continued)

| | | |
|---|---|---|
| 33. | Mr. Yi Wang | Vice Chair, National Expert Committee on Climate Change;Professor of School of Public Administration, University of Chinese Academy of Sciences;Member, Standing Committee of the National People's Congress of China |
| 34. | Mr. Tianyi Wang | Professor, Hong Kong University of Science and Technology (Guangzhou) |
| 35. | Ms. Marjorie Yang | Chairman, Esquel Group |

## International Members

| | | |
|---|---|---|
| 1. | Mr. Steven Guilbeault | **Executive Vice Chairperson**<br>Minister, Environment and Climate Change Canada |
| 2. | Mr. Achim Steiner | **Vice Chairperson**<br>Administrator, The United Nations Development Programme |
| 3. | Ms. Inger Andersen | **Vice Chairperson**<br>Executive Director, The United Nations Environment Programme |
| 4. | Ms. Kristin Halvorsen | **Vice Chairperson**<br>Director, CICERO Center for International Climate and Environmental Research; Former Minister of Finance and Education of Norway; Former Deputy Prime Minister of Norway |
| 5. | Mr. Scott Vaughan | **International Chief Advisor of CCICED**<br>Senior Fellow of International Institute for Sustainable Development |
| 6. | Mr. Danny Alexander | Vice President, Policy and Strategy, The Asian Infrastructure Investment Bank |
| 7. | Ms. Grethel Aguilar | Director General of the International Union for Conservation of Nature |
| 8. | Mr. Per Ängquist | Director General, Swedish Chemicals Agency |
| 9. | Mr. Peter Bakker | President and CEO, World Business Council for Sustainable Development |
| 10. | Mr. Manish Bapna | President and Chief Executive Officer, the Natural Resources Defense Council |
| 11. | Mr. Børge Brende | President, World Economic Forum |
| 12. | Mr. Jerry Brown | Chair, California-China Climate Institute |
| 13. | Mr. Francesco La Camera | Director-General, the International Renewable Energy Agency |
| 14. | Mr. Tomas Anker Christensen | Climate Ambassador of Denmark |
| 15. | Mr. Srun Darith | Secretary of State, Ministry of Environment, Cambodia |

(continued)

(continued)

| | | |
|---|---|---|
| 16. | Mr. Aniruddha (Ani) Dasgupta | President and CEO of World Resources Institute |
| 17. | Mr. John J. DeGioia | President, Georgetown University |
| 18. | Mr. Jos Delbeke | The first EIB Chair on Climate Policy and International Carbon Markets |
| 19. | Ms. Patricia Espinosa | CEO and Founding Partner of Onepoint5, Former Executive Secretary of the United Nations Framework Convention on Climate Change (UNFCCC) |
| 20. | Ms. Patricia Fuller | President and CEO, International Institute for Sustainable Development |
| 21. | Mr. Jan Hendrik Dronkers | Secretary-General, the Ministry of Infrastructure and Water Management, the Netherlands |
| 22. | Ms. Kate Hampton | CEO, Children's Investment Fund Foundation |
| 23. | Mr. Arthur Hanson | Senior Advisor and Former President of International Institute for Sustainable Development |
| 24. | Mr. Hal Harvey | Founder, Energy Innovation |
| 25. | Mr. Pete Harrison | Executive Vice President, Regions, Environmental Defense Fund |
| 26. | Mr. Stephen Heintz | President and CEO, the Rockefeller Brothers Fund |
| 27. | Ms. Naoko Ishii | Professor, Executive Vice President and Director of the Center for Global Commons, University of Tokyo; Former CEO and Chair, Global Environment Facility |
| 28. | Mr. Marco Lambertini | Nature Positive Initiative Convener |
| 29. | Mr. Stanley Loh | Permanent Secretary, Ministry of Sustainability and the Environment, Singapore |
| 30. | Mr. Michael McElroy | Gilbert Butler Professor of Environmental Studies, Harvard University |
| 31. | Mr. Dirk Messner | President, German Federal Environment Agency |
| 32. | Mr. Hideki Minamikawa | President, Japan Environmental Sanitation Center |
| 33. | Ms. Jennifer Morris | Chief Executive Officer, The Nature Conservancy |
| 34. | Mr. Scott Morris | Vice-President (East Asia, Southeast Asia and the Pacific), Asian Development Bank |
| 35. | Mr. Gerd Müller | Director General of the United Nations Industrial Development Organization |
| 36. | Mr. Bruno Oberle | President of the World Resources Forum, Chair of International Resources Panel |
| 37. | Mr. Jonathan Pershing | Program Director of Environment, William and Flora Hewlett Foundation |
| 38. | Mr. Carlos Manuel Rodriguez | CEO and Chairperson of the Global Environment Facility; Former Environment and Energy Minister, Costa Rica |

(continued)

(continued)

| | | |
|---|---|---|
| 39. | Ms. Kirsten Schuijt | Director General, World Wide Fund for Nature |
| 40. | Mr. Erik Solheim | Senior Advisor, World Resources Institute |
| 41. | Mr. Andrew Steer | President and CEO, Bezos Earth Fund |
| 42. | Mr. Sukanto Tanoto | Founder and Chairman, Royal Golden Eagle |
| 43. | Mr. James Thornton | Founder and President, ClientEarth |
| 44. | Ms. Anna Toni | National Secretary for Climate Change at the Ministry of Environment and Climate Change, Brazil |
| 45. | Ms. Nomfundo Tshabalala | Director-General of the Department of Forestry, Fisheries, and the Environment, Republic of South Africa |
| 46. | Ms. Laurence Tubiana | CEO, European Climate Foundation |
| 47. | Ms. Jo Tyndall | Director, Environment Directorate, Organization for Economic Co-operation and Development |
| 48. | Ms. Christie Ulman | President, Sequoia Climate Foundation |
| 49. | Mr. Kurt Vandenberghe | Director-General, Directorate-General Climate Action (DG CLIMA) |
| 50. | Mr. Juergen Voegele | Vice President for Sustainable Development, the World Bank |
| 51. | Mr. Jan-Gunnar Winther | Pro-rector for research and development,The Arctic University of Norway;Specialist Director, Norwegian Polar Institute |
| 52. | Mr. Zhang Hongjun | Partner, Holland and Knight LLP;Board Chair, Energy Foundation China |

# CCICED Phase VII Special Advisors (as of February 2025)

## Chinese Special Advisors

| | | |
|---|---|---|
| 1. | Mr. Yongjun Ji | Director General of Department of Asian & African Affairs, The Chinese People's Association for Friendship with Foreign Countries; Secretary-General, China-Africa People's Association for Friendship |
| 2. | Mr. Yong Zhang | Executive Vice President, China COSCO Shipping Corporation Limited |
| 3. | Ms. Wenling Chen | Chief Economist, China Center for International Economic Exchanges |
| 4. | Mr. Yansheng Zhang | Chief Researcher of China Center for International Economic Exchanges |

(continued)

(continued)

| | | |
|---|---|---|
| 5. | Mr. Haisheng Li | The Chief Scientist of Chinese Research Academy of Environmental Sciences; Former President of China Research Academy of Environmental Sciences |
| 6. | Mr. Jing Guo | President, BRI International Green Development Coalition |
| 7. | Mr. Heng Zhou | Former Director General, Department of International Cooperation, China Meteorological Administration |
| 8. | Mr. Yanfei Ye | Former Senior Inspectorate Advisor(DG level), Policy Research Bureau of National Financial Regulatory Administration |
| 9. | Mr. Baolin Hu | Former Deputy Director General, Three Gorges Project Construction Commission of the State Council |
| 10. | Mr. Panmao Zhai | Co-chair, Intergovernmental Panel on Climate Change (IPCC)Working Group I Co-chair; Chief Scientist and Principle Investigator of Chinese Academy of Meteorological Sciences |
| 11. | Mr. Yongsheng Zhang | Director-General and Research Fellow, Research Institute for Eco-civilization, Chinese Academy of Social Sciences |
| 12. | Mr. Jun Ma | Chairman, Green Finance Committee of China Society for Finance and Banking; President, The Institute of Finance and Sustainability (IFS) |
| 13. | Mr. Xiaojiang Li | Professor and former President, China Academy of Urban Planning and Design |
| 14. | Mr. Ping Yu | Former President of China Council for the Promotion of International Trade |
| 15. | Mr. Qi Zhai | Executive Secretary General, China Business Council for Sustainable Development |
| 16. | Mr. Jie Tang | Director of the Chinese University of Hong Kong (Shenzhen), and former vice mayor of Shenzhen |
| 17. | Ms. Qing Hu | Professor of the School of Environmental Sciences and Engineering and Director of the Engineering Innovation Center (Beijing) at Southern University of Science and Technology |
| 18. | Mr. Lin Xu | Chairman of China-US Green Fund, Committee Member of Demonstration Area in the Yangtze River Delta on Ecologically Friendly Development, Former Director General of the Department of Fiscal and Financial Affairs and the Department of Development Planning, NDRC |
| 19. | Mr. Zhenguo Li | Founder and President of LONGi |
| 20. | Mr. Yusuo Wang | Founder and Chairman of the Board of ENN Group |

## International Special Advisors

| | | |
|---|---|---|
| 1. | Mr. Knut Halvor Alfsen | Former Director, Center for International Climate and Environmental Research Oslo |
| 2. | Mr. Kuat Akizhanov | Deputy Director, Central Asia Regional Economic Cooperation Institute |

(continued)

(continued)

| | | |
|---|---|---|
| 3. | Mr. Howard Bamsey | Honorary Professor, School of Regulation and Global Governance of Australian National University; Former Chair, Global Water Partnership; Former Executive Direcotor of GCF |
| 4. | Mr. Dimitri de Boer | Regional Director of Programmes for Asia & Chief Representative of China, ClientEarth |
| 5. | Mr. Guillermo Castilleja | Senior Advisor, Global Alliance for the Future of Food |
| 6. | Ms. Galit Cohen | Director, Program on Climate Change and National Security, Israel Institute for National Security Studies (INSS) |
| 7. | Mr. Kevin P. Gallagher | Director, Global Development Policy Center, Boston University |
| 8. | Ms. Shenyu G. Belsky | China Program Director/China Chief Representative, Rockefeller Brothers Fund |
| 9. | Mr. Stephan Contius | Special Advisor on SDGs at the Foundations Platform F20 |
| 10. | Mr. Mark Halle | Former European Representative and Director for Trade and Investment, International Institute for Sustainable Development |
| 11. | Ms. Ilka Hirt | Deputy Director-General for International Policy, German Federal Ministry for the Environment, Nature Conservation, Nuclear Safety and Consumer Protection (BMUV) |
| 12. | Ms. Jeanne-Marie Huddleston | Director General of Bilateral Affairs and Trade, the International Affairs Branch of Environment and Climate Change Canada |
| 13. | Ms. Bernice Lee | Research Director, Futures, Chatham House-Royal Institute of International Affairs |
| 14. | Mr. Hongpeng Lei | Global Director, Climate, Children's Investment Fund Foundation |
| 15. | Mr. Jian Liu | Director of the Early Warning and Assessment Division, United Nations Environment Programme |
| 16. | Mr. Sze Ping Lo | Program Director, China and Southeast Asia, Sequoia Climate Foundation |
| 17. | Mr. Zafar Makhmudov | Executive Director, the Regional Environmental Centre for Central Asia |
| 18. | Mr. Hans Mommaas | Director-General, PBL Netherlands Environmental Assessment Agency |
| 19. | Ms. Neo Gim Huay | Managing Director, World Economic Forum |
| 20. | Ms. Mary Nichols | Vice Chair, California-China Climate Institute |
| 21. | Ms. Oyun Sanjaasuren | Director, External Affairs, Green Climate Fund |
| 22. | Ms. Catherine Stewart | Canada's Ambassador for Climate Change |
| 23. | Mr. Ismo Tiainen | Director-general, Administration and International Affair, Ministry of the Environment, the Republic of Finland |
| 24. | Mr. Ji Zou | CEO & President of Energy Foundation China |
| 25. | Ms. Meng Fei | China Program Senior Director, Energy Innovation |